AF785305

Catalysis on Zeolites and Zeolite-Like Materials, 3rd Edition

Catalysis on Zeolites and Zeolite-Like Materials, 3rd Edition

Guest Editor

Wladimir Reschetilowski

MDPI

Basel • Beijing • Wuhan • Barcelona • Belgrade • Novi Sad • Cluj • Manchester

Guest Editor
Wladimir Reschetilowski
Fakultät Chemie und
Lebensmittelchemie
Technische Universität Dresden
Dresden
Germany

Editorial Office
MDPI AG
Grosspeteranlage 5
4052 Basel, Switzerland

This is a reprint of the Special Issue, published open access by the journal *Catalysts* (ISSN 2073-4344), freely accessible at: https://www.mdpi.com/journal/catalysts/special_issues/8W7QCK98W2.

For citation purposes, cite each article independently as indicated on the article page online and as indicated below:

Lastname, A.A.; Lastname, B.B. Article Title. *Journal Name* **Year**, *Volume Number*, Page Range.

ISBN 978-3-7258-7000-4 (Hbk)
ISBN 978-3-7258-7001-1 (PDF)
https://doi.org/10.3390/books978-3-7258-7001-1

© 2026 by the authors. Articles in this reprint are Open Access and distributed under the Creative Commons Attribution (CC BY) license. The reprint as a whole is distributed by MDPI under the terms and conditions of the Creative Commons Attribution-NonCommercial-NoDerivs (CC BY-NC-ND) license (https://creativecommons.org/licenses/by-nc-nd/4.0/).

Contents

catalysts MDPI

Editorial

Catalysis on Zeolites and Zeolite-like Materials, 3rd Edition

Wladimir Reschetilowski

Fakultät Chemie und Lebensmittelchemie, Technische Universität Dresden, Helmholtzstraße 14, 01069 Dresden, Germany; wladimir.reschetilowski@tu-dresden.de

The scientific and practical interest in catalysts based on zeolites and zeolite-like materials continues to be strong; therefore, this Special Issue of *Catalysts*, "Catalysis on Zeolites and Zeolite-Like Materials, 3rd Edition," continues the successful publication of the preceding two volumes. These materials have regular pore systems with cavities and channels in molecular dimensions, granting them unique properties such as high thermal stability, intrinsic acidity, and well-defined, shape-selective pore structures. A growing number of researchers are investigating the key catalytic properties of zeolites and related materials, and they are additionally being used commercially in an expanding range of fields. Ongoing research on the synthesis of novel structures and on the modification of existing zeolite-based and zeolite-type catalysts continues to advance, opening up new possibilities for innovative and environmentally friendly chemical processes. Beyond their traditional use in the petrochemical industry, zeolites are gaining importance in sustainable catalysis, and are beginning to play a key role in conversion of biomass or biomass-derived compounds into green fuels and other basic chemicals. In addition, the zeolites are often combined with photo(electro)catalytic components to further increase catalytic efficiency.

In this context, the reviews by Contributions 1 and 2 presented in this Special Issue are of particular interest. Contribution 1 demonstrates the potential of zeolite-supported TiO_2 for environmental applications, and highlights a sustainable approach for removing pollutants from the environment. The combination of the regular pore structure, high surface area, and adsorption capacity of various zeolites with the photocatalytic properties of TiO_2 gives rise to synergistic effects. These effects enable the effective degradation of a wide range of organic pollutants and hazardous substances found in wastewater. Contribution 2 discusses the mechanisms of photocatalytic and photothermal catalytic substances. It summarizes recent progress in zeolite-based catalysts for treatment of environmental pollutants through photocatalytic or photothermal catalytic processes. In addition, Contribution 3 describes the synthesis of the core–shell-structured microspheres composed of SiO_2@Fe(III)-based metal–organic frameworks. These materials enable efficient Fenton-like degradation of dyes, achieving degradation efficiencies of up to 96%. These catalysts can also be easily recovered by filtration and washing, and maintain high catalytic activity over six consecutive cycles in photo-Fenton wastewater purification processes. In Contribution 4, the conversion of glucose to levulinic acid, catalyzed by synergistic Cr/HZSM-5 in a GVL/H_2O biphasic system, was investigated. The study focuses on the effects of the solvent GVL (γ-valerolactone) using kinetic analysis. It shows that the presence of GVL reduces the activation energy for the conversion of glucose to levulinic acid, thereby improving the glucose dehydration process.

The review by Contribution 5 presents the recent progress in and strategies for designing zeolite-based catalysts for the hydroformylation of olefins. The catalytic properties are defined by the characteristics of the zeolite channels, the method used to deposit the active metal, the size of metal particles, and the microenvironment around them. Notably, exceptionally high activity in the hydroformylation of styrene was achieved with an MFI-type zeolite-based catalyst. This system also showed the highest reported *n*-regioselectivity in the hydroformylation of linear α-olefins among phosphorous-free catalytic systems—values close to 100%. For propylene hydroformylation, the best performance was obtained using a Rh-loaded MEL zeolite-based catalyst.

In Contribution 6, the hierarchical structuring of BEA-type zeolite was investigated using two different approaches. In addition, niobium pentoxide at different loadings was impregnated into the hierarchical materials. Both treatments increased the SiO_2/Al_2O_3 ratio and produced crystals with domains of approximately equal size. The hierarchical methods produced secondary mesopores and reduced the micropore volume of the treated zeolites. These structural changes had different effects on the dehydration of ethanol and propanol. Contribution 7 reports on a multifunctional Na-promoted FeCo-modified H-ZSM-5 zeolite catalyst for CO_2 hydrogenation to light aromatics via a modified Fischer–Tropsch synthesis pathway. This study systematically varies the Si/Al ratio of acidic zeolite and investigates its crucial role in influencing selectivity toward light aromatics. The results provide guidance for the rational design of bifunctional catalysts with targeted, shape-selective functions. In Contribution 8, the dimethyl ether-to-olefin (DTO) reaction was studied on ferrierite zeolites loaded with different amounts of phosphorus. The goal was to maximize the *n*-butene yield by optimizing the amount of phosphorus loading. Multiple regression analysis of the data obtained revealed that the ratio of strong-to-weak acid sites and the total pore volume correlate strongly with the *n*-butene yield. Based on these correlations, the DTO reaction mechanism was discussed using the dual-cycle reaction model. Finally, Contribution 9 describes the use of different synthesized FER zeolites as catalysts for *n*-butene skeletal isomerization. FER-type zeolites were successfully prepared either with pyrrolidine as an organic template or without pyrrolidine, using Na-form or H-form seeds. The seed-derived FER samples showed similar topological and morphological properties compared to the pyrrolidine-derived sample, as well as similar total acid densities. The study demonstrates that combining pyrrolidine-containing seeds with an optimal aluminum distribution in the FER framework enhances the performance in the zeolite in terms of the skeletal isomerization of *n*-butene.

Acknowledgments: As Guest Editor of this Special Issue, I would like to thank all authors for submitting their high-quality reviews and articles and wish all contributors continued success.

Conflicts of Interest: The author declares no conflicts of interest.

List of Contributions:

1. Armaković, S.J.; Armaković, S. Zeolite-Supported TiO_2 for Enhanced Photocatalytic Performance in Environmental Applications: A Review. *Catalysts* **2025**, *15*, 174. https://doi.org/10.3390/catal15020174.
2. Zhang, S.; Xu, L.; Xu, J.; Shen, B. A Mini-Review of Recent Progress in Zeolite-Based Catalysts for Photocatalytic or Photothermal Environmental Pollutant Treatment. *Catalysts* **2025**, *15*, 158. https://doi.org/10.3390/catal15020158.
3. Liu, K.; Zhu, Y.; Cheng, T.; Liu, G.; Tan, C. SiO_2@Fe(III)-Based Metal–Organic Framework Core–Shell Microspheres for Water-Purification-Based Photo-Fenton Processes. *Catalysts* **2025**, *15*, 23. https://doi.org/10.3390/catal15010023.
4. Wu, H.; Zhang, R.; Li, J.; Chang, J.; Liu, Z.; Chen, J.; Xiong, J.; Qiao, Y.; Yu, Z.; Lu, X. Experimental and Kinetic Studies on the Conversion of Glucose to Levulinic Acid Catalyzed by Synergistic

Cr/HZSM-5 in GVL/H_2O Biphasic System. *Catalysts* **2025**, *15*, 162. https://doi.org/10.3390/catal15020162.
5. Nerasheva, M.V.; Gorbunov, D.N. Recent Progress and Strategies on the Design of Zeolite-Based Catalysts for Hydroformylation of Olefins. *Catalysts* **2024**, *14*, 942. https://doi.org/10.3390/catal14120942.
6. da Silva Valadares, D.; de Carvalho, W.H.R.; Fonseca, A.L.F.; de França Machado, G.; Silva, M.R.; Campos, P.T.A.; Dias, J.A.; Dias, S.C.L. Different Routes for the Hierarchization of *BEA Zeolite, Followed by Impregnation with Niobium and Application in Ethanol and 1-Propanol Dehydration. *Catalysts* **2025**, *15*, 340. https://doi.org/10.3390/catal15040340.
7. Wang, S.; Sun, Y.; Lin, S.; Bian, Z.; Han, Y.; Bi, X.; Zhang, Z.; Liu, X.; Liu, D.; Wang, Y.; et al. Modified Fischer–Tropsch Pathway for CO_2 Hydrogenation to Aromatics: Impact of Si/Al Ratio of H-ZSM-5 Zeolite on Light Aromatics Selectivity. *Catalysts* **2025**, *15*, 557. https://doi.org/10.3390/catal15060557.
8. Hanaoka, T.; Aoyagi, M.; Edashige, Y. Synthesis of n-Butene via Dimethyl Ether-to-Olefin Reaction over P-Loaded Ferrierite Zeolites. *Catalysts* **2024**, *14*, 902. https://doi.org/10.3390/catal14120902.
9. Fan, J.; Zhu, X.; Yang, F.; Xu, Y.; Chen, L. Seeds Combining Pyrrolidine Control the Framework Al Distribution of FER Zeolite to Enhance Its Performance in the Skeletal Isomerization of *n*-Butene. *Catalysts* **2024**, *14*, 697. https://doi.org/10.3390/catal14100697.

Disclaimer/Publisher's Note: The statements, opinions and data contained in all publications are solely those of the individual author(s) and contributor(s) and not of MDPI and/or the editor(s). MDPI and/or the editor(s) disclaim responsibility for any injury to people or property resulting from any ideas, methods, instructions or products referred to in the content.

https://doi.org/10.3390/catal16010004

catalysts

MDPI

Article

Seeds Combining Pyrrolidine Control the Framework Al Distribution of FER Zeolite to Enhance Its Performance in the Skeletal Isomerization of *n*-Butene

Jinlong Fan [1,2], Xuedong Zhu [1,*], Fan Yang [1], Yarong Xu [2] and Lantian Chen [2]

1 State Key Laboratory of Chemical Engineering, East China University of Science and Technology, Shanghai 200237, China; fanjlws@petrochina.com.cn (J.F.)
2 Research Institute of Urumqi Petrochemical Company, Petrochina Co., Ltd., Urumqi 830019, China; xuyrws@petrochina.com.cn (Y.X.); chenltws@petrochina.com.cn (L.C.)
* Correspondence: xdzhu@ecust.edu.cn; Tel.: +86-21-64251005

Abstract: **FER** zeolites have a unique framework structure and acid distribution, and are widely studied as a catalyst for reactions such as *n*-butene skeletal isomerization and dimethyl ether carbonylation. The Brönsted acid site (BAS) located in the 10-member ring (10-MR) of **FER** zeolites serves as the active site for the isomerization reaction of skeletal *n*-butene to produce isobutene. This study prepared five types of **FER** zeolites using different methods: using pyrrolidine (PY) alone as a template; using Na-form **FER** as seeds (SN) or H-form **FER** as seeds without organic structure directing agents (OSDAs); and combining the seeds of SN or SH with PY as OSDAs. The differences in the structure and acid distribution of the five zeolites were investigated, as well as their catalytic performance for the skeletal isomerization of *n*-butene. Experiments and characterization results showed that under hydrothermal synthesis conditions, the FER-PY+SH zeolites synthesized by using both H-form zeolites seeds and pyridine exhibited the highest aluminum concentrations at T1 and T3 sites, along with the greatest BAS located in the 10-MR. This unique composition contributed to the highest selectivity of isobutene. The FER-PY+SH catalyst was continuously used for 720 h at 350 °C, 0.1 MPa, and an *n*-butene mass space velocity of 2.0 h^{-1} for three cycles of 2160 h. During this period, the conversion of *n*-butene was over 39%, while the selectivity of isobutene exceeded 95%. The FER-PY+SH catalyst exhibited excellent stability and activity.

Keywords: **FER** zeolites; seed; pyrrolidine; Al distribution; *n*-butene; skeletal isomerization

1. Introduction

Isobutene is an important basic organic chemical raw material, which can be used to produce many high value-added products, including butyl rubber, polyisobutylene, methyl tert-butyl ether (MTBE), and methyl methacrylate. One of the primary industrial methods of production involves the skeletal isomerization of *n*-butene to generate isobutene [1–3]. In the past few decades, extensive research has focused on the production of isobutene through the skeleton isomerization of *n*-butene, utilizing various catalysts such as phosphoric acid, metal halides, aluminum oxide, phosphated and halogenated aluminum oxide, and zeolite molecular sieves [4,5]. The pore structure of **FER** zeolite includes two vertically intersecting two-dimensional pore systems: an 8-member ring (8-MR) pore (0.35 × 0.48 nm) along the [010] direction, a 10-member ring (10-MR) pore (0.43 × 0.55 nm) along the [001] direction, and a ferrierite cage with a diameter of approximately 0.6–0.7 nm formed by the intersection of a 6-member ring (6-MR) and 8-MR [6,7]. **FER** zeolite is currently the highest-performing *n*-butene skeleton isomerization catalyst and was successfully applied in industrial applications in the 1990s [8]. In the following decades, extensive research was conducted on the reaction mechanism of the skeletal isomerization of *n*-butene and on the synthesis and post-treatment modification of high-performance **FER** zeolites [9–11].

Numerous studies have demonstrated that the Brönsted acid site (BAS) located on the 10-MR in **FER** zeolite acts as a selective active site for catalyzing the skeleton isomerization of *n*-butene to produce isobutene. In contrast, the Lewis acid site (LAS) and BAS located on the 8-MR can exacerbate the occurrence of side reactions such as butene polymerization and cracking [12–16]. As is well known, zeolites are crystal compounds composed of TO4 (T atoms are generally Si or Al) tetrahedra according to certain rules. Their acidity mainly arises from the introduction of protons, which balance the negative charge of aluminum atoms in the framework of zeolites. It can be seen that the catalytic performance of zeolites is closely linked to the strength and position of acid sites. The strength and position of these acid sites are directly influenced by the distribution of Al atoms within the zeolite framework [17–20]. According to relevant information from the International Zeolite Association (IZA), the framework T atom of **FER** zeolites exhibits four distinct positions, namely the T1/T3 position, which can be connected to a 10-MR, and the T2/T4 position, which is solely connected through an 8-MR (note: labels of T atoms have been unified according to the **FER** framework in the IZA). Xiong et al. [21–23] used characterization methods such as MAS NMR and X-ray emission spectroscopy to study the distribution of framework aluminum atoms in **FER** zeolites. The results showed that **FER** zeolites induced by different organic structural directing agents (OSDAs) had different distributions of framework aluminum. Moreover, they combined MAS NMR spectroscopy and density functional theory to calculate the chemical shifts of aluminum located at positions T1/T2/T3/T4 in **FER** zeolites, which were 61, 51, 53, and 58 ppm, respectively.

In the synthesis process of zeolites, OSDAs play a crucial role in balancing the negative charge introduced by the aluminum atoms in the framework. Consequently, these OSDAs significantly influence the distribution of aluminum in zeolites. Pinar et al. [14,24,25] synthesized **FER** zeolites using a fluorinated system without the use of inorganic cations, using a combination of 1-benzoyl-1-methylpyrrolidinium, trimethylamine (TMA), pyridine, pyrrolidine (PY), and several organic template agents. The investigations indicate that **FER** zeolites induced by different organic structure directing agents have different skeletal aluminum distributions and exhibit different performances in *n*-butene skeletal isomerization reactions. Due to the different sizes of template molecules and their interactions with skeleton atoms, **FER** zeolites with different framework aluminum distributions can be obtained. When larger size OSDAs are used, a longer crystallization time is required to increase the distribution probability of aluminum atoms in the 10-MR, enhance the BAS of the 10-MR, and improve the control of the activity and selectivity of *n*-butene skeleton isomerization. Conversely, this tends to lead to side reactions such as butene polymerization and cracking, and accelerate catalyst deactivation. In addition, the studies of Leshkov [26] and Chu [27] also indicate that when different sizes of cyclic or chain amines are used as OSDAs, **FER** zeolites with different 10-MR and 8-MR aluminum distributions can also be obtained. Numerous studies have shown that the distribution of aluminum in the framework of zeolites is not only influenced by OSDAs, but also by the synthesis conditions of zeolites [28] and the selection of silicon and aluminum sources [29,30], solvents [31], inorganic cations [32–34], and post-treatment [35–37]. During the seed-induced synthesis of zeolites, the seed can play a role in reducing or avoiding the use of OSDAs, shortening crystallization time, avoiding impurities, and regulating the crystal morphology of target zeolites [38–40]. Xiao et al. [41] believe that due to the multiple roles of the seed in inducing the synthesis of zeolites, the seed can be regarded as the third type of structure directing agents, in addition to inorganic cations and OSDAs. Zhang et al. [42] synthesized **FER** zeolite with good crystallinity, large surface area, uniform micropores, and abundant acidic sites using RUB-37 as the crystal seed without OSDAs. Studies by Ham et al. [43–45] showed a change in the distribution of framework aluminum atoms in seed-induced synthesized **FER** zeolites as compared to **FER** zeolites as the seed. The higher BAS content of the 8-MR than the 10-MR resulted in better dimethyl ether carbonylation activity and stability of the seed-induced **FER** zeolites than the seed. In addition, Nishitoba [30] found that when using TMA as an OSDA to synthesize **CHA** zeolites, the addition of **FAU** zeolites as crystal seeds and aluminum sources could regulate the distribution of aluminum atoms in the framework of **CHA** zeolites. These studies indicate that the distribution of aluminum in the zeolite framework can be modulated using the seed induction methods.

In this work, the synthesis of **FER** zeolites under different conditions was investigated as follows: using only pyrrolidine in OSDA synthesis; using only Na-form seed (SN)-induced synthesis; using only H-form seed (SH)-induced synthesis; and combining SN or SH + PY. The distribution of aluminum atoms in the framework of **FER** zeolites synthesized by different methods and their catalytic performance for *n*-butene skeleton isomerization reaction were investigated. The obtained catalyst of **FER** zeolites by inducing H-form seeds + PY as OSDAs has the highest distribution of 10-MR framework aluminum, which enhanced the acidic strength in the 10-MR and significantly improved the catalytic activity and selectivity of **FER** zeolites for the *n*-butene skeleton isomerization reaction. A new method for regulating the distribution of the framework of aluminum and acid in the **FER** zeolites was proposed.

2. Results and Discussion

2.1. Structural and Textural Properties

As shown in Figure 1, XRD patterns of FER-PY, FER-SN, FER-SH, FER-PY+SN, and FER-PY+SH exhibit typical diffraction peaks at 9.3°, 22.3°, 23.5°, 24.3°, 25.2°, 25.7°, and 28.5° out of the (200), (321), (330), (112), (040), (202), and (312) crystal planes in the **FER** topology (PDF #44-0104), respectively, indicating the successful formation of the **FER** zeolite without detectable impurities [22,23]. The relative crystallinity (RC) was calculated by integrating the areas of diffraction peaks within 9.3-28.5°, and the sum of the areas of FER-PY was set as 100% for reference. As listed in Table 1, the calculated RC values are comparable, measuring 101%, 99%, 100%, and 101% for FER-SN, FER-SH, FER-PY+SN, and FER-PY+SH, respectively. The above results indicate that several templates, such as PY, SN, SH, PY+SN, and PY+SH, can induce the synthesis of well-crystallized **FER** zeolites. The composition information is listed in Table 1, where the SiO_2 to Al_2O_3 ratios (SARs) of the five samples are within 24.6–25.6. Additionally, the Na_2O contents are all below 43 mg/g, suggesting that all FER-X samples have similar compositions and acidic densities. The calculated texture parameters are shown in Table 1; in line with the XRD observations, the specific surfaces and pore volumes remain relatively uniform, for instance, the values of S_{BET} average around 430 m^2/g, while V_{total} and V_{micro} are all about 0.27 cm^3/g and 0.14 cm^3/g, respectively.

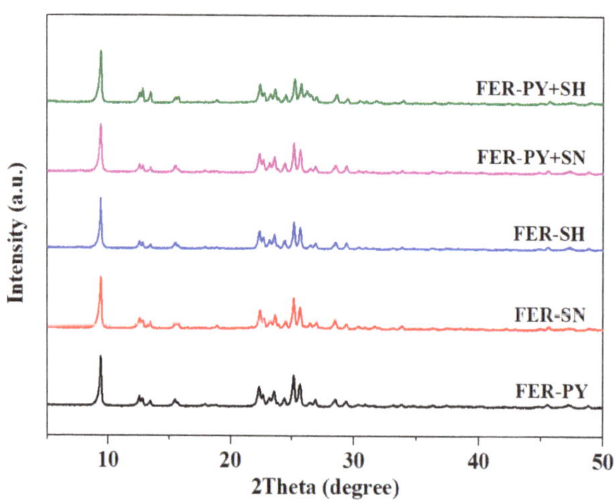

Figure 1. XRD patterns of various FER-X zeolites.

Table 1. Textural properties of various FER-X zeolites.

Sample	Na_2O (mg/g)	SARs	RC (%)	Surface Area (m^2/g)		Pore Volume (cm^3/g)	
				Total	Micro	Total	Micro
FER-PY	40	25.6	100	431	388	0.27	0.14
FER-SN	39	25.3	101	428	386	0.27	0.14
FER-SH	36	24.6	99	422	380	0.26	0.13
FER-PY+SN	42	25.3	100	430	378	0.28	0.14
FER-PY+SH	41	24.8	101	432	390	0.28	0.15

The SEM images (Figure 2) present the morphologies of the obtained FER-X zeolites. All samples are composed of hexagonal flake-like crystals with smooth surfaces; the average size of each flake is ~1 μm in length, ~500 nm in width, and 20~50 nm in thickness. Combined XRD and SEM analyses indicate that their morphology and structure are identical.

Figure 2. SEM images of various FER-X zeolites. (**a**) FER-PY, (**b**) FER-SN, (**c**) FER-SH, (**d**) FER-PY+SN, (**e**) FER-PY+SH.

2.2. Acidity Characterization

The acidic properties were characterized using NH_3-TPD and Py-IR. As illustrated in Figure 3, two NH_3 desorption peaks are observed at 195 °C and 450 °C (Figure 3a), corresponding to the desorption of NH_3 molecules that interact with the weak acids and strong acids, respectively. The positions of these peaks provide insight into acid strength, indicating that the weak and strong acid sites in FER-X samples exhibit similar acidic densities. Figure 3b–f shows the Py-IR spectrum was extracted at 150 °C, 250 °C, and 350 °C. The IR adsorption peak at 1450 cm^{-1} corresponds to the C-N stretching of pyridine adsorbed on LAS, while the peak at 1490 cm^{-1} is attributed to the C-H vibration of pyridine on both BAS and LAS. The peak at 1540 cm^{-1} is due to the N-H stretching of pyridinium cations generated by protonation of pyridine on BAS. As shown in Figure 3b–f, with the increase in desorption temperature, both BAS and LAS in the sample decrease. When the desorption temperature reaches 350 °C, LAS decreases to a very low level, indicating a scarcity of strong LAS in the FER-X samples. In contrast, the decrease in BAS is considerably smaller, indicating a stronger acidity of BAS present in the FER-X samples.

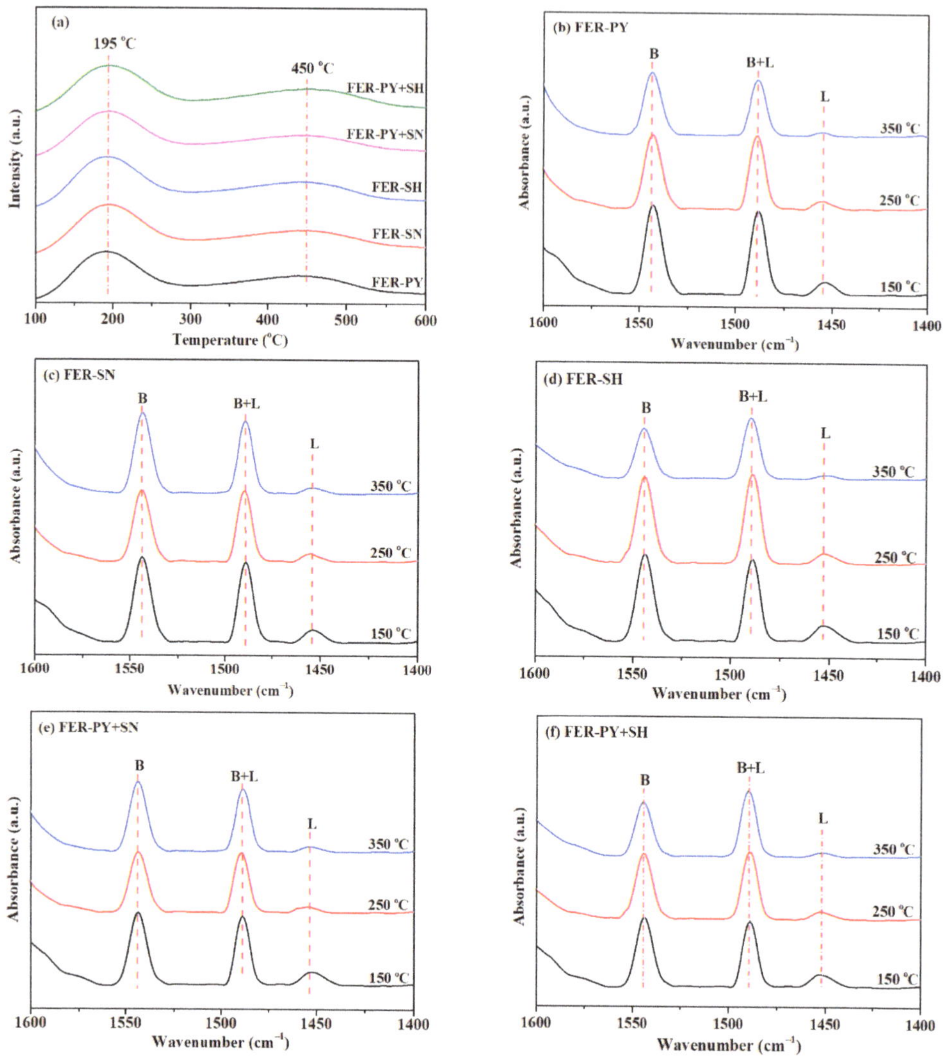

Figure 3. NH_3-TPD and Py-IR profiles of various FER-X zeolites. (**a**) NH_3-TPD profiles of FER-X zeolites, (**b**) Py-IR profiles of FER-PY, (**c**) Py-IR profiles of FER-SN, (**d**) Py-IR profiles of FER-SH, (**e**) Py-IR profiles of FER-PY+SN, (**f**) Py-IR profiles of FER-PY+SH.

The quantitative analysis of NH_3-TPD and Py-IR experiments is presented in Table 2. Notably, NH_3-TPD provides an overall assessment of the acidity of the samples, as the kinetic diameter of NH_3 (3.7 Å) is smaller than the 8-MR and 10-MR apertures in the **FER** structure. In contrast, Py-IR could only detect the acid sites located in 10-MR due to its larger kinetic diameter (5.7 Å for pyridine vs. 4.8 Å for 8-MR). Consequently, NH_3-TPD can be used to characterize the total acidity of FER-X zeolites, and Py-IR can be used to characterize the acidity of the 10-MR of FERX-zeolites (see the caption below Table 2 for more details). From the NH_3-TPD profiles, all samples have a similar acid distribution and acid content, with acid densities around approximately 1.3 mmol/g for the five samples, consistent with their structural and compositional properties. However, more BAS can be identified over Py-IR for the FER-X samples (0.29 mmol/g for FER-PY and 0.28–0.34 mmol/g for others), especially for FER-PY+SH, whose value is 0.34 mmol/g. In addition, the BAS of desorption temperature as measured from Py-IR ($B_{10\text{-}350}$) is also shown in Table 2; this value is 0.20 mmol/g for FER-PY and increases monotonically to 0.32 mmol/g for FER-PY+SH. The differing acid distributions may arise from the distinct nucleation and crystal growth processes of zeolites synthesized using seed crystals, or a combination of seed crystals and organic templates, compared to those formed solely with organic templates. The different $B_{10\text{-}350}$ provides direct evidence that utilizing seeds or employing a combination of seeds and organic templates will alter the acid distribution in **FER** zeolite.

Table 2. Amounts of acid in the various FER-X samples.

Sample	Acidity by NH_3-TPD [a] (mmol/g)			Acidity by Py-IR [b] (mmol/g)			
	Weak	Strong	Total	Brønsted	Lewis	Total	$B_{10\text{-}350}$ [c]
FER-PY	0.79	0.56	1.35	0.29	0.03	0.32	0.20
FER-SN	0.78	0.55	1.33	0.28	0.03	0.31	0.24
FER-SH	0.75	0.53	1.28	0.28	0.03	0.31	0.25
FER-PY+SN	0.76	0.52	1.28	0.30	0.03	0.33	0.28
FER-PY+SH	0.81	0.56	1.37	0.34	0.02	0.35	0.32

[a] Acid sites on FER-X and their amounts were measured by NH_3-TPD analysis, and the quantities of weak and strong acid sites were calculated by using an integrated area at the maximum desorption temperatures of NH_3 at 195 °C and 450 °C, respectively. [b] Amounts of BAS and LAS were measured by Py-IR spectroscopy at a desorption temperature of 150 °C and 350 °C, and the number of acid sites was calculated using the well-known formula of $BAS = 1.88A_BR^2/W$, $LAS = 1.42A_LR^2/W$, where A_B and A_L are the integrated absorbance peak areas of BAS and LAS located at 1550 and 1450 cm^{-1}, respectively, R is the radius of a catalyst pellet of FER-X, cm, and W is the weight of catalyst, g [46]. [c] The number of BAS located in 10-MR at 350 °C was calculated by conducting Py-IR measurements at a desorption temperature of 350 °C.

2.3. ^{29}Si and ^{27}Al MAS NMR

The ^{29}Si MAS NMR spectra show closely similar features (Figure 4), exhibiting four sub-peaks from the curve at approximately −100, −107, −112, and −116 ppm, respectively. The peaks at −100 and −107 ppm correspond to the Si atoms bonded to two Al atoms (Si(2Al)) and one Al atom (Si(1Al)), respectively, while the peaks at −112 and −116 ppm can be attributed to the Si atoms that are not adjacent any Al atoms (Si(0Al)), and silicon connecting to four Si tetrahedrons in the **FER** framework [21,47]. The weaker signal at −100 ppm of all samples indicates a lower prevalence of Al–O–Si–O–Al species on FER-X zeolites.

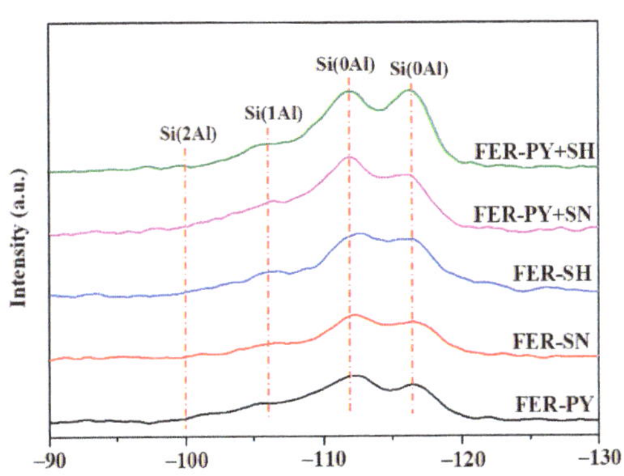

Figure 4. ^{29}Si MAS NMR spectra of various FER-X zeolites.

The ^{27}Al MAS NMR spectra of various FER-X zeolites are shown in Figure 5. As shown in Figure 5a, all of the FER-X samples exhibit a prominent resonance peak at 55 ppm corresponding to tetrahedrally coordinated Al species, while no signal peak appears at 0 ppm. This indicates that the aluminum atoms in zeolites samples are located on the framework of FER-X zeolites, and there is no non-framework aluminum present [42,43,48]. Previous studies have shown that aluminum atoms exhibit different NMR chemical shifts depending on their positions of T1, T2, T3, and T4 within the **FER** framework [21,23,47], and the ^{27}Al MAS NMR spectrum at 55 ppm is deconvolved into four peaks located at 61, 51, 53, and 58 ppm, respectively, as illustrated in Figure 5b–f. The Al population at T1–T4 sites in the FER-X samples was quantified from the corresponding ^{27}Al NMR spectra (Figure 5b–f) and Table 3. The FER-PY sample exhibited a minimal fraction of Al atoms on the T1 + T3 sites (30%). In comparison, the distribution proportion of aluminum on the T1 + T3 positions in FER-SN, FER-SH, and FER-PY+SN gradually increased to 41%, 44%, and 50% respectively, while the distribution proportion of aluminum at T1 + T3 positions on FER-PY+SN reached the highest value of 60%. Notably, each of the five samples contained a small amount of Al on the T1 site (0~12%).

Table 3. Aluminum distribution obtained from the curve fitting of ^{27}Al MAS NMR spectra of various FER-X zeolites.

Sample	Aluminum Distribution (%)				
	T1	T2	T3	T4	T1 + T3
FER-PY	0	30	30	39	30
FER-SN	8	19	33	40	41
FER-SH	10	25	34	31	44
FER-PY+SN	12	21	38	29	50
FER-PY+SH	6	15	54	25	60

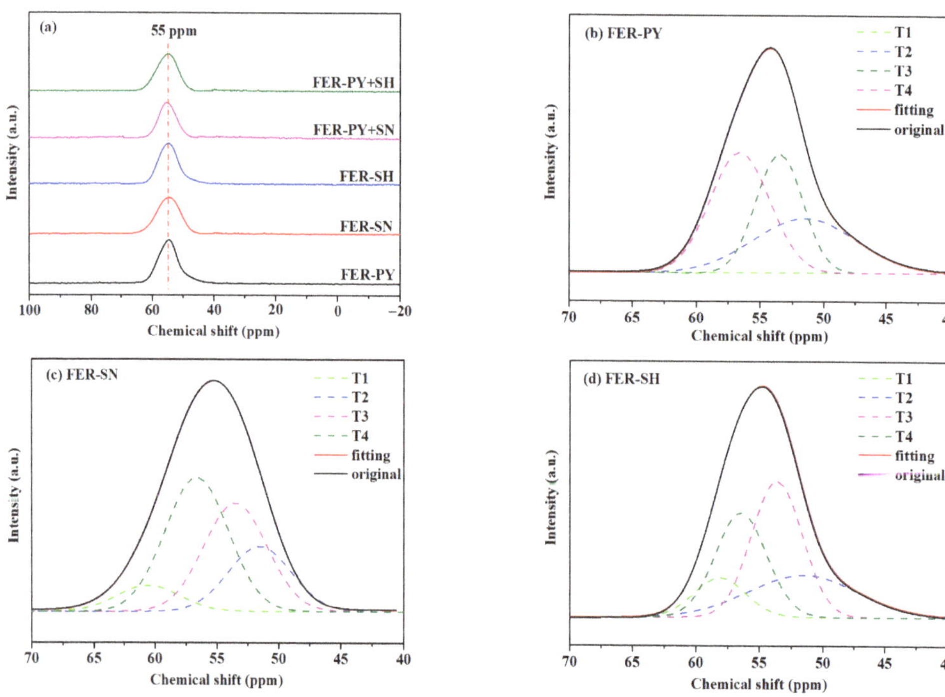

Figure 5. *Cont.*

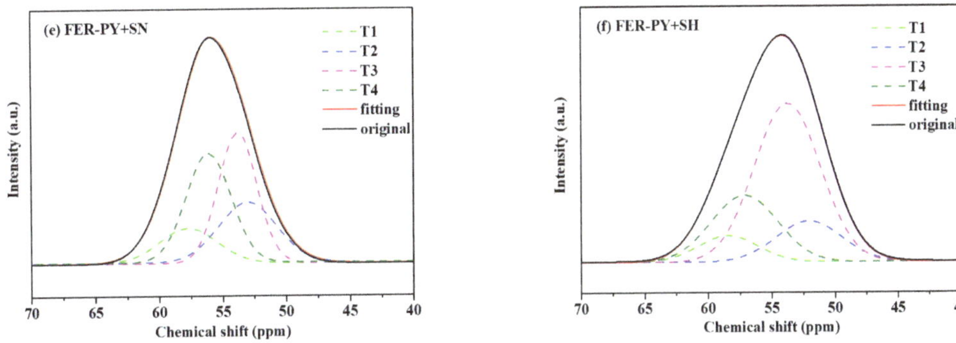

Figure 5. ^{27}Al MAS NMR spectra of various FER-X zeolites and Al distribution obtained from the curve deconvolved of FER-X zeolites. (**a**) ^{27}Al MAS NMR spectra of FER-X zeolites, (**b**) deconvolved of FER-PY, (**c**) deconvolved of FER-SN, (**d**) deconvolved of FER-SH, (**e**) deconvolved of FER-PY+SN, (**f**) deconvolved of FER-PY+SH.

In summary, using PY as an OSDA to induce the synthesis of FER-PY zeolites as a reference, all samples possessed similar compositional and textural properties, are total acid densities were closely aligned. Nevertheless, the combined use of pyrrolidine and H-form **FER** seeds renders a distinctive acid distribution, with a greater number of strong acid sites located in the 10-MR. To elucidate the relationship between the aluminum distribution and the acid distributions of FER-X zeolites, Figure 6 correlates the aluminum distribution data at T1 + T3 sites of different samples with $B_{10\text{-}350}$. A linear relationship can be observed between T1 + T3 and $B_{10\text{-}350}$, implying the acid distributions positively correlate with the quality of the aluminum distribution on T1 + T3. Several factors may contribute to this result. Firstly, the use of seed crystals to facilitate **FER** zeolite synthesis accelerates the nucleation and crystallization processes, potentially leading to a higher distribution of aluminum within the 10-MR. Secondly, the lower sodium content in H-form **FER** seeds allows for larger species to balance the framework charge, thereby increasing the likelihood of aluminum being positioned at T1 + T3 sites in the 10-MR. Moreover, this effect is more pronounced in the presence of PY as OSDAs alongside H-form **FER** seeds.

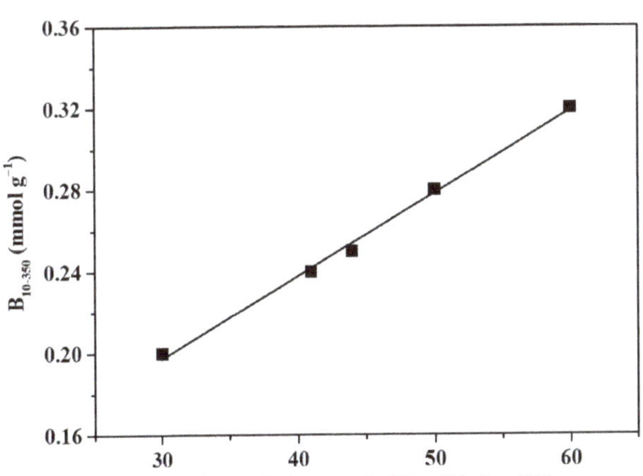

Figure 6. The influence of the Al populations on the T1 + T3 sites on the $B_{10\text{-}350}$ of various FER-X zeolites.

2.4. Catalytic Performance

The catalytic performances of all FER-X samples in n-butene skeletal isomerization are shown in Figure 7; the C4 hydrocarbons produced by fluidized catalytic cracking in refineries react with methanol to produce MTBE after removing isobutene (post-MTBE C4) and are used as feed.

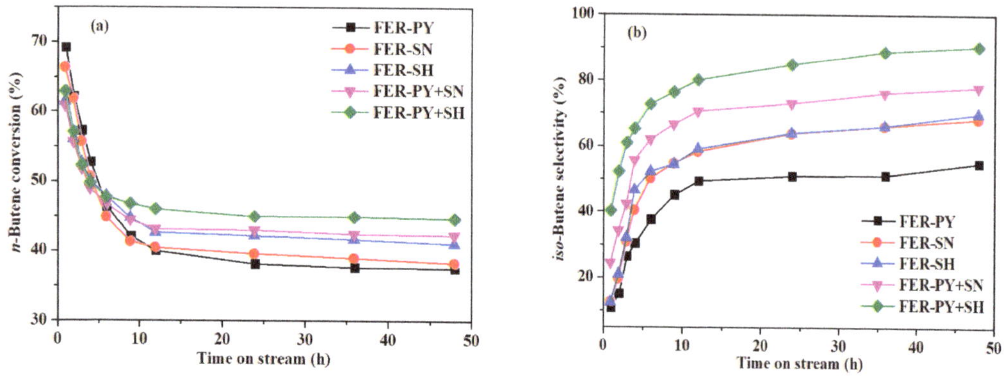

Figure 7. Catalytic performance of various FER-X zeolites for skeletal isomerization of n-butene with post-MTBE C4 as feed. (**a**) n-butene conversion of FER-X zeolites within time on stream, (**b**) *iso*-butene selectivity of FER-X zeolites within time on stream. Reaction conditions: T = 350 °C; p = 0.1 MPa; $WHSV_{n\text{-butene}} = 2.0\ h^{-1}$.

Figure 7a illustrates the relationship between the n-butene conversion and time on stream for FER-X catalyzed n-butene skeletal isomerization. As can be seen from the figure, FER-X catalysts exhibit the same trend, that is, with the extension of time on stream, the conversion initially decreases rapidly and then gradually stabilizes. Both FER-PY and FER-SN showed the best initial activities, achieving n-butene conversions exceeding 65%, while the initial conversions of the other samples fall within the range of 60% to 63%. This may be mainly due to the presence of more framework aluminum at the T2 + T4 sites in FER-PY and FER-SN, which resulted in more side reactions such as polymerization and cracking, leading to a higher conversion of n-butene. As the reaction proceeded, all conversions dropped rapidly during the first 12 h and then stabilized, by approximately 45% after running for 48 h.

In contrast, the selectivity of isobutene showed a completely reversed trend from that in Figure 7b. In the initial stage, the selectivity over FER-PY, FER-SN, and FER-SH was below 20%, while FER-PY+SN and FER-PY+SH recorded 25% and 40%, respectively. With the extension of time on stream, the selectivity over FER-PY gradually increased to ~55% after 48 h. FER-SN and FER-SH both presented a sharp increase in the first 12 h and then kept steady, with the selectivity reaching ~63% at 48 h. The selectivity over FER-PY+SN and FER-PY+SH also presented a sharp increase in the first 12 h, leveling off at about 78% and 90% after 48 h, respectively.

The experimental results in Figure 7 are mainly due to the different acid distributions of the FER-X catalysts. It is generally believed that the skeleton isomerization of n-butene to produce isobutene is carried out through the joint action of the monomolecular mechanism and the bimolecular mechanism. The BAS located in the 10-MR of **FER** zeolites is the active site for the selective produce of isobutene from n-butene [48–50]. In the monomolecular mechanism, n-butene is first adsorbed and protonated into a carbenium cation on the acid site, after which it undergoes skeleton isomerization and proton feedback to afford isobutene. The bimolecular mechanism mainly involves polymerization and isomerization to generate various long-chain carbocations, which subsequently crack to produce isobutene and by-products such as propylene. In the initial stage of this reaction, the strong acid sites of **FER** zeolite were well retained, resulting in the predominance of the bimolecular mechanism, which was followed by cracking and the generation of by-products. After the initial stage, carbon deposition began to cover the non-selective acid sites, leading to a decrease in catalyst activity while suppressing undesired side reactions. Previous investigations have proved that the BAS in 10-MR of **FER** zeolites is accountable for the skeleton isomerization of *n*-butene, while BAS in the 8-MR and LAS can exacerbate side reactions such as polymerization and cracking, despite their higher activity [51,52]. Therefore, FER-PY+SH demonstrates improved selectivity and isobutene yields owing to its higher $B_{10\text{-}350}$ concentration. As shown in Figure 8, the yield of isobutene increases with the increase in $B_{10\text{-}350}$ and Al distribution on the T1 + T3.

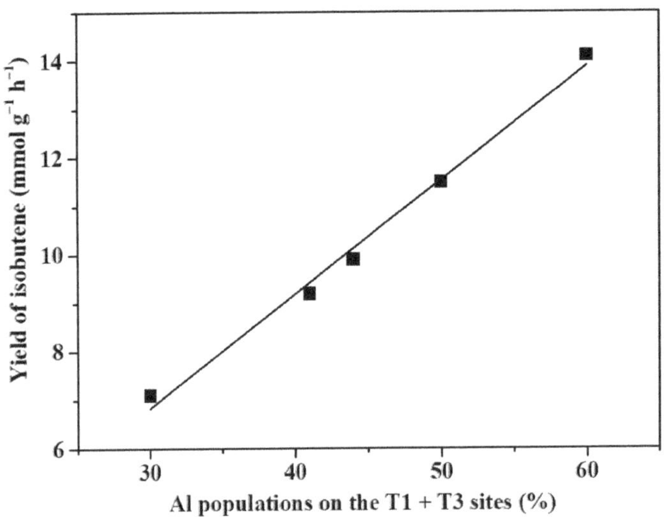

Figure 8. Relationship between yield of isobutene and Al populations on the T1 + T3 sites.

2.5. *Catalyst Stability*

Figure 9 displays the long-term stability tests using FER-PY+SH as candidate catalysts. The catalytic results were obtained with three cycles of regeneration (2160 h). The one-way life of the catalyst reaches 720 h, the conversion of *n*-butene (x_n) is greater than 39%, the selectivity of isobutene (s_i) is greater than 95%, and the yield of isobutene is greater than 37.5%. After 720 h of continuous reaction, the conversion of *n*-butene was below 39%, and catalyst regeneration was carried out at this time. After stopping the reaction, all deactivated catalysts were placed in a muffle furnace and the temperature was raised at a rate of 1 °C/min from 30 °C to 500 °C, and calcined at 500 °C for 12 h to complete the regeneration of the deactivated catalysts. Interestingly, after 720 h of one-way operation, the activity of FER-PY+SH in skeletal isomerization of *n*-butene was fully recovered after regeneration. This finding substantiates that FER-PY+SH can be effectively reused, thereby extending its lifespan beyond most reported catalysts in the skeletal isomerization of *n*-butene. The catalyst of FER-PY+SH exhibits superior catalytic activity due to its unique distribution of framework aluminum and enhanced acidic strength in the 10-MR.

In conclusion, the preparation of FER-PY+SH zeolites using a combination of H-form seeds and pyrrolidine enriches the number of BASs in the 10-MR, while improving the selectivity of isobutene and effectively suppressing the side reactions such as polymerization and cracking. The FER-PY+SH samples thus demonstrate excellent isobutene selectivity and yields in the skeletal isomerization of *n*-butene.

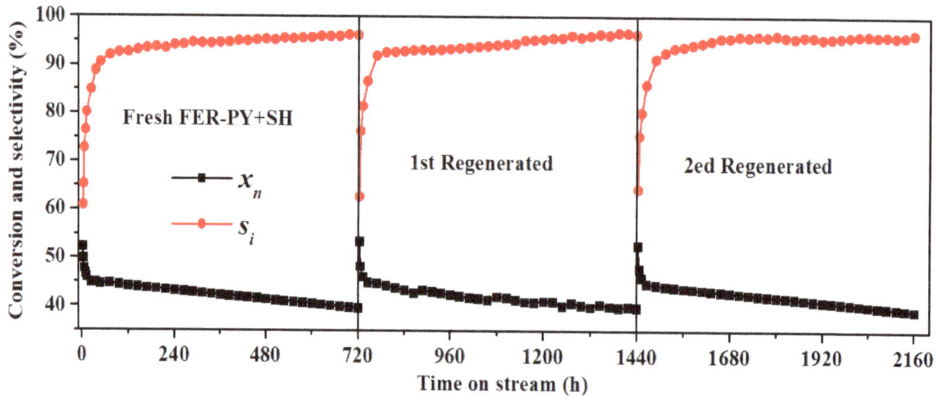

Figure 9. Stability of FER-PY+SH for skeletal isomerization of *n*-butene. Reaction conditions: T = 350 °C; p = 0.1 MPa; $WHSV_{n\text{-butene}}$ = 2.0 h^{-1}.

3. Experimental

3.1. Materials

The reagents used for the synthesis of **FER** zeolite were silica sol (SS3015, 30% SiO_2, 0.02%NaO_2, Shandong Baite New Material Co., Ltd. Linyi, Shandong, China), sodium metaaluminate ($NaAlO_2$, 99.0%, Tianjin Kermel Chemical Reagent Co., Ltd. Tianjin, China), sodium hydroxide (NaOH, 96.0%, Tianjin Kermel Chemical Reagent Co., Ltd. Tianjin, China), pyrrolidine (PY) (C_4H_9N, 99%, Tianjin Kermel Chemical Reagent Co., Ltd. Tianjin, China), ammonium chloride (NH_4Cl, 99%, Tianjin Kermel Chemical Reagent Co., Ltd. Tianjin, China), and deionized water.

3.2. Preparation of FER-X Zeolites

3.2.1. Preparation of Seeds

Typically, deionized water and NaOH aqueous solution were slowly added to silica sol (SiO_2, 30 wt%), and stirred vigorously for 1 h. Then, $NaAlO_2$ aqueous solution and PY were added, followed by stirring for 2 h to form a homogeneous suspension with a molar composition of 0.1 Na_2O:0.033 Al_2O_3:1 SiO_2:25 H_2O:0.22 PY. The suspension was transferred into a Teflon-lined stainless-steel autoclave and crystallized at 160 °C under a rotation of 200 rpm for 72 h. After the hydrothermal synthesis, the product was washed with distilled water until it reached a neutral pH. It was then dried at 120 °C overnight and calcinated at 550 °C in air for 6 h to obtain Na-form **FER**. The Na-form **FER** zeolites were subsequently converted into NH_4-form via ion exchange with 1 mol/L NH_4Cl solution at 80 °C for 2 h; this process was repeated twice. Then, the resulting white slurry was filtered and dried at 120 °C for 6 h. Finally, H-form **FER** zeolite catalysts were obtained by calcination of NH_4-form **FER** zeolites in air at 500 °C for 4 h. These Na-form **FER** and H-form **FER** were used as the seeds, and the seeds' abbreviations are SN and SH respectively.

3.2.2. Preparation of Seed-Derived FER-X Zeolites

The seed-derived **FER** zeolites were prepared as fellows: organic structure directing agents were not used, while the remaining synthetic processes were kept identical to the procedures described above. Finally, the SN and SH were added to form a homogeneous suspension with a molar composition of 0.1 Na_2O:0.033 Al_2O_3:1 SiO_2:25 H_2O: 15% wt seeds. The prepared mother gel was transferred into a Teflon-lined stainless-steel autoclave and crystallized at 160 °C under a rotation of 200 rpm for 72 h, followed by centrifuging, rinsing, and drying overnight at 120 °C to afford Na-form **FER**.

The method of synthesizing **FER** zeolites using both PY and SN or SH as seeds simultaneously is similar to the synthesis methods of the two aforementioned **FER** zeolites. Finally, the homogeneous suspension with a molar composition of 0.1 Na_2O:0.033 Al_2O_3:1 SiO_2:25 H_2O: 0.12 PY:15% wt SN or SH was transferred into a Teflon-lined stainless-steel reactor, and then hydrothermal synthesis was conducted for 72 h at 160 °C with 200 rpm. The white gel was obtained after several washing steps and dried at 120 °C for 6 h, and further calcined at 550 °C in air for 6 h to afford Na-form **FER**.

3.2.3. Preparation of H-form **FER** Zeolites

All of the Na-form **FER** zeolites were transformed into NH_4-form via ion exchange with 1 mol/L NH_4Cl solution at 80 °C for 2 h. This process was repeated twice. Then, the resulting white slurry was filtered and dried at 120 °C for 6 h. Finally, H-form **FER** zeolite catalysts were obtained by calcining of NH_4-form **FER** zeolites in air at 500 °C for 4 h. The H-form **FER** zeolite prepared using PY was labeled as FER-PY, the H-form **FER** zeolite prepared by Na-form **FER** zeolites as seeds induction was denoted as FER-SN, the H-form **FER** zeolite prepared by H-form **FER** zeolites as seeds induction was denoted as FER-SH, and the H-form **FER** zeolite prepared with PY and Na-form **FER** zeolites or H-form **FER** zeolites as seeds simultaneously was denoted as FER-PY+SN and FER-PY+SH, respectively. Finally, the H-form **FER** zeolite was compressed and formed to obtain a catalyst for the skeletal isomerization of *n*-butene.

3.3. Characterization of Catalysts

X-ray diffraction (XRD) was used to measure the crystallization and phase of the **FER** catalysts performed with a Rigaku D/max 2550 VB/PC diffractometer apparatus with Cu Ka radiation (λ = 0.15406 nm) at a voltage and current of 40 kV and 40 mA, respectively. The 2θ scanning range was from 5° to 40° with a scan rate of 4°/min and the step size of 0.02°. The R.C.s of the catalysts were obtained by calculating the sum of the peak intensity at $2\theta = 9.3 \pm 0.1°$, $22.3 \pm 0.1°$, $23.5 \pm 0.1°$, $24.3 \pm 0.1°$, $25.2 \pm 0.1°$, $25.7 \pm 0.1°$, and $28.5 \pm 0.1°$, and the R.C.s of FER-PY was set as 100%.

The scanning electron microscopy (SEM) images of the samples were analyzed using a Nava Nano SEM 450 microscope to confirm the crystal morphology and size. The sodium content and SiO_2/Al_2O_3 molar ratio of the samples were determined by X-ray fluorescence (XRF) analysis using a ZSX Primus II instrument (Rigaku, Tokyo, Japan) operated at 60 kV and 150 mA. The textural characterization of the **FER** catalysts was performed by N_2 adsorption–desorption at −196 °C with a US Micromeritics ASAP 2020 V3.00 H nitrogen adsorption analyzer (Atlanta, GA, USA). The **FER** catalysts were outgassed at 350 °C for 5 h under the vacuum of 10^{-3} Pa to remove moisture and volatile impurities before the adsorption measurement. The specific surface areas were obtained by the Brunauer–Emmett–Teller (BET) method and the pore volume was calculated by N_2 adsorption at a relative pressure of 0.99.

The temperature-programmed desorption of ammonium (NH_3-TPD) with a Micromeritics chemisorb 2720 (Atlanta, GA, USA) instrument equipped with a thermal conductivity detector was used to study the acid properties of the catalysts. Typically, 150 mg of the sample was first preheated under a He stream of 30 mL·min^{-1} at 550 °C for 1 h, then purged in a 5% NH_3-He flow of 30 mL/min at 100 °C for 40 min. Subsequently, the sample was treated under a He flow of 30 mL/min at 100 °C for 60 min to remove the physically adsorbed NH_3. Finally, the samples were heated from 100 °C to 550 °C using a temperature ramp of 10 °C/min, and the quantity of NH_3 desorbed was measured by a TCD detector.

Pyridine adsorption Fourier transform infrared (Py-IR) was used to determine B acid and L acid concentrations using a Bruker Tensor 27 infrared spectrometer. Approximately, the sample powder (30 mg) was formed into a self-supporting wafer (dimeter 20mm) and then placed inside the reaction tank. The sample was degassed under vacuum at 500 °C for 2 h under high vacuum conditions of 10^{-3} Pa, and then the sample was saturated with the probe molecule pyridine at 40 °C for 30 min, and subsequently, the physisorbed molecules were removed at 150 °C for 30 min. IR spectra were recorded at 350 °C. The adsorption peaks located at 1540 cm^{-1} and 1450 cm^{-1} were assigned to BAS and LAS, respectively. The concentration of acid sites was calculated from the IR results using the extinction coefficient and integrated intensity of the desorption peak and the sample weight reported by Emeis [46].

Magic-angle-spinning nuclear-magnetic-resonance (MAS NMR) spectra of ^{27}Al and ^{29}Si were collected on a Bruker Avance III 600 MHz Wide Bore spectrometer (Bruker, Rheinstetten, Germany). The single-pulse sequence was adopted with a 10° pulse and a delay time of 0.3 s. The chemical shifts for ^{27}Al and ^{29}Si were calibrated by referring to $AlCl_3$ and tetramethylsilane, respectively. The ^{27}Al MAS NMR spectra were deconvoluted using the mixed Gaussian–Lorentzian equation.

3.4. Catalyst Evaluation

The skeletal isomerization of *n*-butene was conducted under atmospheric pressure in a fixed-bed stainless-steel reactor unit, with the tubular reactor having 12 mm inner diameter and 800 mm length. Firstly, 3 g of the FER-X catalyst (20–40 mesh) was placed in the middle of reactor, the catalyst was pretreated under a N_2 flow at a heating rate of 10 °C/min, and the system was purged for 1 h. The C4 hydrocarbons produced by fluidized catalytic cracking in refineries react with methanol to produce MTBE after removing isobutene (post-MTBE C4) and were used as reaction materials. The feed stream of post-MTBE C4 was fed into the reactor by a mass flow controller to start the skeletal isomerization; when heated to 350 °C, the weight hourly space velocity (WHSV) of *n*-butene was 2.0 h^{-1}. The properties of the post-MTBE C4 are listed in Table 4.

Table 4. Composition of post-MTBE C4 from refinery.

Composition	wt/%
propane	0.12
propylene	0.15
isobutane	52.86
butane	10.94
trans-2-butene	13.76
1-butene	12.00
isobutene	0.26
cis-2-butene	9.91

The products were analyzed using an Agilent 7890B gas chromatograph (Santa Clara, CA, USA) equipped with a flame ionization detector and an Agilent-GS-Alumina capillary column (50 m × 0.532 mm). Since the equilibrium of 1-butene, *trans*-2-butene, and *cis*-2-butene was established rapidly through double-bond isomerization at the temperature of the reaction, and all the straight-chain butene isomers were transformed to isobutylene, all of the straight-chain butene was considered as the *n*-butene reactant.

The conversion of *n*-butene (x_n) was defined as:

$$x_n = \frac{m_0 - m_1}{m_0} \times 100\% \tag{1}$$

The selectivity of isobutene (s_i) was defined as:

$$s_i = \frac{m_2 - m_3}{m_0 - m_1} \times 100\% \tag{2}$$

where m_0, m_1, m_2, and m_3 are the weights of *n*-butene in the feed, the weights of *n*-butene in the products, the weights of isobutene in the reaction products, and the weights of isobutene in the feed, respectively.

The production rate of isobutene was defined as the average moles of isobutene produced at the reaction duration of 24–48 h [mmol g^{-1} h^{-1}].

4. Conclusions

FER-type zeolites have been successfully prepared in pyrrolidine as an organic template or without pyrrolidine systems utilizing the Na-form or H-form as the seed. The seed-derived **FER** samples own similar topological and morphological properties compared to the pyrrolidine-derived sample, as well as similar total acid densities. The choice of the H-form seed and pyrrolidine induces more BAS in the 10-MR of the **FER** structure. As the BAS in 8-MR and LAS is responsible for the bimolecular mechanism that tends to cause more carbon depositions and side-products in *n*-butene skeleton isomerization, the FER-PY+SH samples exhibit better isobutene selectivity and higher stability than the counterparts synthesized by Na-form as the normal seed. By optimizing the synthetic parameters, FER-PY+SH could be continuously operated for 720 h at 350 °C and 0.1 MPa for three cycles, and a weight hourly space velocity of *n*-butene of 2.0 h^{-1}, during which the *n*-butene conversion keeps higher than 39%, the selectivity of isobutene is higher 95%, and the isobutene yield is higher than 37.5%.

Author Contributions: J.F.: investigation, data curation, writing—original draft. X.Z.: methodology, conceptualization, supervision, writing—review and editing. F.Y.: methodology, writing—original draft. Y.X.: methodology formal analysis, project administration. L.C.: validation, formal analysis. All authors have read and agreed to the published version of the manuscript.

Funding: This research received the funding of the project by CNPC (2018B-1907).

Data Availability Statement: Data are contained within the article.

Conflicts of Interest: Author Jinlong Fan, Yarong Xu and Lantian Chen were employed by the company of Research Institute of Urumqi Petrochemical Company, Petrochina Company Limited. The remaining authors declare that the research was conducted in the absence of any commercial or financial relationships that could be construed as a potential conflict of interest.

Abbreviations

Content	Abbreviations
Brönsted acid sites	BAS
10-member ring	10-MR
pyrrolidine	PY
Na-form **FER** as seeds	SN
H-form **FER** as seeds	SH
organic structure directing agents	OSDAs
8-member ring	8-MR
6-member ring	6-MR
International Zeolite Association	IZA
methyl tert-butyl ether	MTBE
trimethylamine	TMA
relative crystallinity	RC
SiO_2 to Al_2O_3 ratios	SARs
BAS of desorption temperature as measured from Py-IR	$B_{10\text{-}350}$
C4 hydrocarbons produced by fluidized catalytic cracking in refineries react with methanol to produce MTBE after removing isobutene	post-MTBE C4
weight hourly space velocity	WHSV

References

1. Díaz Velázquez, H.; Likhanova, N.; Aljammal, N.; Verpoort, F.; Martínez-Palou, R. New Insights into the Progress on the Isobutane/Butene Alkylation Reaction and Related Processes for High-Quality Fuel Production. A Critical Review. *Energy Fuels* **2020**, *34*, 15525–15556. [CrossRef]
2. Xu, Q.; Yang, W.; Chen, Z.; Ye, Y.; Luo, Y.; Street, J.; Zhou, H.; Xu, C. Formation and regeneration of shape-selective ZSM-35 catalysts for n-Butene skeletal isomerization to isobutylene. *ACS Omega* **2018**, *3*, 8202–8211. [CrossRef] [PubMed]

3. Corma, A. Inorganic solid acids and their use in acid-catalyzed hydrocarbon reactions. *Chem. Rev.* **1995**, *95*, 559–614. [CrossRef]
4. Houzvicka, J.; Ponec, V. Skeletal isomerization of n -Butenes. *Catal. Reviews. Sci. Eng.* **1997**, *39*, 319–344. [CrossRef]
5. Paul, M.; Naccache, C. Skeletal isomerisation of n-butenes catalyzed by medium-pore zeolites and aluminophosphates. *Adv. Catal.* **2010**, *31*, 505–543.
6. Vaughan, P.A. The crystal structure of the zeolite ferrierite. *Acta Crystallogr.* **1966**, *21*, 983–990. [CrossRef]
7. Xu, H.; Zhu, J.; Zhu, L.; Zhou, E.; Shen, C. Advances in the synthesis of ferrierite zeolite. *Molecules* **2020**, *25*, 3722. [CrossRef]
8. Mooiweer, H.H.; de Jong, K.P.; Kraushaar-Czarnetzki, B.; Stork WH, J.; Krutzen BC, H. Skeletal isomerisation of olefins with the zeolite ferrierite as catalyst. *Stud. Surf. Sci. Catal.* **1994**, *84*, 2327–2334.
9. Guo, G.Q.; Long, Y.C. Advances in the catalysis of skeletal isomerization of n-butene into iso-butene on FER zeolite. *Chem. Ind. Eng. Prog.* **2006**, *6*, 16–19.
10. Liu, W.; Hu, H.; Liu, Y.; Zhang, L.; Xia, C.; Wang, Q.; Ke, M. Distribution of effective ferrierite active sites for skeletal isomerization of n-butene to isobutene. *ChemistrySelect* **2019**, *4*, 7851–7857. [CrossRef]
11. Wang, Y.; Gao, Y.; Chu, W.; Zhao, D.; Chen, F.; Zhu, X.; Li, X.; Liu, S.; Xie, S.; Xu, L. Synthesis and catalytic application of FER zeolites with controllable size. *J. Mater. Chem. A* **2019**, *7*, 7573–7580. [CrossRef]
12. Wichterlová, B.; Žilkova, N.; Uvarova, E.; Čejka, J.; Sarv, P.; Paganini, C.; Lercher, J.A. Effect of Broensted and Lewis sites in ferrierites on skeletal isomerization of n-Butenes. *Appl. Catal. A Gen.* **1999**, *182*, 297–308. [CrossRef]
13. Domokos, L.; Lefferts, L.; Seshan, K.; Lercher, J. The importance of acid site locations for n-Butene skeletal isomerization on ferrierite. *J. Mol. Catal. A Chem.* **2000**, *162*, 147–157. [CrossRef]
14. Pinar, A.; Márquez-Álvarez, C.; Grande-Casas, M.; Pérez-Pariente, J. Template-controlled acidity and catalytic activity of ferrierite crystals. *J. Catal.* **2009**, *263*, 258–265. [CrossRef]
15. Seo, G.; Jeong, H.S.; Hong, S.B.; Uh, Y.S. Skeletal isomerization of 1-Butene over ferrierite and ZSM-5 zeolites: Influence of zeolite acidity. *Catal. Lett.* **1996**, *36*, 249–253. [CrossRef]
16. Van Donk, S.; Bus, E.; Broersma, A.; Bitter, J.H.; de Jong, K.P. Probing the accessible sites for n-Butene skeletal isomerization over aged and selective H-Ferrierite with d3-acetonitrile. *J. Catal.* **2002**, *212*, 86–93. [CrossRef]
17. Corma, A. From Microporous to Mesoporous Molecular Sieve Materials and Their Use in Catalysis. *Chem. Rev.* **1997**, *97*, 2373–2420. [CrossRef]
18. Li, J.; Gao, M.; Yan, W.; Yu, J. Regulation of the Si/Al ratios and Al distributions of zeolites and their impact on properties. *Chem. Sci.* **2023**, *14*, 1935–1959. [CrossRef]
19. Mercedes, B.; Corma, A. What Is Measured When Measuring Acidity in Zeolites with Probe Molecules? *ACS Catal.* **2019**, *9*, 1539–1548.
20. Palčić, A.; Valtchev, V. Analysis and control of acid sites in zeolites. *Appl. Catal. A Gen. Int. J. Devoted Catal. Sci. Its Appl.* **2020**, *606*, 117795. [CrossRef]
21. Xiong, Z.; Qi, G.; Bai, L.; Zhan, E.; Chu, Y.; Xu, J.; Ta, N.; Hao, A.; Deng, F.; Shen, W. Preferential population of Al atoms at the T4 site of ZSM-35 for the carbonylation of dimethylether. *Catal. Sci. Technol.* **2022**, *12*, 4993–4997. [CrossRef]
22. Bohinc, R.; Hoszowska, J.; Dousse, J.-C.; Błachucki, W.; Zeeshan, F.; Kayser, Y.; Nachtegaal, M.; Pinar, A.B.; van Bokhoven, J.A. Distribution of aluminum over different T-sites in ferrierite zeolites studied with aluminum valence to core X-ray emission spectroscopy. *Phys. Chem. Chem. Phys.* **2017**, *19*, 29271–29277. [CrossRef] [PubMed]
23. Pinar, A.B.; Verel, R.; Pérez-Pariente, J.; van Bokhoven, J.A. Direct evidence of the effect of synthesis conditions on aluminum sitting in zeolite ferrierite: A ^{27}Al MQ MAS NMR study. *Microporous Mesoporous Mater.* **2014**, *193*, 111–114. [CrossRef]
24. Pinar, A.B.; Gómez-Hortigüela, L.; McCusker, L.B.; Pérez-Pariente, J. Controlling the Aluminum Distribution in the Zeolite Ferrierite via the Organic Structure Directing Agent. *Chem. Mater.* **2013**, *25*, 3654–3661. [CrossRef]
25. Márquez-Alvarez, C.; Pinar, A.B.; García, R.; Grande-Casas, M.; Pérez-Pariente, J. Influence of Al distribution and defects concentration of ferrierite catalysts synthesised from na-free gels in the skeletal isomerisation of n-Butene. *Top. Catal.* **2009**, *52*, 1281–1291. [CrossRef]
26. Román-Leshkov, Y.; Moliner, M.; Davis, M.E. Impact of Controlling the Site Distribution of Al Atoms on Catalytic Properties in Ferrierite-Type Zeolites. *J. Phys. Chem. C* **2011**, *115*, 1096–1102. [CrossRef]
27. Chu, W.; Liu, X.; Yang, Z.; Nakata, H.; Tan, X.; Liu, X.; Xu, L.; Guo, P.; Li, X.; Zhu, X. Constrained Al sites in FER-type zeolites. *Chin. J. Catal.* **2021**, *42*, 2078–2087. [CrossRef]
28. Kim, S.; Park, G.; Woo, M.H.; Kwak, G.; Kim, S.K. Control of hierarchical structure and framework-Al distribution of ZSM-5 via adjusting crystallization temperature and their effects on methanol conversion. *ACS Catal.* **2019**, *9*, 2880–2892. [CrossRef]
29. Gábová, V.; Dědeček, J.; Čejka, J. Control of Al distribution in ZSM-5 by conditions of zeolite synthesis. *J. Chem. Commun.* **2003**, *10*, 1196–1197. [CrossRef]
30. Nishitoba, T.; Yoshida, N.; Kondo, J.N.; Yokoi, T. Control of Al distribution in the CHA-type aluminosilicate zeolites and its impact on the hydrothermal stability and catalytic properties. *Ind. Eng. Chem. Res* **2018**, *57*, 3914–3922. [CrossRef]
31. Heard, C.J.; Grajciar, L.; Nachtigall, P. The effect of water on the validity of Lwenstein's rule. *Chem. Sci.* **2019**, *10*, 5705–5711. [CrossRef] [PubMed]
32. Di Iorio, J.R.; Gounder, R. Controlling the isolation and pairing of aluminum in Chabazite zeolites using mixtures of organic and inorganic structure-directing agents. *Chem. Mater.* **2016**, *28*, 2236–2247. [CrossRef]

33. Di Iorio, J.R.; Li, S.; Jones, C.B.; Nimlos, C.T.; Wang, Y.; Kunkes, E.; Vattipalli, V.; Prasad, S.; Moini, A.; Schneider, W.F.; et al. Cooperative and competitive occlusion of organic and inorganic structure directing agents within Chabazite zeolites influences their aluminum arrangement. *J. Am. Chem. Soc.* **2020**, *142*, 4807–4819. [CrossRef]
34. Fletcher, R.E.; Ling, S.; Slater, B. Violations of Löwensteins rule in zeolites. *Chem Sci* **2017**, *8*, 7483–7491. [CrossRef]
35. Inagaki, S.; Yamada, N.; Nishii, M.; Nishi, Y.; Kubota, Y. Control of framework Al distribution in ZSM-5 zeolite via post-synthetic $TiCl_4$ treatment. *Microporous Mesoporous Mater.* **2020**, *302*, 1102231–1102239. [CrossRef]
36. Vjunov, A.; Fulton, J.L.; Huthwelker, T.; Pin, S.; Mei, D.; Schenter, G.K.; Govind, N.; Camaioni, D.M.; Hu, J.Z.; Lercher, J.A. Quantitatively Probing the Al Distribution in Zeolites. *Am. Chem. Soc.* **2014**, *136*, 8296–8306. [CrossRef]
37. Holzinger, J.; Beato, P.; Lundegaard, L.F.; Skibsted, J. Distribution of Aluminum over the tetrahedral sites in ZSM-5 zeolites and their evolution after steam treatment. *J. Phys.Chem. C Nanomater. Interfaces* **2018**, *122*, 15595–15613. [CrossRef]
38. Meng, X.J.; Xiao, F.S. Green routes for synthesis of zeolites. *Chem. Reviews.* **2014**, *114*, 1521–1543. [CrossRef]
39. Ye, Z.; Zhang, H.; Zhang, Y.; Tang, Y. Seed-induced synthesis of functional MFI zeolite materials: Method development, crystallization mechanisms, and catalytic properties. *Front. Chem. Sci. Eng.* **2020**, *14*, 143–158. [CrossRef]
40. Zhang, H.; Zhao, Y.; Zhang, H.; Wang, P.; Shi, Z.; Mao, J.; Zhang, Y.; Tang, Y. Tailoring zeolite ZSM-5 crystal morphology porosity through flexible utilization of Silicalite-1 seeds as templates: Unusual crystallization pathways in a heterogeneous system. *Chem. A Eur. J.* **2016**, *22*, 7141–7151. [CrossRef]
41. Ji, Y.; Wang, Y.; Xie, B.; Xiao, F.S. Zeolite Seeds: Third Type of Structure Directing Agents in the Synthesis of Zeolites. *Comments Inorg. Chem.* **2016**, *36*, 1–16. [CrossRef]
42. Zhang, H.; Guo, Q.; Ren, L.; Yang, C.; Zhu, L.; Meng, X.; Li, C.; Xiao, F.-S. Organotemplate-free synthesis of high-silica ferrierite zeolite induced by CDO-structure zeolite building units. *J. Mater. Chem.* **2011**, *21*, 9494–9497. [CrossRef]
43. Ham, H.; Jung, H.S.; Kim, H.S.; Kim, J.; Cho, S.J.; Lee, W.B.; Park, M.J.; Bae, J.W. Gas-phase carbonylation of dimethyl ether on the stable seed derived Ferrierite. *ACS Catal* **2020**, *10*, 5135–5146. [CrossRef]
44. Kim, J.; Ham, H.; Jung, H.S.; Wang, Y.; He, Y.; Tsubaki, N.; Cho, S.J.; Han, G.Y.; Bae, J.W. Dimethyl ether carbonylation to methyl acetate over highly crystalline zeolite-seed derived Ferrierite. *Catal. Sci. Technol.* **2018**, *8*, 3060–3072. [CrossRef]
45. Kwak, S.J.; Kim, H.S.; Park, N.; Park, M.J.; Lee, W.B. Recent progress on Al distribution over zeolite frameworks: Linking theories and experiments. *Korean J. Chem. Eng.* **2021**, *38*, 1117–1128. [CrossRef]
46. Emeis, C.A. Determination of Integrated Molar Extinction Coefficients for Infrared Absorption Bands of Pyridine Adsorbed on Solid Acid Catalysts. *J. Catal.* **1993**, *141*, 347–354. [CrossRef]
47. Guo, Y.; Wang, S.; Geng, R.; Wang, P.; Li, S.; Dong, M.; Qin, Z.; Wang, J.; Fan, W. Enhancement of the dimethyl ether carbonylation activation via regulating acid sites distribution in FER zeolite framework. *iScience* **2023**, *26*, 107748. [CrossRef]
48. Xue, T.; Li, S.S.; Wu, H.H. Surfactant-promoted synthesis of hierarchical zeolite ferrierite nano-sheets. *Microporous Mesoporous Mater.* **2021**, *312*, 1107481–1107488. [CrossRef]
49. Xie, S.; Peng, J.; Xu, L.; Wu, Z.; Wang, Q. Synthesis of ZSM-35 zeolite using cyclohexylamine as organic template and its catalytic performance. *Chin. J. Catal.* **2003**, *24*, 531–534.
50. Houzvicka, J.; Nienhuis, J.G.; Ponec, V. The role of the acid strength of the catalysts in the skeletal isomerisation of *n*-Butene. *Appl. Catal. A Gen.* **1998**, *174*, 207–212. [CrossRef]
51. Xu, W.Q.; Yin, Y.G.; Suib, S.L.; Edwards, J.C.; O'Young, C.L. Modification of non-template synthesized ferrierite/ZSM-35 for n-Butene skeletal isomerization to isobutylene. *J. Catal.* **1996**, *163*, 232–244. [CrossRef]
52. Comelli, R.A. Skeletal Isomerization of Linear Butenes on Boron Promoted Ferrierite: Effect of the Catalyst Preparation Technique. *Catal. Lett.* **2008**, *122*, 302–309. [CrossRef]

Disclaimer/Publisher's Note: The statements, opinions and data contained in all publications are solely those of the individual author(s) and contributor(s) and not of MDPI and/or the editor(s). MDPI and/or the editor(s) disclaim responsibility for any injury to people or property resulting from any ideas, methods, instructions or products referred to in the content.

catalysts

Article

Synthesis of *n*-Butene via Dimethyl Ether-to-Olefin Reaction over P-Loaded Ferrierite Zeolites

Toshiaki Hanaoka [1,*], Masaru Aoyagi [1] and Yusuke Edashige [2]

[1] Organic Materials Diagnosis Group, Research Institute for Sustainable Chemistry, National Institute of Advanced Industrial Science and Technology (AIST), 3-11-32 Kagamiyama, Higashihiroshima 739-0046, Hiroshima, Japan; masaru-aoyagi@aist.go.jp
[2] Faculty of Agriculture, Ehime University, 3-5-7 Tarumi, Matsuyama 790-8566, Ehime, Japan; yedash@hb.tp1.jp
* Correspondence: t.hanaoka@aist.go.jp

Abstract: In the dimethyl ether (DME)-to-olefin (DTO) reaction over 20 types of P-loaded ferrierite zeolites with different P loading amounts, the synthesis of *n*-butenes such as 1-butene, *trans*-2-butene, and *cis*-2-butene was investigated to maximize the *n*-butene yield by optimizing the P loading amount. The zeolites were characterized using X-ray diffractometry (XRD), N_2 adsorption-desorption isotherms, and NH_3 temperature-programmed desorption (NH_3-TPD). Micropore and external surface areas, total pore and micropore volumes, and weak and strong acids affected the DTO reaction's characteristics. The P-loaded ferrierite zeolite with a P loading of 0.3 wt.% calcined at 500 °C exhibited an *n*-butene yield of 35.7 C-mol%, which exceeds the highest yield reported to date (31.2 C-mol%). Multiple regression analysis using the obtained data showed that the strong acid/weak acid ratio and total pore volume had a high correlation with the *n*-butene yield, with a contribution rate of 64.3%. Based on the multiple regression analysis results, the DTO reaction mechanism was discussed based on the proposed reaction model involving the dual-cycle mechanism.

Keywords: dimethyl ether; *n*-butene; P-loaded ferrierite zeolite; weak acid; strong acid; total pore volume; multiple regression analysis

1. Introduction

A circular economy is an economic model that creates added value while reducing resource input and making effective use of stock. For products using fossil resources, it is necessary to minimize resource consumption and suppress waste generation. Because biomass is the only renewable carbonaceous resource, the social implementation of high-value-added bio-based products is expected to contribute to the development of a circular economy [1,2].

Natural rubber is a bio-based rubber material. Synthetic rubber has superior properties compared to natural rubber in terms of oil resistance, weather resistance, and heat resistance. One essential component in the production of various synthetic rubbers such as styrene butadiene rubber and butadiene rubber, as well as polymers and latexes, is 1,3-butadiene (BD). Biomass-derived BD (Bio-BD) is typically synthesized from cellulose and hemicellulose [3–5]. Simulation results of these synthesis processes have revealed technical as well as environmental challenges [6,7]. Lignin contains complex aromatic units and can be converted into various phenolics [8,9]. However, the conversion process to Bio-BD has not yet been reported, and there are issues with its utilization.

Three processes for Bio-BD synthesis from lignin were proposed, and the one with the most promising economic efficiency was selected via simulation (Figure 1) [10]. This process consisted of four steps: gasification, dimethyl ether (DME) synthesis, DME-to-olefin (DTO) reaction, and isomerization/dehydrogenation. In the final step (isomerization/dehydrogenation), the conversion of *iso*-butene, which has a branched structure, to BD is difficult. Therefore, it is necessary to selectively synthesize *n*-butenes such as 1-butene, *trans*-2-butene, and *cis*-2-butene in the DTO reaction. Simulation results have revealed that the technical challenge of this process is improving the *n*-butene yield of the DTO reaction [11].

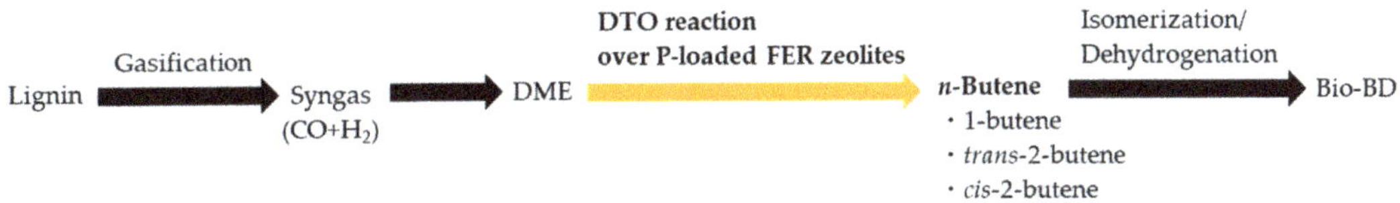

Figure 1. Synthesis process of Bio-BD from lignin.

Zeolite is a typical catalyst used in the DTO reaction. Several synthesis methods suitable for use when propylene and aromatic compounds are the target products have been reported [12–18]. These methods take into account the light olefin synthesis mechanism using kinetic models [12], reaction mechanisms based on the pore structure [13–15], functional groups [16], acid properties [17], and the propylene yield improvement by increasing acid sites using multiple zeolites [18]. Although butene was detected as a byproduct in the DTO reaction over zeolites with the topologies of MFI [19,20], MFI/MTW, MFI/MEL [21], AEI/CHA [22], YFI [23], and CHA [24,25], none of these studies was aimed at *n*-butenes as the main product.

In general, the physicochemical properties of zeolites significantly influence their reaction properties. The addition of heteroatoms to zeolites is an effective approach to changing their properties. The loading of Mg [26], La, Ca [27], Zr [28], Ni [29], Co [30], Ce [31], or P [32,33] onto zeolite results in a hierarchical porous structure and changes in the acid properties, thereby affecting the reaction properties.

The ferrierite (FER) zeolite reportedly exhibited a high *n*-butene yield in the DTO reaction with *n*-butene as the target product [34]. Furthermore, for P-loaded FER zeolites with P loading amounts of 0 and 1.4–12 wt.%, the pore structure and acid properties changed monotonically with increasing P loading amounts [35]. The maximum *n*-butene yield was obtained using FER zeolites without P loading, and the DTO reaction showed a close relationship between the P loading amount and *n*-butene yield (Figure 1) [35]. Zhang et al. reported an optimal Ni loading amount for light olefin selectivity in the DTO reaction over the Ni/SAPO-34 zeolite [29].

In this study, the DTO reaction over P-loaded FER zeolites with P loading amounts of 0–0.8 wt.% was investigated. Our results reveal that there is an optimal P loading amount that leads to the highest *n*-butene yield; a yield which exceeds the highest yield reported to date. Moreover, the DTO reaction mechanism as well as the relationship between the physicochemical properties of the zeolite and *n*-butene yield was discussed.

2. Results and Discussion

Twenty different P-loaded FER zeolites with different P loading amounts (0, 0.1, 0.3, 0.6, and 0.8 wt.%) calcined at different temperatures (450, 500, 550, and 600 °C) were prepared using the impregnation method. Henceforth, the expression for P-loaded FER zeolite will be P(P loading amount)_calcination temperature. For example, if the P loading amount is 0.3 wt.% and the calcination temperature is 500 °C, the representation will be P(0.3)_500.

2.1. Characterization

The P-loaded FER zeolites were characterized using X-ray diffractometry (XRD), N_2 adsorption/desorption isotherms, and NH_3 temperature-programmed desorption (NH_3-TPD) analysis.

Figure 2a shows the XRD patterns of the P-loaded FER zeolites with different P loading amounts at a calcination temperature of 500 °C. In all the cases, diffraction peaks were observed at $2\theta = 9.3° \pm 0.1°$ [200], $22.3° \pm 0.1°$ [420], $23.5° \pm 0.1°$ [330], $25.2° \pm 0.1°$

[040], and 25.7° ± 0.1° [202], which are characteristic of FER [36,37]. This indicates that phosphoric acid treatment did not damage the crystal structure of the FER zeolite. However, no distinct peaks related to P were observed. Van der Bij et al. reported that amorphous $AlPO_4$ is formed by the reaction of FER with phosphoric acid [38].

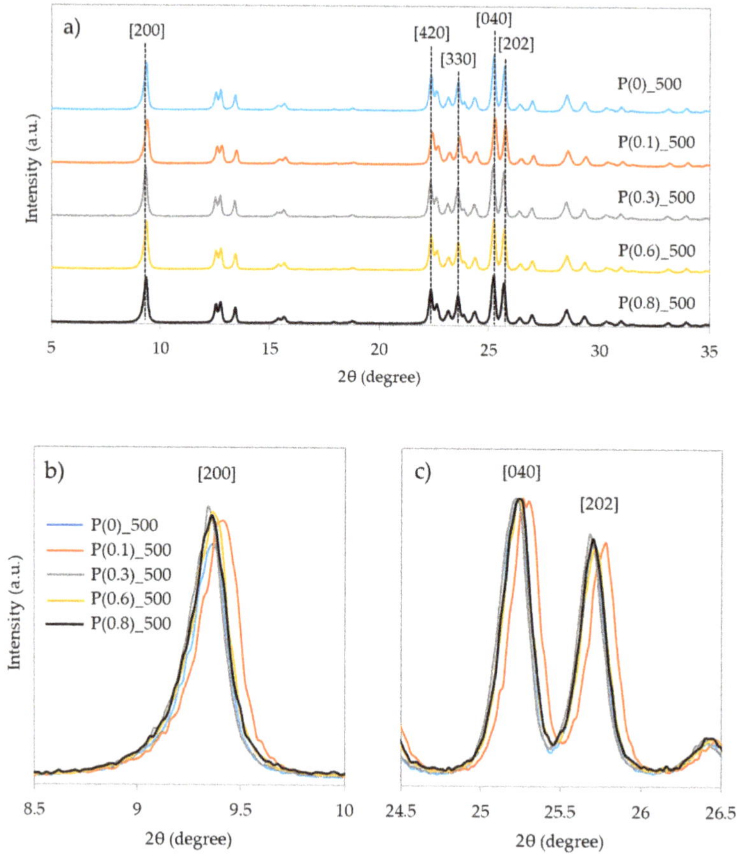

Figure 2. (**a**) XRD patterns of the P-modified FER zeolites calcined at 500 °C and the zoomed regions at 2θ = (**b**) 8.5°–10° and (**c**) 24.5°–26.5°.

In all the as-prepared zeolites, the peak corresponding to 2θ = 25.2° ± 0.1° [040] showed a relatively higher intensity. All XRD patterns were normalized based on the intensity of this peak (Figure 2b,c). The intensity of the peaks corresponding to 2θ = 9.3° ± 0.1° [200] and 25.7° ± 0.1° [202] showed the maximum value for P(0.3)_500, i.e., the minimum value at half-value width. This indicates that the P loading amount influences the crystal diameter of the FER zeolite. Soh et al. reported the optimal P loading amount at which the crystal diameter exhibited the maximum value for the P-loaded H-Y zeolite [39]. Interestingly, the crystal size increased when the P loading amount was increased to 0.3 wt.%, above which the crystal size decreased with increasing P loading amount. In the impregnation method, the zeolite is exposed to phosphoric acid, which results in the dissolution of some of the crystals. This suggests that an increase in the P loading amount increases the acid-to-zeolite weight ratio, which in turn increases the amount of dissolved crystal, thereby resulting in a smaller crystal size.

Similarly, for the P-loaded FER zeolites calcined at 450 and 550 °C, the P loading amount, at which the crystal diameter attained its maximum value, was determined

(Figures S1 and S2, respectively). In contrast, the intensities of the peaks observed at $2\theta = 9.3° \pm 0.1°$ [200] and $25.7° \pm 0.1°$ [202] were almost independent of the P loading amount at 600 °C (Figure S3).

The N_2 adsorption-desorption isotherms for the P-loaded FER zeolites calcined at 500 °C are shown in Figure 3. All of the isotherms were classified as type I based on the IUPAC classification. In all of the samples, adsorption was detected in the low relative pressure region of $p/p_0 < 0.01$, indicating pore filling of the N_2 molecules, and no hysteresis was observed in the desorption curves. These results indicate the absence of mesopores [40]. The N_2 adsorption–desorption isotherms of the P-loaded FER zeolites calcined at 450, 550, and 600 °C showed the same trends as those observed for the sample calcined at 500 °C (Figures S4, S5 and S6, respectively).

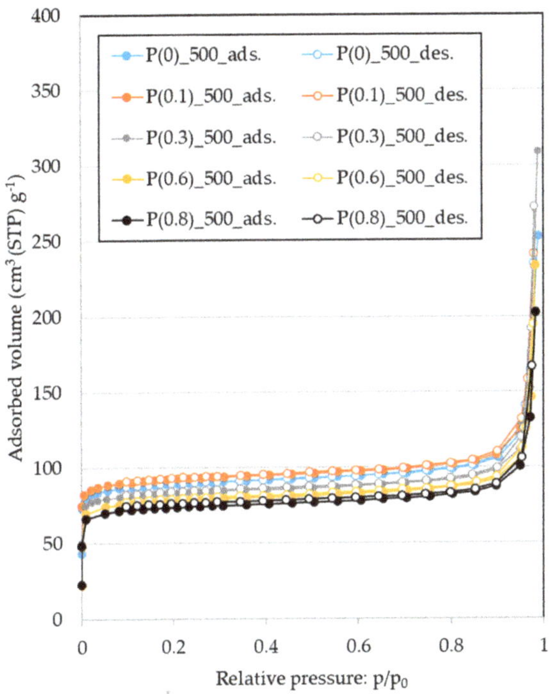

Figure 3. N_2 adsorption-desorption isotherms of the P-modified FER zeolites calcined at 500 °C.

The structural parameters (Table 1) of the P-loaded FER zeolites were obtained from the N_2 adsorption-desorption isotherms. At a calcination temperature of 500 °C, the total and micropore surface areas and the total pore volume were maximized. At other calcination temperatures, most of the structural parameters showed the same trends. Wang et al. reported that in the case of P-loaded beta zeolites, phosphoric acid primarily interacts with Al_2O_3 and destroys the original crystal structure [41]. Van der Bij et al. reported that P acts as a glue, promoting the aggregation of zeolite particles [42]. The increase in the P loading amount to 0.3 wt.% possibly led to the generation of new micropores, owing to an increase in the amount of Al eluted from inside the micropores and the formation of macropores owing to zeolite particle aggregation, thereby increasing both the surface area and pore volume. However, excessive P loading may have led to pore blockage, resulting in a decrease in these parameters.

Table 1. Pore parameters for the P-loaded FER zeolites calcined at different temperatures.

Catalyst	Surface Area ($m^2\ g^{-1}$)			Pore Volume ($cm^3\ g^{-1}$)	
	Total [1]	External [2]	Micro [2]	Total [1]	Micro [2]
P(0)_450	301	26	365	0.47	0.11
P(0.1)_450	304	29	366	0.53	0.11
P(0.3)_450	253	31	258	0.34	0.09
P(0.6)_450	154	29	156	0.36	0.05
P(0.8)_450	57	26	16	0.32	0.01
P(0)_500	352	27	442	0.39	0.13
P(0.1)_500	408	26	461	0.41	0.13
P(0.3)_500	333	23	416	0.48	0.12
P(0.6)_500	307	23	381	0.36	0.11
P(0.8)_500	291	22	363	0.31	0.11
P(0)_550	336	28	416	0.39	0.12
P(0.1)_550	356	24	446	0.44	0.13
P(0.3)_550	307	25	377	0.55	0.11
P(0.6)_550	286	25	348	0.42	0.10
P(0.8)_550	316	24	386	0.51	0.12
P(0)_600	363	22	457	0.38	0.13
P(0.1)_600	362	25	455	0.39	0.13
P(0.3)_600	349	27	435	0.46	0.13
P(0.6)_600	340	26	422	0.53	0.12
P(0.8)_600	319	25	395	0.49	0.11
S.D [3]	76	2	107	0.07	0.03

[1] Calculated using the BET method, [2] calculated using the t-plot method, [3] standard deviations.

The NH_3-TPD profiles of the P-loaded FER zeolites calcined at 500 °C are shown in Figure 4. Two distinct peaks corresponding to the weak and strong acid sites were observed. The maximum intensities of the low-temperature and high-temperature peaks were observed in the temperature ranges of 196–208 and 467–481 °C, respectively, for all loadings. The desorption maxima of both peaks shifted to lower temperatures with increasing P loading amount. These results indicate that as the P loading amount increases, the number of acid sites that interact relatively strongly with NH_3 molecules decreases for both acid sites. A similar trend was observed in the NH_3-TPD profiles of the P-loaded FER zeolites calcined at 450, 550, and 600 °C (Figures S7, S8 and S9, respectively).

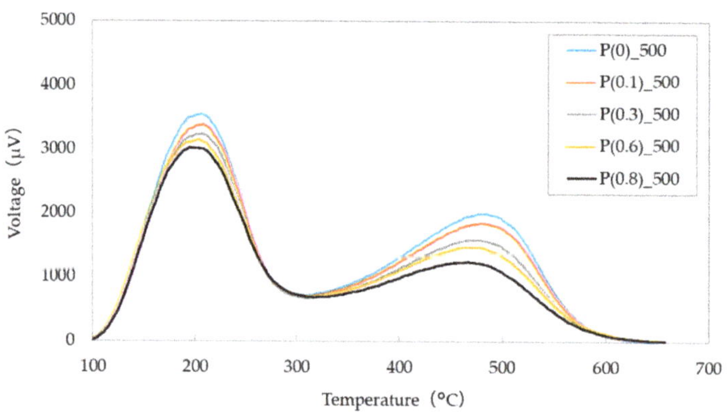

Figure 4. NH_3-TPD profiles of the P-modified FER zeolites calcined at 500 °C.

The number of weak and strong acid sites evaluated based on the peak area is shown in Table 2. Ammonia adsorbs onto strong acid sites in the form of ammonium cations (protonic acidity). Conversely, it adsorbs onto the weak acid sites via hydrogen-coordinated bonds [43]. Therefore, weak acids with Lewis acidity and strong acids with Brønsted acidity could be considered briefly. At calcination temperatures of 450 and 500 °C, the number of both acid sites decreased with increasing P loading amount (Table 2). Conversely, at calcination temperatures of 550 and 600 °C, the number of both acid sites was almost the same when the P loading amounts were 0 and 0.1 wt.%, and decreased for a higher P loading amount. Xue et al. proposed a surface structure model for phosphoric acid-modified zeolites and suggested the presence of a non-participating P-OH group on the zeolite surface due to the condensation with hydroxyl groups of phosphoric acid and the Al-OH and Si-OH groups [44]. The acid strength of P-OH is lower than that of Al-OH; therefore, the acid properties for P-loaded zeolites decreases [39]. This finding is consistent with the fact that, in this study, the number of relatively strong acid sites decreased with increases in the P loading amount.

Table 2. Measured acidities of the P-modified FER zeolites calcined at different temperatures.

Catalyst	Weak (mmol g^{-1})	Strong (mmol g^{-1})	Total (mmol g^{-1})	Strong/Weak (-)
P(0)_450	1.60	1.54	3.14	0.96
P(0.1)_450	1.47	1.49	2.96	1.01
P(0.3)_450	1.38	1.30	2.68	0.94
P(0.6)_450	1.35	1.24	2.59	0.92
P(0.8)_450	1.31	1.19	2.50	0.91
P(0)_500	1.41	1.37	2.78	0.98
P(0.1)_500	1.33	1.28	2.61	0.96
P(0.3)_500	1.29	1.15	2.43	0.89
P(0.6)_500	1.27	1.11	2.39	0.87
P(0.8)_500	1.20	0.99	2.19	0.88
P(0)_550	1.48	1.33	2.81	0.89
P(0.1)_550	1.49	1.39	2.89	0.93
P(0.3)_550	1.36	1.09	2.45	0.80
P(0.6)_550	1.29	1.02	2.30	0.79
P(0.8)_550	1.18	0.95	2.12	0.81
P(0)_600	1.25	0.98	2.23	0.78
P(0.1)_600	1.25	0.99	2.24	0.79
P(0.3)_600	1.20	0.93	2.13	0.77
P(0.6)_600	1.16	0.84	2.00	0.72
P(0.8)_600	0.71	0.56	1.27	0.78
S.D [1]	0.18	0.23	0.41	0.08

[1] standard deviations.

Py-IR spectra of P-supported FER zeolites calcined at 500 °C show large peak areas of 1543 cm^{-1}, indicating a lot of Brønsted (B) acid sites, compared to Lewis (L) acid sites due to small peaks of 1455 cm^{-1} (Figure S10). This result suggests that B acid sites affect the DTO reaction.

2.2. Relationship Between P Loading Amount and DTO Reaction

The DTO reaction was performed using 20 different P-loaded FER zeolites, and the corresponding results are presented in Table 3. The DME conversion rate and product yield ranged from 2.4 to 95.5% and 0.2 to 35.7 C-mol%, respectively, and were highly dependent on the P loading amount and calcination temperature. In particular, the P(0.3)_500 exhibited the highest *n*-butene yield (35.7 C-mol%), which exceeds the previously reported value of 31.2 C-mol% [35]. At each calcination temperature, there was an optimal P loading amount at which the DME conversion rate and *n*-butene yield exhibited the maximum values. Wang et al. reported an optimal amount of P that resulted in the highest conversion in the synthesis of polyoxymethylene dimethyl ether using P-loaded beta zeolites [41].

Excess P covers the active site, possibly hindering the contact between the substrate and active site. For hydrogen production from $NaBH_4$ using Co-loaded zeolites, the Co loading amount that resulted in the highest H_2 yield has been previously reported [30,45]. As the Co loading amount increases, the number of Co particles that are easily accessible to the substrate also increases in the cage. However, excessive Co loading blocks the entrance of the cage, indicating that there is an optimal Co loading amount [45].

Table 3. Catalytic performance of the P-modified FER zeolites in the DTO reaction.

Catalyst	DME Conv. (C-mol%)	Product Yield (C-mol%)										
		C_1–C_3	$C_2^=$	$C_3^=$	MeOH	C_4 [1]	i-$C_4^=$	BD	n-$C_4^=$ [2]	C_5	C_6	C_7
P(0)_450	75.3	2.6	2.4	5.1	8.4	0.1	1.3	0.8	26.1	21.3	6.3	0.8
P(0.1)_450	95.5	4.0	4.6	7.7	4.6	0.2	3.4	0.8	33.6	28.7	7.2	0.9
P(0.3)_450	81.8	4.5	6.7	12.1	0.5	0.2	3.0	0.1	34.1	17.5	3.0	0.3
P(0.6)_450	39.9	3.6	5.3	5.7	3.6	0.1	0.1	0.2	17.4	3.4	0.5	0.0
P(0.8)_450	27.7	3.8	6.1	4.8	0.3	0.1	0.0	0.1	11.1	1.3	0.1	0.0
P(0)_500	50.4	2.7	5.8	12.9	0.2	0.3	6.1	0.0	11.1	9.8	1.1	0.4
P(0.1)_500	88.1	4.7	5.4	9.2	1.1	0.4	7.3	0.1	23.8	29.6	5.5	0.9
P(0.3)_500	90.5	5.7	8.9	15.0	0.2	0.1	4.7	0.0	35.7	17.2	2.5	0.3
P(0.6)_500	35.3	3.6	3.9	4.1	5.2	0.1	0.0	0.2	14.1	3.5	0.5	0.0
P(0.8)_500	3.4	1.2	0.5	0.1	1.3	0.0	0.0	0.0	0.2	0.0	0.0	0.0
P(0)_550	14.9	3.1	1.7	3.4	0.2	0.4	1.5	0.0	2.9	1.1	0.4	0.2
P(0.1)_550	72.5	3.7	3.1	4.4	12.9	0.2	4.0	0.4	19.1	19.4	4.4	0.9
P(0.3)_550	36.0	4.9	4.7	9.0	0.2	0.4	4.0	0.2	7.9	4.1	0.6	0.0
P(0.6)_550	2.9	1.3	0.3	0.4	0.2	0.0	0.1	0.0	0.5	0.1	0.0	0.0
P(0.8)_550	2.4	1.4	0.3	0.2	0.2	0.0	0.0	0.0	0.3	0.0	0.0	0.0
P(0)_600	3.4	1.6	0.3	0.3	0.2	0.1	0.2	0.0	0.5	0.1	0.0	0.0
P(0.1)_600	12.8	3.4	1.5	2.7	0.2	0.4	1.2	0.0	2.3	0.8	0.3	0.1
P(0.3)_600	3.4	1.7	0.2	0.2	0.3	0.1	0.1	0.0	0.6	0.2	0.0	0.0
P(0.6)_600	2.8	1.3	0.3	0.3	0.2	0.0	0.1	0.0	0.4	0.1	0.0	0.0
P(0.8)_600	5.8	1.1	0.3	0.2	2.5	0.0	0.0	0.0	1.2	0.3	0.1	0.0

[1] *n*-butane + *iso*-butane, [2] 1-butene + *trans*-2-butene + *cis*-2-butene.

In this study, the DME conversion rate and *n*-butene yield decreased monotonically with increasing calcination temperature at different P loading amounts (0, 0.1, 0.6, and 0.8 wt.%). However, when P loading was 0.3 wt.%, the highest DME conversion rate and *n*-butene yield were obtained at a calcination temperature of 500 °C. Both the characterization and DTO reaction results show that the reaction properties depend on the pore structure and the acidity.

2.3. Multiple Regression Analysis

Statistical analysis of the physicochemical properties of zeolites and the DTO reaction characteristics will add useful information to the discussion of the reaction mechanism. In this study, multiple regression analysis was performed using a stepwise method to examine the physical properties of the P-loaded FER zeolites, which have a high correlation to the *n*-butene yield. We considered the values of seven physical properties: micropore surface area (x_1), external surface area (x_2), total pore volume (x_3), micropore volume (x_4), weak acid (x_5), strong acid (x_6), and strong acid/weak acid (x_7, Table 2) as explanatory variables and calculated the correlation coefficient (r_{ij}) with respect to the *n*-butene yield (y) (Table S1). The absolute values of the seven correlation coefficients were in the following order: x_7 (0.766) > x_6 (0.692) > x_5 (0.525) > x_2 (0.470) > x_1 (0.171) > x_4 (0.141) > x_3 (0.037).

Multiple regression analysis requires selecting an appropriate combination of explanatory variables. The tolerance ($1 - r_{ij}^2$) of the explanatory variables x_1 and x_4 was 0.009 (Table S2), which is less than the critical value (0.1). Multiple regression analysis using the combination of x_1 and x_4 leads to multicollinearity, resulting in a decrease in correlation accuracy. Therefore, other combinations of explanatory variables were employed in the multiple regression analysis.

The multiple regression analysis results, including the explanatory variables, residual sum of squares (RSS), F-ratio, regression equation, and contribution rate are shown in Table 4. The RSS is expressed as shown in Equation (1).

$$\mathrm{RSS} = \sum_{i=1}^{20} \varepsilon_i^2 = \sum_{i=1}^{20} \{y_i - a_0 - (a_1x_{1i} + a_2x_{2i} + a_3x_{3i} + a_4x_{4i} + a_5x_{5i} + a_6x_{6i} + a_7x_{7i})\}^2, \quad (1)$$

Table 4. Multiple regression analysis.

Trial	Explanatory Variable	RSS	F-Ratio	Regression Equation	Contribution Rate (%)
1	None	6012			
2	x_7	1266	67.5	$y = -89.7 + 117.2x_7$	58.6
3	x_7, x_1 x_7, x_2 x_7, x_3 x_7, x_4 x_7, x_5 x_7, x_6	1265 1221 1092 1262 1260 1258	7.8	$y = -116.2 + 42.6x_3 + 126.7x_7$	64.3
4	x_7, x_3, x_1 x_7, x_3, x_2 x_7, x_3, x_4 x_7, x_3, x_5 x_7, x_3, x_6	1070 1059 1081 1090 1091	0.5	$y = -124.3 + 0.6x_2 + 41.3x_3 + 117.8x_7$	65.4
5	$x_7, x_3 - x_7$ $x_7, x_3 - x_3$	2385 1266	2.7 20.1	$y = 9.4 + 6.5x_3$ $y = -89.7 + 117.2x_7$	0.1 58.6

Here, ε_i is the error, a_0 is the intercept, a_1–a_7 are partial regression coefficients, and x_1–x_7 are the experimental data. The RSS is derived from the error between the data and the regression equation, and a small value is desirable.

The F-ratio represents the rate of decrease in the RSS when explanatory variables are added, and it is expressed as shown in Equation (2).

$$\mathrm{F-ratio} = \frac{\mathrm{RSS_q} - \mathrm{RSS_{q+1}}}{\mathrm{RSS_{q+1}}}((\text{Number of experimental data}(=20) - (q+1) - 1), \quad (2)$$

Here, q represents the number of explanatory variables. When the F-ratio is 2 or more, the rate of decrease in the RSS is high, indicating that the addition of the explanatory variables is effective.

The F-ratio is also used to determine whether variables already included in the regression equation are necessary. If 'q' explanatory variables are already in the regression equation and one of them is removed, the F-ratio represents the rate of increase in RSS and is expressed as shown in Equation (3).

$$\mathrm{F-ratio} = \frac{\mathrm{RSS_q} - \mathrm{RSS_{q-1}}}{\mathrm{RSS_q}}((\text{Number of experimental data}(=20) - q - 1), \quad (3)$$

If this value is 2 or more, the RSS increase rate is high and the explanatory variable cannot be removed.

Multiple regression analysis can determine the partial regression coefficients (a_1–a_7) that minimize the RSS in Equation (1). The regression equation is obtained as shown below [Equation (4)].

$$y = a_0 + a_1x_1 + a_2x_2 + a_3x_3 + a_4x_4 + a_5x_5 + a_6x_6 + a_7x_7 \quad (4)$$

When no explanatory variables are used (trial 1), the RSS is the sum of the squares of the *n*-butene yield (6012). After the addition of x_7, which has the highest correlation coefficient to y (trial 2), the RSS decreased to 1266. Because the F-ratio was 2 or greater (67.5), x_7 was adopted as an explanatory variable. The RSS was calculated when x_1, x_2, x_3, x_4, x_5, and x_6 were added to x_7 (trial 3). When x_3 was added, the RSS was the smallest (1092). Because the F-ratio was 7.8 (>2), we adopted x_3. The RSS was calculated when x_1, x_2, x_4, x_5, and x_6 were added to x_7 and x_3 (trial 4). The RSS was the lowest (1059) when x_2 was added. However, because the F-ratio was less than 2 (0.5), x_2 was not adopted.

We also considered whether the selected x_7 and x_3 values are useless variables. When one of the variables was removed (one at a time) from the regression equation involving the variables x_7 and x_3, the RSS values were 2385 and 1266, respectively (trial 5), while the F-ratios were 2.7 and 20.1, respectively, both of which are greater than 2. This indicates that neither x_7 nor x_3 can be removed. The multiple regression analysis including x_7 and x_3 ($y = -116.2 + 42.6x_3 + 126.7x_7$, Trial 2) had no variables to add or delete, and hence the multiple regression analysis was terminated. When x_7 (strong acid/weak acid) and x_3 (total pore volume) were used as the explanatory variables, the contribution rate of the regression equation was 64.3% compared to that when only x_7 was used (Trials 1 and 5), and the *n*-butene yield (y) was 5.7% higher.

Multiple regression analysis without x_7 yielded a contribution rate of 59.1% (Table S3). Thus, the adoption of x_7 as an explanatory variable was effective and could successfully explain the 5.2% higher *n*-butene yield.

In the previous paper [35], the pore structure and acid properties of the P-loaded FER zeolites were not optimized, and the multiple regression analysis showed that micropore surface area and strong acid sites had the highest correlation with the *n*-butene yield, leading to the contribution rate of 53.7%. In this paper, the pore structure and acid properties, that is, the P loading amount, were optimized, and the multiple regression analysis showed that the strong acid/weak acid ratio and total pore volume had the highest correlation, leading to the contribution rate of 64.3%, an increase of 10.6% compared to the previous study.

2.4. DTO Reaction Mechanism

A dual-cycle mechanism has been proposed for the DTO reaction [46,47]. We proposed a reaction model involving the dual-cycle mechanism to discuss the DTO reaction mechanism, while considering the results of the multiple regression analysis, as shown in Figure 5.

The reaction pathways include the formation of methoxy groups on the catalyst surface (reactions 1 and 2), direct formation of ethylene and propylene (reactions 3–5), methylation within the olefinic cycle (reactions 6–10), formation of methylcyclopentenyl cations, benzene (C6) production from propylene and hexene via cyclization of CP^+ (reactions 11–14), methylation within the aromatic cycle (reaction 15), and cracking (reactions 16 and 17). In the previous report [35], the proposed model contained only aromatic and olefinic cycles. In the paper, reactions 15–17, included in the aromatic cycle, are the same as in the previous report [35]. In contrast, reactions 1–5 were proposed, and in the olefinic cycle, heptene production was added, and further alkylation or cracking was not proposed.

In the initial stage of the DTO reaction, DME reacts with the hydroxyl groups on the catalyst surface, resulting in the formation of methoxy groups (reactions 1 and 2). Subsequently, the methoxy group and DME, that is, two methoxy groups, react to yield ethylene and propylene (reactions 3–5). Sun et al. reported the formation of active ethylene species on acidic sites based on nuclear magnetic resonance (NMR) spectroscopic measurements during the early stages of the MeOH-to-olefin (MTO) reaction [48].

In the olefinic cycle, methylation is promoted by the reactions of ethylene and propylene with the methoxy group (reactions 6–10). Sun et al. reported that the strong acid site is the active site for benzene alkylation involving β-zeolites [49]. Hill et al. reported that alkylation in the DTO reaction originates from the MeOH/DME complex formed on the

B acid site [50]. Song et al. reported an increase in selectivity for light olefins, with an increase in B acid sites in the MeOH-to-propylene reaction over P-loaded ZSM-5 [51]. In this study, the methoxy group derived from strong acid and B acid site would be involved in methylation.

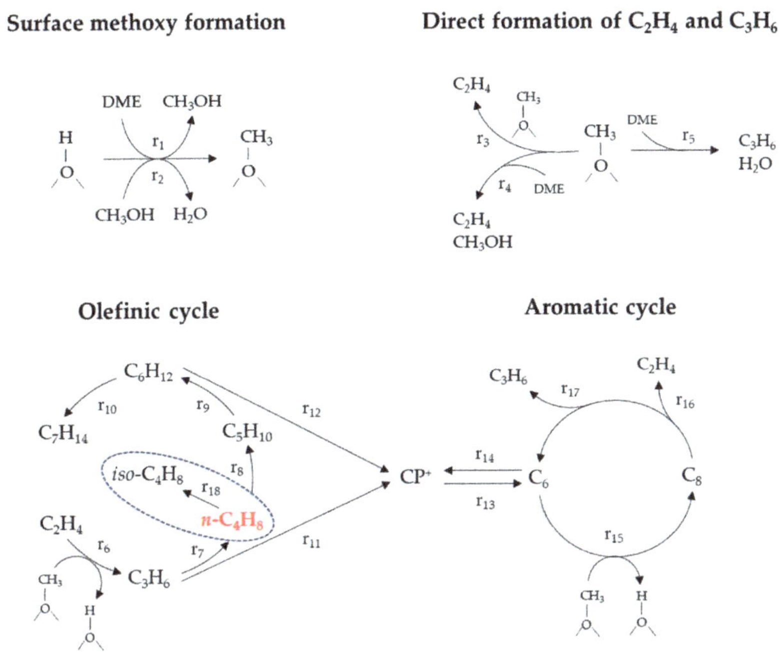

Figure 5. Proposed DTO reaction model involving the dual-cycle mechanism.

The activation energy of olefin methylation in the DTO reaction is, in order, ethylene > propylene > butene > pentene, indicating that the synthesis of relatively long hydrocarbons is advantageous [46]. This suggests that the selective formation of butene requires not only the promotion of propylene methylation but also the suppression of butene methylation via steric hindrance. The total pore volume exhibited its maximum value at an optimal P loading amount (Table 1). The increase in the total pore volume was due to the increase in macropores owing to P acting as a glue and the aggregation of the primary particles of zeolite, indicating that the distance between the primary particles became shorter. At any given calcination temperature, the *n*-butene yield increased as the P loading amount increased from 0 to 0.3 wt.% (Table 3). Thus, ethylene and propylene flowing out of the primary particles may have had easier access to the methoxy groups in adjacent particles, resulting in further methylation. Conversely, excessive P loading could have resulted in low *n*-butene yield due to pore blockage and the inhibition of the access of heptene to the B acid site, because no products with eight or more carbon numbers were detected (Table 3, Figure 5).

The morphology was investigated using an atomic resolution TEM (Figure S11), and we will be able to discuss the relationship between the existence state of P and the DTO reaction using atomic resolution TEM images in near future.

Propylene and hexene in the olefinic cycle produce benzene as a hydrocarbon pool via the methylcyclopentenyl cations (CP^+) (reactions 11–14) [47,52]. In the aromatic cycle, C6 reacts with a methoxy group, producing an aromatic compound with eight carbon atoms (C8) (reaction 15). The detailed reactions in the aromatic cycle were mentioned in the previous study [35]. Subsequently, cracking produces ethylene and propylene, and C6 is again formed (reactions 16 and 17). In this study, weak acid in the micropores probably suppressed the formation of CP^+ from propylene and hexene (reactions 11 and 12). The low ethylene and propylene yields in the DTO reaction (Table 3) are consistent with the suppression of alkylation and cracking (reactions 15–17) in the aromatic cycle.

Ling et al. reported that in dodecylbenzene synthesis over P-loaded MCM-22, a P loading of 0.6 wt.% showed the highest yield [53]. The NH_3-TPD profiles showed that as the P loading increased, the amount of weak acid increased and that of strong acid decreased [53], thereby supporting the fact that strong acid/weak acid influences the reaction characteristics. Magomedova et al. reported that there are desirable strong/weak acids for the selective synthesis of ethylene and propylene in the DTO reaction over ZSM-5 [54]. In this study, both the desired ratio of strong acid (which promotes methylation) to weak acid (which suppresses cyclization) and the total pore volume (which suppresses butene methylation) resulted in a high *n*-butene yield.

In this study, the production of *iso*-butene was suppressed (Table 3, reaction 18 in Figure 5). Hill et al. reported that the reaction rate constant for the alkylation of *iso*-butene is at least 40 times higher than those for 1-butene, *trans*-2-butene, and *cis*-2-butene [50]. This indicates that *iso*-butene production was suppressed owing to *n*-butene isomerization in this study. Kitev et al. reported that the B acid site of FER zeolite is the active site of isomerization in the skeletal isomerization reaction of 1-butene, and that *iso*-butene selectivity decreased with increasing micropore/mesopore ratios [55]. This indicates supports that the presence of micropores in FER zeolites makes it difficult for *n*-butene to access B acid sites, resulting in steric hindrance to *iso*-butene formation (r_{18}). The fact that the presence of mesopores was not confirmed (Table 1) is consistent with the fact that *iso*-butene production from *n*-butene is suppressed.

Accordingly, the DTO reaction over P-loaded FER zeolites was explained using the proposed reaction model involving the dual-cycle mechanism.

3. Materials and Methods

3.1. Catalyst Preparation and Characterization

Commercially available FER zeolite (720NHA, Tosoh Co., Ltd., Tokyo, Japan) with SiO_2/Al_2O_3 (mol/mol)) = 18.5, cation type: NH_4^+, crystal size $\leq$ 1 μm, and particle size = 6 μm was used in this study. The P-loaded FER zeolites with different P loading amounts (0, 0.1, 0.3, 0.6, and 0.8 wt.%) were prepared via the impregnation method. The calcination temperatures were 450, 500, 550, and 600 °C.

P-loaded FER zeolites were characterized as follows.

The XRD patterns were obtained using an X-ray diffractometer (RINT-TTR III, Rigaku Corporation, Tokyo, Japan). The target, voltage, current, and wavelength were Cu, 50 kV, 300 mA, and 0.15418 nm, respectively.

The structural parameters were measured using a specific surface area/pore distribution measuring device (BELSORP-mini II, MicrotracBEL Cort., Osaka, Japan).

NH_3-TPD was performed using an auto-chemisorption system (BELCAT-B, MicrotracBEL Corp., Osaka, Japan) equipped with a thermal conductivity detector (TCD).

The detailed procedures were mentioned in the previous paper [35].

3.2. DTO Reaction

A fixed-bed flow reactor was used for the DTO reaction [10]. Zeolite powder (0.2 g) was employed in each run. The detailed procedure for the DTO reaction and estimation method for the catalytic behavior were both mentioned in the previous paper [35].

4. Conclusions

DTO reactions were performed using P-loaded FER zeolites with different loading amounts calcined at different temperatures. The structural factors of zeolite that influenced the *n*-butene yield were clarified using multiple regression analysis. The following conclusions were drawn.

1. P(0.3)_500 exhibited the highest *n*-butene yield (35.7 C-mol%), which exceeded the previously reported highest yield (31.2 C-mol%);
2. Multiple regression analysis results showed that *n*-butene yield had a high correlation with strong acid/weak acid and total pore volume, and was explained by these two explanatory variables based on a contribution rate of 64.3%;
3. The DTO reaction over the P-loaded FER zeolites was explained using the proposed reaction model involving the dual-cycle mechanism.

Supplementary Materials: The following supporting information can be downloaded at: https://www.mdpi.com/article/10.3390/catal14120902/s1. Figure S1: (a) XRD patterns of the P-modified FER zeolites calcined at 450 °C and the zoomed regions from 2θ = (b) 8.5 to 10 and (c) 24.5 to 26.5; Figure S2: (a) XRD patterns of the P-modified FER zeolites calcined at 550 °C and the zoomed regions from 2θ = (b) 8.5 to 10 and (c) 24.5 to 26.5; Figure S3: (a) XRD patterns of the P-modified FER zeolites calcined at 600 °C and the zoomed regions from 2θ = (b) 8.5 to 10 and (c) 24.5 to 26.5; Figure S4: N_2 adsorption-desorption isotherms of the P-modified FER zeolites calcined at 450 °C; Figure S5: N_2 adsorption-desorption isotherms of the P-modified FER zeolites calcined at 550 °C; Figure S6: N_2 adsorption-desorption isotherms of the P-modified FER zeolites calcined at 600 °C; Figure S7: NH_3-TPD profiles of the P-modified FER zeolites calcined at 450 °C; Figure S8: NH_3-TPD profiles of the P-modified FER zeolites calcined at 550 °C; Figure S9: NH_3-TPD profiles of the P-modified FER zeolites calcined at 600 °C; Figure S10. Py-IR spectra of the P-modified FER zeolites calcined at 500 °C; Figure S11: STEM image of P(0.8)_500; Table S1: Supplementary table for multiple regression analysis; Table S2. Tolerance between the selected explanatory variables; Table S3. Multiple regression analysis without adding x_7 (strong acid/weak acid).

Author Contributions: Data curation, T.H.; writing—original draft, T.H.; writing—review and editing, M.A. and Y.E.; project supervision, Y.E. The manuscript has been read and revised by all the authors before submission. All authors have read and agreed to the published version of the manuscript.

Funding: This work was supported by the Fundamental Research Fund of the National Institute of Advanced Industrial Science and Technology (AIST), Japan, and did not receive any specific grant from funding agencies in the public, commercial, or not-for-profit sectors.

Data Availability Statement: Data is contained within the article or Supplementary Materials.

Acknowledgments: The authors are grateful for the fundamental grant from AIST. The authors are also thankful for experimental and technical assistance from Tadahiro Fujitani, Naoto Kamiuchi, Koki Sawa, and Yuko Tsuji.

Conflicts of Interest: The authors declare no conflicts of interest.

References

1. Sherwood, J. The significance of biomass in a circular economy. *Bioresour. Technol.* **2020**, *300*, 122755. [CrossRef] [PubMed]
2. Kumar, B.; Verma, P. Biomass-based biorefineries: An important architype towards a circular economy. *Fuel* **2021**, *288*, 119622. [CrossRef]
3. Cai, D.; Zhu, Q.; Chen, C.; Hu, S.; Qin, P.; Wang, B.; Tan, T. Fermentation-pervaporation-catalysis integration process for bio-butadiene production using sweet sorghum juice as feedstock. *J. Taiwan Inst. Chem. Eng.* **2018**, *82*, 137–143. [CrossRef]
4. Rodriguez, A.C.; Sad, M.E.; Padró, C.L. Acid site requirement and reaction pathway for selective bio-butadiene synthesis by 1,3-butanediol dehydration. *Appl. Catal. A-Gen.* **2023**, *664*, 119349. [CrossRef]
5. Abdelrahman, O.A.; Park, D.S.; Vinter, K.P.; Spanjers, C.S.; Ren, L.; Cho, H.J.; Vlachos, D.G.; Fan, W.; Tsapatsis, M.; Dauenhauer, P.J. Biomass-derived butadiene by dehydra-decyclization of tetrahydrofuran. *ACS Sustain. Chem. Eng.* **2017**, *5*, 3732–3736. [CrossRef]
6. Camacho, C.E.C.; Alonso-Fariñas, B.; Perales, A.L.V.; Vidal-Barrero, F.; Ollero, P. Techno-economic and Life-Cycle Assessment of One-Step Production of 1,3-Butadiene from Bioethanol using Reaction Data under Industrial Operating Conditions. *ACS Sustain. Chem. Eng.* **2020**, *8*, 10201–10211. [CrossRef]
7. Kuznetsov, A.; Kumar, G.; Ardagh, A.; Tsapatsis, M.; Zhang, Q.; Dauenhauer, P.J. On the economics and process design of renewable butadiene from biomass-derived furfural. *ACS Sustain. Chem. Eng.* **2020**, *8*, 3273–3282. [CrossRef]
8. Hu, R.; Zhan, J.; Zhao, Y.; Xu, X.; Luo, G.; Fan, J.; Clark, J.H.; Zhang, S. Bio-based platform chemicals synthesized from lignin biorefinery. *Green Chem.* **2023**, *25*, 8970–9000. [CrossRef]

9. Martínková, L.; Grulich, M.; Pátek, M.; Křístková, B.; Winkler, M. Bio-Based Valorization of Lignin-Derived Phenolic Compounds: A Review. *Biomolecules* **2023**, *13*, 717. [CrossRef]
10. Hanaoka, T.; Fujimoto, S.; Kihara, H. Improvement of the 1,3-butadiene production process from lignin—A comparison with the gasification power generation process. *Renew. Energy* **2019**, *135*, 1303–1313. [CrossRef]
11. Hanaoka, T.; Fujimoto, S.; Kihara, H. Evaluation of *n*-butene synthesis from dimethyl ether in the production of 1,3-butadiene from lignin: A techno-economic analysis. *Renew. Energy* **2021**, *163*, 964–973. [CrossRef]
12. Khadzhiev, S.N.; Magomedova, M.V.; Peresypkina, E.G. Kinetic models of methanol and dimethyl ether conversion to olefins over zeolite catalysts (Review). *Petrol. Chem.* **2015**, *55*, 503–521. [CrossRef]
13. Han, Q.; Park, S.; Inagaki, S.; Kubota, Y. Selective production of light olefins over MSE-type zeolite catalyst. *J. Jpn. Pet. Inst.* **2017**, *60*, 288–300. [CrossRef]
14. Park, S.; Watanabe, Y.; Nishita, Y.; Fukuoka, T.; Inagaki, S.; Kubota, Y. Catalytic conversion dimethyl ether into propylene over MCM-68 zeolite. *J. Catal.* **2014**, *319*, 265–273. [CrossRef]
15. Kubota, Y.; Inagaki, S. High-performance catalysts with MSE-type zeolite framework. *Top. Catal.* **2015**, *58*, 480–493. [CrossRef]
16. Inagaki, S.; Park, S.; Yamazaki, H.; Kondo, J.N.; Kubota, Y. Investigation of the acidic nature of MCM-68 zeolite based on the adsorption of CO and bulky probe molecules. *Microporous Mesoporous Mater.* **2018**, *272*, 16–23. [CrossRef]
17. Al-Dughaither, A.S.; de Lasa, H. Neat dimethyl ether conversion to olefins (DTO) over HZSM-5: Effect of SiO_2/Al_2O_3 on porosity, surface chemistry, and reactivity. *Fuel* **2014**, *138*, 52–64. [CrossRef]
18. Zhao, D.; Zhang, Y.; Li, Z.; Wang, Y.; Yu, J. Synthesis of AEI/CHA intergrowth zeolites by dual templates and their catalytic performance for dimethyl ether to olefins. *Chem. Eng. J.* **2017**, *323*, 295–303. [CrossRef]
19. Giglio, E.; Ferrarelli, G.; Salomone, F.; Corrao, E.; Migliori, M.; Bensaid, S.; Pirone, R.; Giordano, G. Tailoring the acidity of ZSM-5 via surface passivation: Catalytic assessment on dimethyl ether to olefin (DTO) process. *Fuel* **2024**, *362*, 130559. [CrossRef]
20. Ortega, C.; Hessel, V.; Kolb, G. Dimethyl ether to hydrocarbons over ZSM-5: Kinetic study in an external recycle reactor. *Chem. Eng. J.* **2018**, *354*, 21–34. [CrossRef]
21. Magomedova, M.V.; Starozhitskaya, A.V.; Davidov, I.A.; Tsaplin, D.E.; Maximov, A.L. Dimethyl Ether to Olefins on Hybrid Intergrowth Structure Zeolites. *Catalysts* **2023**, *13*, 570. [CrossRef]
22. Zhao, D.; Zhang, Y.; Li, Z.; Wang, Y.; Yu, J. Synthesis of SAPO-18/34 intergrowth zeolites and their enhanced stability for dimethyl ether to olefins. *RSC Adv.* **2017**, *7*, 939–946. [CrossRef]
23. Liu, Q.; Yoshida, Y.; Nakazawa, N.; Inagaki, S.; Kubota, Y. The synthesis of YNU-5 zeolite and its application to the catalysis in the dimethyl ether-to-olefin reaction. *Materials* **2020**, *13*, 2030. [CrossRef]
24. Zhu, J.; Cui, Y.; Nawaz, Z.; Wang, Y.; Wei, F. In situ Synthesis of SAPO-34 Zeolites in Kaolin Microspheres for a Fluidized Methanol or Dimethyl Ether to Olefins Process. *Chin. J. Chem. Eng.* **2010**, *18*, 979–987. [CrossRef]
25. Haas, A.; Hauber, C.; Kirchmann, M. Time-Resolved Product Analysis of Dimethyl Ether-to-Olefins Conversion on SAPO-34. *ACS Catal.* **2019**, *9*, 5679–5691. [CrossRef]
26. Kolesnichenko, N.V.; Konnov, S.V.; Pavlov, V.S.; Yashina, O.V.; Ezhova, N.N.; Khadzhiev, S.N. Dimethyl Ether to Olefins Conversion in a Slurry Reactor: Effects of the Size of Particles and the Textural and Acidic Properties of the MFI-Type Zeolite. *Pet. Chem.* **2017**, *57*, 29–37. [CrossRef]
27. Bakare, I.A.; Muraza, O.; Sanhoob, M.A.; Miyake, K.; Hirota, Y.; Yamani, Z.H.; Nishiyama, N. Dimethyl ether-to-olefins over aluminum rich ZSM-5: The role of Ca and La as modifiers. *Fuel* **2018**, *211*, 18–26. [CrossRef]
28. Crodero-Lanzac, T.; Bodríguez-Cano, M.A.; Palomoto, J.; Valero-Romero, M.J.; Aguayo, A.T.; Bilbao, J.; Rodríguez-Mirasol, J.; Cordero, T. Binderless ZrO_2/HZSM-5 fibrillar composites by electrospinning as catalysts for the dimethyl ether-to-olefins process. *Microporous Mesoporous Mater.* **2022**, *342*, 112102. [CrossRef]
29. Zhang, S.; Chen, W.; Yang, L.; Xie, T.; Li, W.; Yu, D.; Fang, Y. Effect of morphology and acidity control of Ni-SAPO-34 zeolite on catalytic performance of dimethyl ether to olefins. *J. Solid State Chem.* **2021**, *303*, 122503. [CrossRef]
30. Saka, C.; Eygi, M.S.; Balbay, A. CoB doped acid modified zeolite catalyst for enhanced hydrogen release from sodium borohydride hydrolysis. *Int. J. Hydrogen Energy* **2020**, *45*, 15086–15099. [CrossRef]
31. Han, Q.; Enoeda, K.; Inagaki, S.; Kubota, Y. Catalytic performance of Ce-modified MCM-68 zeolite in the dimethyl ether-to-olefin reaction: Impact of high calcination temperature. *Chem. Lett.* **2017**, *46*, 1434–1437. [CrossRef]
32. Omata, K.; Yamaguchi, Y.; Watanabe, Y.; Kodama, K.; Yamada, M. Artificial neural network (ANN)-aided optimization of ZSM-5 catalyst for the dimethyl ether to olefin (DTO) reaction from neat dimethyl ether (DME). *Ind. Eng. Chem. Res.* **2009**, *48*, 6256–6261. [CrossRef]
33. Park, S.; Inagaki, S.; Kubota, Y. Selective formation of light olefins from dimethyl ether over MCM-68 modified with phosphate species. *Catal. Today* **2016**, *265*, 218–224. [CrossRef]
34. Hanaoka, T.; Aoyagi, M.; Edashige, Y. *n*-Butene synthesis in the dimethyl ether-to-olefin reaction over zeolites. *Catalysts* **2021**, *11*, 743. [CrossRef]
35. Hanaoka, T.; Aoyagi, M.; Edashige, Y. Improvement of *n*-Butene Yield in Dimethyl Ether-to-Olefin Reaction Using Ferrierite Zeolite Catalysts. *Catalysts* **2023**, *13*, 1040. [CrossRef]
36. Wei, P.; Zhu, X.; Wang, Y.; Chu, W.; Xie, S.; Yang, Z.; Liu, Z.; Li, Z.; Xu, L. Rapid synthesis of ferrierite zeolite through microwave assisted organic template free route. *Microporous Mesoporous Mater.* **2019**, *279*, 220–227. [CrossRef]

37. Berdiell, I.C.; Braghin, G.B.; Cordero-Lanzac, T.; Beato, P.; Lundegaard, L.F.; Wragg, D.; Bordiga, S.; Svelle, S. Tracking Structural Deactivation of H-Ferrierite Zeolite Catalyst During MTH with XRD. *Top. Catal.* **2023**, *66*, 1418–1426. [CrossRef]
38. van der Bij, H.E.; Cicmil, D.; Wang, J.; Meirer, F.; de Groot, F.M.F.; Weckhuysen, B.M. Aluminum-Phosphate Binder Formation in Zeolites as Probed with X-ray Absorption Microscopy. *J. Am. Chem. Soc.* **2014**, *136*, 17774–17787. [CrossRef]
39. Soh, J.C.; Chong, S.L.; Hossain, S.S.; Cheng, C.K. Catalytic ethylene production from ethanol dehydration over non-modified and phosphoric acid modified Zeolite H-Y (80) catalysts. *Fuel Process. Technol.* **2017**, *158*, 85–95. [CrossRef]
40. Kamimura, Y.; Kowenje, C.; Yamanaka, K.; Itabashi, K.; Endo, A.; Okubo, T. Synthesis of hydrophobic siliceous ferrierite by using pyridine and sodium fluoride. *Microporous Mesoporous Mater.* **2013**, *181*, 154–159. [CrossRef]
41. Wang, B.; Yan, X.; Zhang, H. Interaction between phosphorus and zeolite/binder: A realumination study on beta zeolites. *Microporous Mesoporous Mater.* **2021**, *312*, 110735. [CrossRef]
42. van der Bij, H.E.; Aramburo, L.R.; Arstad, B.; Dynes, J.J.; Wang, J.; Weckhuysen, B.M. Phosphatation of zeolite H-ZSM-5: A combined microscopy and spectroscopy study. *Chem. Phys. Chem.* **2014**, *15*, 283–292. [CrossRef] [PubMed]
43. Al-Dughaither, A.S.; de Lasa, H. HZSM-5 zeolites with different SiO_2/Al_2O_3 ratios. Characterization and NH_3 desorption kinetics. *Ind. Eng. Chem. Res.* **2014**, *53*, 15303–15316. [CrossRef]
44. Xue, N.; Chen, X.; Nie, L.; Guo, X.; Ding, W.; Chen, Y.; Gu, M.; Xie, Z. Understanding the enhancement of catalytic performance for olefin cracking: Hydrothermally stable acids in P/HZSM-5. *J. Catal.* **2007**, *248*, 20–28. [CrossRef]
45. Rakap, M.; Özkar, S. Intrazeolite cobalt(0) nanoclusters as low-cost and reusable catalyst for hydrogen generation from the hydrolysis of sodium borohydride. *Appl. Catal. B-Environ.* **2009**, *91*, 21–29. [CrossRef]
46. Lee, D.; Kim, J.J.; Ali, M.; Choung, J.W.; Lee, W.B.; Bae, J.W.; Park, M.J. Mechanistic kinetic modeling for catalytic conversion of DME to gasoline-range hydrocarbons over nanostructured ZSM-5. *Catal. Sci. Technol.* **2022**, *12*, 4798–4810. [CrossRef]
47. Magomedova, M.; Starozhiskaya, A.; Davidov, I.; Maximov, A.; Kravtsov, M. Dual-cycle mechanism based kinetic model for DME-to-olefin synthesis on HZSM-5-type catalysts. *Catalysts* **2021**, *11*, 1459. [CrossRef]
48. Sun, T.; Chen, W.; Xu, S.; Zheng, A.; Wu, X.; Zeng, S.; Wang, N.; Meng, X.; Wei, Y.; Liu, Z. The first carbon-carbon bond formation mechanism in methanol-to-hydrocarbons process over chabazite zeolite. *Chem* **2021**, *7*, 2415–2428. [CrossRef]
49. Sun, X.; Wang, Q.; Xu, L.; Liu, S. The role of acid strength of zeolites in liquid-phase alkylation of benzene with ethylene. *Catal. Lett.* **2004**, *94*, 75–79. [CrossRef]
50. Hill, I.M.; Ng, Y.S.; Bhan, A. Kinetics of butene isomer methylation with dimethyl ether over zeolite catalysts. *ACS Catal.* **2012**, *2*, 1742–1748. [CrossRef]
51. Song, Y.; Zhang, L.L.; Li, G.D.; Shang, Y.S.; Zhao, X.M.; Ma, T.; Zhang, L.M.; Zhai, Y.L.; Gong, Y.J.; Xu, J.; et al. ZSM-5 extrudates modified with phosphorus as a super effective MTP catalyst: Impact of the acidity on binder. *Fuel Process. Technol.* **2017**, *168*, 105–115. [CrossRef]
52. Hu, M.; Wang, C.; Gao, X.; Chu, Y.; Qi, G.; Wang, Q.; Xu, G.; Xu, J.; Deng, F. Establishing a link between the dual cycles in methanol-to-olefins conversion on H-ZSM-5: Aromatization of cycloalkenes. *ACS Catal.* **2020**, *10*, 4299–4305. [CrossRef]
53. Tian, L.; Li, J.; Li, Y.; Chen, B. Synthesis of dodecylbenzene with benzene and 1-dodecene over MCM-22 zeolite modified with phosphorus. *Chin. J. Catal.* **2008**, *29*, 889–894. [CrossRef]
54. Magomedova, M.; Galanova, E.; Davidov, I.; Afokin, M.; Maximov, A. Dimethyl ether to olefins over modified ZSM-5 based catalysts stabilized by hydrothermal treatment. *Catalysts* **2019**, *9*, 485. [CrossRef]
55. Khitev, Y.P.; Ivanova, I.I.; Kolyagin, Y.G.; Ponomareva, O.A. Skeletal isomerization of 1-butene over micro/mesoporous materials based on FER zeolite. *Appl. Catal. A-Gen.* **2012**, *441–442*, 123–135. [CrossRef]

Disclaimer/Publisher's Note: The statements, opinions and data contained in all publications are solely those of the individual author(s) and contributor(s) and not of MDPI and/or the editor(s). MDPI and/or the editor(s) disclaim responsibility for any injury to people or property resulting from any ideas, methods, instructions or products referred to in the content.

catalysts

MDPI

Review

Recent Progress and Strategies on the Design of Zeolite-Based Catalysts for Hydroformylation of Olefins

Maria V. Nenasheva and Dmitry N. Gorbunov *

Department of Chemistry, Lomonosov Moscow State University, Moscow 119991, Russia; mariia.nenasheva@chemistry.msu.ru
* Correspondence: gorbunovdn@petrol.chem.msu.ru

Abstract: Due to their unique structural and mechanical characteristics, zeolites are of great interest as carriers for the design of catalysts. This review is focused on the progress in the development of new zeolite-based catalysts for hydroformylation. The procedures of materials synthesis and modification, the methods of metal introduction, and the effect of the catalyst structure on the activity, selectivity, and stability of the hydroformylation catalysts are discussed. This review highlights the role of up-to-date analytical techniques in the development of fine-tuned active and selective hydroformylation catalysts. The greatest achievements over the past five years considered in detail in this review demonstrate a high potential of zeolite-based hydroformylation catalysts. Future prospects and possible directions of upcoming studies are regarded.

Keywords: zeolites; hydroformylation; olefin; catalyst separation

1. Introduction

Hydroformylation (Figure 1), the catalytic interaction of syngas with olefins to form aldehydes, represents a pivotal industrial process that constitutes a critical link in the production chain for a diverse range of valuable products, including plasticizers, surfactants, solvents, cosmetics, and pharmaceuticals [1].

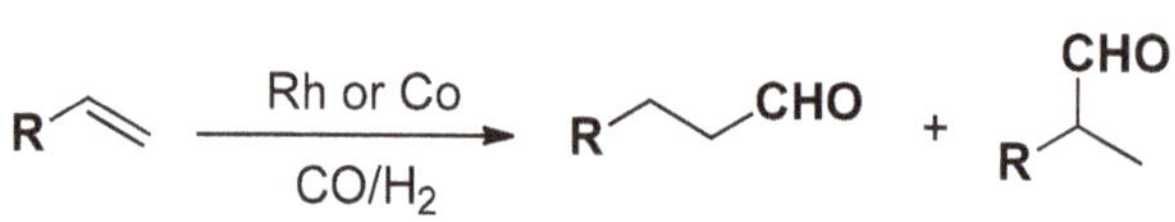

Figure 1. Hydroformylation reaction.

The global production of aldehydes by this reaction exceeds 10 million tons per annum. The initial catalysts employed in hydroformylation, which remain in use in certain processes, were cobalt complexes. However, subsequent research revealed that rhodium compounds exhibited considerably higher activity in this reaction. The difference in the activity of cobalt and rhodium complexes is estimated to be approximately three orders of magnitude [2]. The hydroformylation reaction typically results in the formation of linear and branched aldehydes, with the ratio dependent on the reaction conditions, the employed catalytic system, and electronic effects in the substrate molecule. In the majority of cases, linear aldehydes are regarded as being of greater value due to their higher reactivity, which facilitates further modification, as well as the higher biodegradability of their derivatives. One effective and frequently employed method to enhance the regioselectivity of rhodium-catalyzed hydroformylation on linear aldehyde is the incorporation of phosphine ligands into the catalytic system [3].

One of the most fundamental issues associated with industrial hydroformylation is the complexity of the process of heterogenization, given that under high syngas pressure, active metals are prone to leaching from solid carriers [4]. Presently, the majority of facilities operate under conditions of homogeneous catalysis, wherein the catalysts are hydridocarbonyl complexes of cobalt and rhodium in a dissolved state. This complicates the stage of separation of products from the catalyst, which necessitates the utilization of specialized facilities and a considerable expenditure of energy [5,6] (Table 1).

Table 1. Typical reaction conditions and approaches to catalyst recycling in industrial hydroformylation.

Process	Catalyst Type	Conditions	Products	Separation Procedure
BASF, Exxon	Cobalt carbonyl unmodified	25–30 MPa 130–190 °C	Butanals, higher aldehydes	Decomposition of the catalyst followed by Co extraction
Shell	Cobalt carbonyl modified with phosphine ligand	4–8 MPa 150–190 °C	Higher alcohols	Distillation of products with the addition of high-boiling components
UCC	Rh/PPh_3	1.5–2 MPa 85–115 °C	Lower aldehydes	Catalyst and excess ligand are dissolved in high-boiling dimeric and trimeric oxo-synthesis products, aldehydes evaporate during gas circulation and then condensed and removed
Ruhrchemie/ Rhone–Poulenc	Rh/water soluble phosphine TPPTS	1–10 MPa 50–130 °C	Lower aldehydes	Phase separation

The high cost of the rhodium compounds, which permit hydroformylation under milder conditions, implies particularly stringent requirements for efficient catalyst separation for reusability with minimal losses. This problem is especially crucial for the hydroformylation of higher olefinic substrates (>C_6) as, in this case, distillation of the high-boiling products is limited by the decomposition of Rh complexes at high temperatures. A substantial body of research is dedicated to the advancement of active and stable solid hydroformylation catalysts [7–10]. This might provide a simpler and more efficient type of technological scheme with a fixed bed reactor, which contains catalyst pellets with a gaseous or liquid substrate flowing through them, transforming into the desirable products. So, such catalysts can be universally applied for the hydroformylation of lower and higher olefins.

Over the past few decades, considerable advancements have been made in this field. The catalysts currently under development can be divided by the type of carrier into two principal groups: those based on polymers and those based on inorganic carriers, including those with grafted low- and macromolecular organic fragments [11]. Our research group has also published a number of papers on hybrid polymer-inorganic catalysts [12–14]. A comprehensive account of the advancements in the development of polymer-based hydroformylation catalysts can be found in reviews [15,16]. Notable are the catalysts based on POPs (porous organic polymers), developed by Chinese scientists. POPs are synthesized by polymerizing phosphine monomers modified with vinyl groups, after which rhodium is deposited on them from solutions of its complexes [17]. The catalysts are distinguished by their high stability and regioselectivity, which can be attributed to the robust binding of rhodium by phosphine fragments. A pilot plant for the hydroformylation of ethylene has recently been constructed, which employs a catalyst of this type. The plant has a capacity of approximately 50,000 tons per year and has been in continuous operation for over two years [18]. This is the first documented example of a solid hydroformylation catalyst functioning at such a scale. One disadvantage of this catalyst is the relatively complex synthesis procedure, which necessitates the use of toxic and readily oxidizable phosphorus (III) compounds. It has been shown that catalysts based on phosphorus-free polymers are, in general, significantly less stable [15,19]. Therefore, now that the principal applicability of solid catalysts to industrial hydroformylation has been demonstrated, studies of the applicability of other cheaper and more environmentally friendly types of heterogeneous catalysts appear to be even more pertinent.

2. Zeolite-Based Catalysts for the Hydroformylation of Olefins

Zeolites are porous crystalline aluminosilicates characterized by an ordered structure, high surface, regulated acidity, as well as mechanical strength and good chemical and thermal resistance. All these properties determine the high practical significance of zeolites as catalysts or their components and adsorbents [20,21]. Recently, zeolites have also been considered as promising materials for the synthesis of catalysts for a wide range of energy conversion, energy storage, and chemical synthesis processes [22–24]. Due to the chemical composition and geometry of channels and pores, zeolites ensure the thermal stability of catalysts and prevent the migration of metals. The ordered structure of the zeolites provides control over the processes of substrate adsorption, formation of intermediates, and removal of products from the reaction zone, which determine the overall activity and selectivity of the process [25–28]. In this regard, there is also great interest in the development of zeolite-based hydroformylation catalysts.

The earliest documented accounts of rhodium catalysts for hydroformylation based on zeolites can be traced back to the 1970s–1980s [29–31]. However, despite initial efforts, the desired level of stability and selectivity for n-aldehydes could not be attained. The formation of aldehydes with a normal structure was observed to occur at a rate that was approximately equal to or slightly higher than that of branched aldehydes. Additionally, the mass fraction of rhodium in the catalyst was observed to decrease with repeated use. As anticipated, an enhancement in regioselectivity was observed following the modification of the catalyst with a phosphine ligand [32]. However, this did not result in an increase in stability. In 2000, the authors of [33] developed a sufficiently stable hydroformylation catalyst based on NaY zeolite. This was achieved by boiling a $RhCl_3$ solution with NaY zeolite, followed by the addition of $NaBH_4$, formaldehyde, and triphenylphosphine to yield the $HRh(CO)(PPh_3)_3$ complex encapsulated in the zeolite. The *n:iso* ratio in the hydroformylation of hexene-1 was 1.6, which is not an exceptionally high value for a phosphine-containing system. Nevertheless, the catalyst was successfully used in 10 consecutive reactions without a reduction in the hydroformylation rate. In the study in [34], prior to rhodium deposition from the $HRh(CO)(PPh_3)_3$ solution, the zeolite Y surface was modified with phosphotungstic acid, resulting in a notable enhancement in the regioselectivity of the hydroformylation of linear olefins C_6–C_{10} in comparison to the homogeneous system with $HRh(CO)(PPh_3)_3$ (the *n:iso* ratio increased by 1.7–2.5-times at high conversions).

In the preceding five years, studies employing fundamentally different methodologies for the synthesis of catalysts have emerged. Approaches to the synthesis or modification of zeolite materials were developed, for example, by creating a secondary (hierarchical) structure, forming silanol nests, and placing heteroatoms in channels, both metal and non-metal. The presence of such structural features, along with the selected method of placing the active metal, most often Rh, made it possible to create a specific microenvironment around the metal center, providing the necessary steric and electronic effects. Interestingly, depending on the method of synthesis of a zeolite-based catalyst, not only the activity and stability but also the regioselectivity of the reaction largely vary. It is often individual for a certain type of substrate, which is associated with the sieve effect, i.e., the correspondence between the size of the substrate molecule and the shape selectivity of the catalyst. The key achievements in this sphere emerging in the last 5 years are summarized in Table 2. Below is a discussion of the most important issues.

Table 2. The most recent developments and studies of zeolite-based hydroformylation catalysts.

Structural Type	Method of Rh Introducing	Reactor Type	Conditions	Substrate	Yield of Aldehydes or TOF and *n:iso*	Stability	Ref.
MFI, with introdused P-atoms within the channels	Addition of $RhCl_3$/ethylenediamine complex in gel before hydrothermal synthesis	Flow	GSHV 8000 h^{-1}, 1.0 MPa, 90 °C	ethylene	70 h^{-1}, 95% selectivity to propanal	At least 100 min without loss in activity and selectivity at 90 °C	[35]
MFI-type silicalite	Addition of $RhCl_3$/ethylenediamine complex in gel before hydrothermal synthesis	Batch	5.5 MPa, 85 °C	propylene	7328 h^{-1} 5.0	200 h without loss in activity and selectivity [2]	[36]
MFI-type silicalite with hollow structure contain micro- and mesopores	Impregnation with $RhCl_3$ solution under ultrasonic conditions	Batch	5.0 MPa, 100 °C	hexene-1	3461 h^{-1} 0.7	5 cycles (12 h in total) with aldehyde yield >85%, but some decrease in activity was detected	[37]
MFI-type silicalite with core-shell structure	Previously formed Rh nanoparticles assisted by PVP were added to gel before hydrothermal synthesis	Batch	5.0 MPa, 90 °C	hexene-1	350 h^{-1} 2.5	3 cycles (15 h in total) with no detectable Rh leaching observed	[38]
MFI-type silicalite	Addition of $RhCl_3$ in gel before hydrothermal synthesis	Batch	5.0 MPa, 80 °C	decene-1	500 h^{-1}, 6.65	7 cycles (84 h in total) with gradual decrease in activity [1]	[39]
MFI	Addition of $RhCl_3$/ethylenediamine complex in gel before hydrothermal synthesis	Batch	2.0 MPa, 80 °C	decene-1	217 h^{-1} >300	8 cycles (40 h in total) without loss in activity and selectivity [3]	[40]
MFI-type silicalite with silanol nests d within crystals	Impregnation from $RhCl_3$ solution followed by readuction of Rh in H_2 flow	Batch	3.0 MPa, 110 °C	styrene	50,000 h^{-1} [4] 1.1	5 cycles (12.5 h in total) without loss in activity and selectivity	[41]
Dealuminated BEA	Grafting Co and Rh into silanol nests by impregnation	Flow	0.1 MPa, 130 °C	ethylene	153 h^{-1}	200 h without loss in activity	[42]
MEL	Addition of $RhCl_3$/ethylenediamine complex in gel before hydrothermal synthesis followed by oxidative disruption of Rh clusters	Batch	4.0 MPa, 90 °C	propylene	6500 h^{-1} 99	6 cycles (3 h in total) without loss in activity and selectivity, also catalyst was stable in 40-h test on stream	[43]
Zeolite NaX	Addition of $RhCl_3$/ethylenediamine complex in gel before hydrothermal synthesis	Batch	5.0 MPa, 100 °C	hexene-1	3047 h^{-1} 0.73	5 cycles (10 h in total) with aldehyde yield >85%, but some decrease in activity was detected	[44]
Zeolite Y	$RhCl_3$ was mixed with AEPTS [5], this solution was added to gel before hydrothermal synthesis	Batch	6.0 MPa, 60 °C	hexene-1	6567 h^{-1} 1.2	5 cycles (30 h in total) with no detectable Rh leaching observed	[45]

[1] hexene-1 was used as a substrate. [2] Catalyst 3 g, propylene:CO:H_2 = 1:1:1, 1 MPa, 15 mL min^{-1}, 140 °C, TOF 2.5 h^{-1}. [3] After each run, the catalyst was reduced in H_2 flow at elevated temperature. [4] Calculated based on metal dispersion. [5] 3-[2-(2-Aminoethylamino)ethylamino]propyltrimethoxysilane.

In [39], the synthesis of silicon material, referred to as "seeds", was initially conducted using tetraethoxysilane (TEOS) and tetrapropylammonium hydroxide (TPAOH), which were treated with a solution of (3-mercaptopropyl)trimethoxysilane (MPTMS). Subsequently, rhodium was introduced to the surface of the "seeds" from a $RhCl_3$ solution, after which they were mixed with supplementary quantities of TEOS and TPAOH for a hydrothermal synthesis. The procedure was repeated using particles of the resulting catalyst as "seeds" for the second-order structure (Rh_2O_3@S-1-II) (Figure 2). The Rh_2O_3@S-1-II catalyst exhibited a markedly high regioselectivity for the phosphorus-free system in the hydroformylation of decene-1. At an aldehyde yield of 94%, the *n:iso* ratio was 6.65, whereas for both octene-1 and dodecene-1, the value was lower [39]. It would appear that the structure of the carrier plays a pivotal role in the case of decene-1, allowing it to enter microchannels but hindering its isomerization and formation of branched intermediates. In contrast, for olefins of smaller size, these processes are possible, whereas dodecene-1 is mainly hydroformylated on the outer surface of the catalyst. The catalyst could be employed on multiple occasions; however, its efficacy diminished gradually due to the leaching of rhodium.

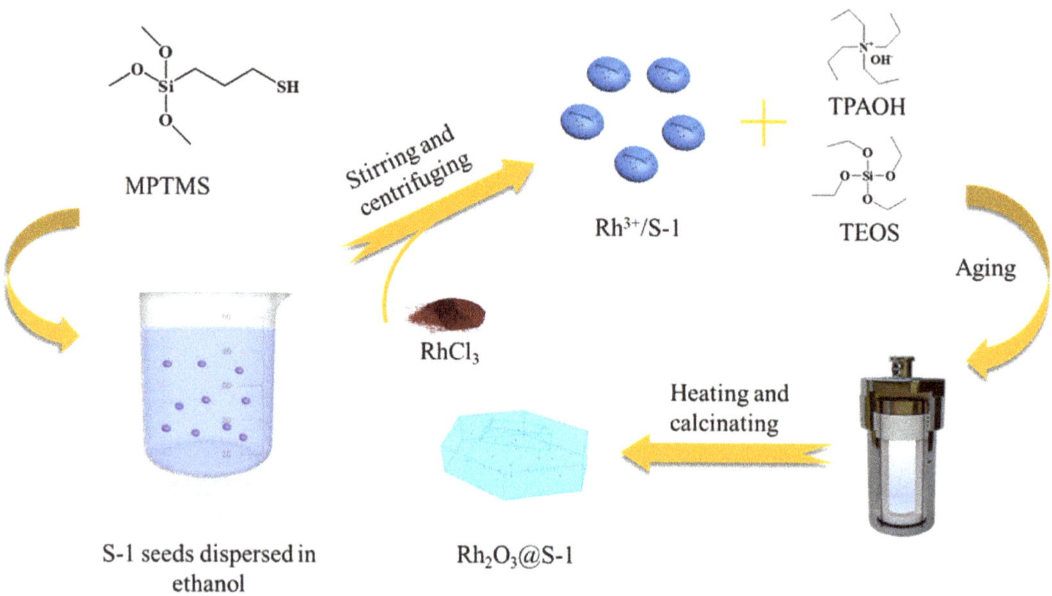

Figure 2. The synthesis of core–shell Rh_2O_3@S-1 catalysts [39].

In [36], a more straightforward approach was employed: the complex [Rh($NH_2CH_2CH_2$ CH_2NH_2)$_3$]Cl_3 was introduced to a solution of TEOS and TPAOH following a six-hour stirring period, prior to the hydrothermal synthesis stage. Also, potassium hydroxide was added to the mixture. After the hydrothermal synthesis, the rhodium was reduced with hydrogen. The procedure allowed for the production of a catalyst with significantly enhanced stability compared to the conventional impregnation method. The catalyst demonstrated sustained activity for 200 h in the hydroformylation of propylene, with a TOF value of 7328 h^{-1}, representing a record value for inorganic hydroformylation catalysts at the time. The introduction of KOH during synthesis resulted in a notable enhancement in the selectivity of the hydroformylation of long-chain olefins on *n*-aldehydes. The authors of [36] ascribe this to a decrease in the acidity of the surface of the material and a deceleration in isomerization.

In a study published in 2024 [40], a series of catalysts based on MFI zeolite with various Rh content were prepared using a similar technique with a modified order of reagent addition. The addition of KOH was made at an earlier stage of the synthesis of zeolite, specifically when preparing the initial mixture of TPAOH template and TEOS. The HAADF-STEM data indicate that rhodium is present within the catalyst structure in the form of clusters with a diameter < 1 nm, which are localized within the zeolite channels. Additionally, X-ray absorption spectroscopy (XAS) was employed to detect Rh_n clusters (n < 10) directly under reaction conditions. Among the Rh-MFI catalysts with various Rh contents, the catalyst containing 0.22% Rh wt (Rh-MFI-0.22) displayed the highest activity in the hydroformylation of decene-1 (Figure 3a). It also contained the highest fraction of 0.3–0.6 nm Rh species (Figure 3b) [40]. So, one can conclude that these species are most active in the target reaction.

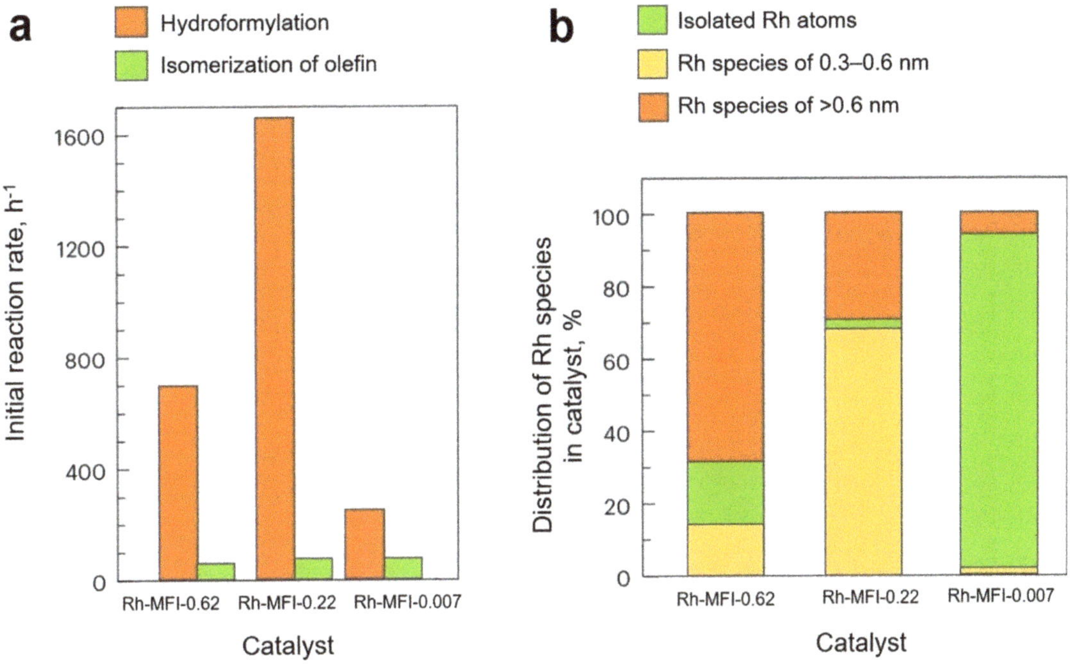

Figure 3. (**a**) The normalized initial reaction rates for the production of aldehydes and iso-olefins on various Rh catalysts at conversion levels below 15%; (**b**) the percentage of different types of Rh species in three representative Rh-MFI catalysts: Rh-MFI-0.62, Rh-MFI-0.22 and Rh-MFI-0.007 [40].

The resulting catalyst exhibited an exceptionally high regioselectivity in the hydroformylation of long-chain olefins; at the conversion approaching quantitation, the formation of n-aldehyde from decene-1 occurred with 100% regioselectivity. While the authors of [40] noted the presence of trace amounts of branched aldehyde, its concentration was found to be more than 300-times lower. The regioselectivity remained at a high level throughout eight consecutive catalytic experiments conducted at high conversions. The TOF value at low conversions was estimated to be 217 h^{-1}. High *n:iso* ratios were demonstrated in the hydroformylation of C_6–C_{11} olefins. Additionally, in this study, catalysts based on other zeolites were obtained. The catalysts Rh-MEL, Rh-MOR, Rh-FER, Rh-TON, Rh-*BEA and Rh-MWW were also tested in hydroformylation. The catalytic results showed that zeolites with three-dimensional pore structures (MFI, MEL and *BEA) are effective supports for Rh-catalyzed hydroformylation of linear α-olefins, whereas the zeolites with two-dimensional (MWW and FER) or one-dimensional (TON and MOR) pore structures exhibit relatively low activities. However, for the structures other than MFI, the amounts of linear and branched aldehydes were comparable (*n:iso* < 10). It is noteworthy that the MFI-based catalyst obtained through the classical impregnation method exhibited reduced regioselectivity compared to the catalyst obtained through the introduction of a rhodium complex with ethylenediamine at the stage of zeolite synthesis; however, it demonstrated a considerably higher regioselectivity compared to catalysts based on other zeolites. This suggests that the selectivity is largely influenced by the structural characteristics of the zeolite material.

In [35], a P-containing Rh-MFI catalyst was obtained through a similar procedure but without the addition of alkali during the synthesis. A portion of TPAOH was replaced by tetraethylphosphonium hydroxide (Rh-(O)-P-MFI-calred). After calcination of the catalyst, the phosphorus present in it was observed to undergo oxidation, resulting in the formation of phosphate groups, as evidenced by ^{31}P NMR data. An in situ XAS study revealed the formation of isolated Rh^{3+} centers under reaction conditions as a result of the oxidative disruption of Rh clusters (Figure 4). In the case of the catalyst obtained with tetraethylphosphonium hydroxide, more such centers are formed than in its phosphorus-free analogue (Rh-MFI-calred). Extensive in situ XAS and IR-MS studies of both phosphorus-containing and phosphorus-free reduced and non-reduced (Rh-MFI-cal) catalysts led the authors of [35] to the conclusion that isolated Rh^{3+} centers are preferred catalytic precursors for ethylene hydroformylation under the condition applied (50–90 °C, 1.0 MPa), in contrast to the classical point of view that lower Rh oxidation states make more effective precursors. So, the stabilization of isolated Rh^{3+} in the Rh-(O)-P structure is supposed to be responsible for the enhanced activity of the phosphorus-containing catalyst, as well as for the preservation of the hydroformylation activity at higher temperatures (200 °C). In this study, TOF values of approximately 30–100 h^{-1} were obtained in the continuous flow hydroformylation of ethylene under comparatively mild conditions (50–90 °C, 1.0 MPa) (Table 3).

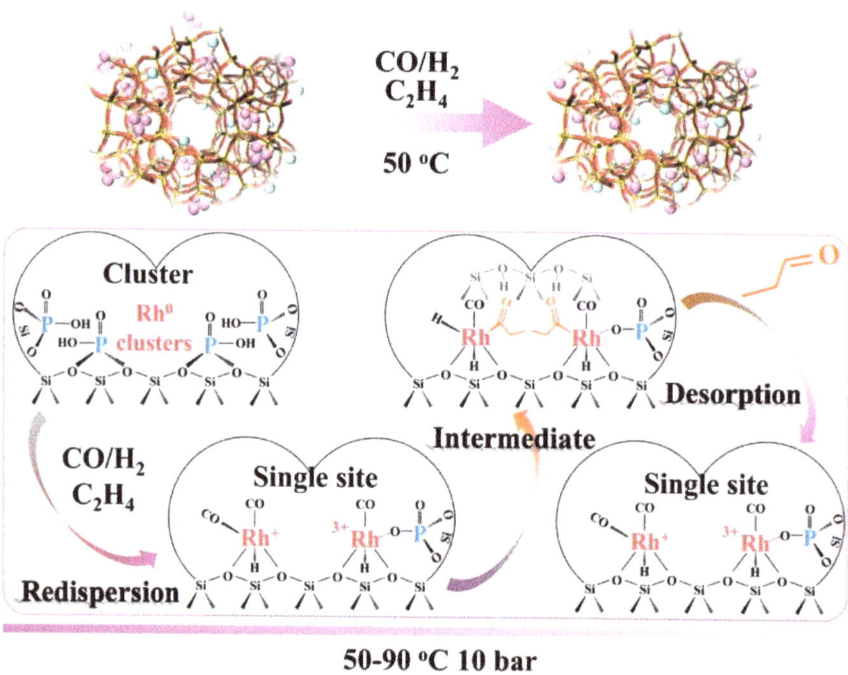

Figure 4. Dynamic behavior of Rh-(O)-P-MFI-calred catalyst under ethylene hydroformylation conditions [35].

The activity of the Rh-OD@MEL catalyst in the hydroformylation of propylene was investigated in a paper published in 2024 [43]. Its precursor material, Rh@MEL, was synthesized using a procedure similar to that described in [40] but with tetrabutylammonium hydroxide employed as a templating agent. It was then exposed to carbon monoxide for the oxidative disruption of rhodium clusters, which resulted in the formation of the Rh-OD@MEL catalyst. The XAS data corroborate the hypothesis that Rh-OD@MEL contains rhodium predominantly in smaller clusters compared to Rh@MEL. Additionally, an increase in the average oxidation degree of Rh was detected. Based on EXAFS, STEM and IR spectroscopy data, the authors of [43] proposed that on the surface of Rh-OD@MEL clusters, rhodium forms Rh(I)- gem-dicarbonyl complexes that are active in hydroformylation. In the hydroformylation of propylene, Rh-OD@MEL demonstrated exceptional selectivity for butanal (*n:iso* > 90), remarkable activity (TOF 6500 h^{-1}), and impressive stability when employed in six consecutive hydroformylation experiments.

Table 3. Catalytic performance of Rh-MFI-cal, Rh-MFI-calred, and Rh-(O)-P-MFI-calred [35].

Catalyst	Reaction Temperature, °C	Propanal TOF, h^{-1}	Selectivity to Propanal, %
Rh-MFI-cal	50	31.0	95
	90	98.7	92
Rh-MFI-calred	50	30.5	91
	90	58.8	96
Rh-(O)-P-MFI-calred	50	45.9	95
	90	70.0	95

Reaction conditions: m(catalyst) = 100 mg, P = 1.0 MPa, $CO/H_2/C_2H_4/N_2$ molar ratio 1:1:1:0.5, total flow rate 35 mL/min, GSHV = 8000 h^{-1}.

In [44], Rh@NaX was synthesized from the gel obtained from silica gel powder, $NaAlO_2$, and NaOH, to which a solution of $RhCl_3$ and ethylendiamine was added before crystallization, followed by calcination. Also, the Rh/NaX catalyst with a similar Rh content was prepared via the impregnation of NaX with $RhCl_3$ solution for the comparison. XPS studies showed that, while in Rh/NaX, Rh^{+3} is the only Rh oxidation stage, in Rh@NaX, along with Rh^{+3}, $Rh^{\delta+}$ species were formed with a lower binding energy ($0 < \delta+ < 1$). The two catalysts demonstrated close TOF values (3047 and 2631 h^{-1}, for Rh@NaX and Rh/NaX, respectively) and similar regioselectivity (*n:iso* = 0.73) in hexene-1 hydroformylation; however, the stability of Rh@NaX was higher and Rh leaching was lower. Also, in contrast with Rh/NaX, it was able to provide an aldehyde yield of >85% in five consecutive runs. Nevertheless, for both catalysts, a decrease in the activity was detected. The low regioselectivity of Rh@NaX indicates that, in the case of NaX, the zeolite structure has no positive effect. Predictably, higher regioselectivity (*n:iso* = 2.59) and activity (TOF = 6588 h^{-1}) were reached after the treatment of Rh@NaX with PPh_3, but no stability tests for Rh@NaX/PPh_3, crucial for phosphine-containing catalysts, were performed [44].

A comparison between three methods of Rh introduction into the S-1 structure was conducted in [37]. For the catalysts obtained different Rh deposition procedures (Table 4), TOF values in hexene-1 hydroformylation were determined to be 3471 h^{-1}, 3461 h^{-1}, and 1725 h^{-1}, and *n:iso* ratios were in the 0.7–0.8 range.

Table 4. Catalytic performance of Rh/S-1, Rh@S-1, and Rh@H-S-1 [37].

Catalyst	Preparation Method	Yield of C_7 Aldehydes, %	*n:iso*	TOF, h^{-1}
Rh/S-1	Impregnation of S-1 with $RhCl_3$ solution under ultrasonic conditions	81.6	0.71	3471
Rh@S-1	$RhCl_3$/ethylenediamine added to the synthesis gel of S-1 before its crystallizations	95.2	0.79	1725
Rh@H-S-1	Rh/S-1 mixed with TPAOH and subjected to further crystallization to obtain mesoporous network	96.0	0.74	3461

Reaction conditions: m (hexene-1) = 1 g, P = 5 MPa, CO/H_2 = 1, toluene = 4.5 mL, m (catalyst) = 0.1 g, T = 100 °C, t = 2 h.

The lowest TOF of Rh@S-1 was explained by diffusion restrictions in this catalyst; at longer reaction times, it demonstrated aldehyde yields comparable to that obtained with two other catalysts. As in the previous study, XPS data showed that Rh@S-1 and Rh@H-S-1, in contrast to Rh/S-1, contained Rh not only in the +3 oxidation state but also at a lower one, δ+ (0 < δ+ < +1). Its proportion increases under reaction conditions. The stability of the performance of the two encapsulated catalysts through the five recycling experiments was also proved to be better than that of Rh/S-1. So, Rh@H-S-1 provided the best combination of activity and stability of the three. After recycling, no Rh signals were observed in Rh@H-S-1 XPS, detecting that hydroformylation takes place on the internal surface. The authors of [37] did not report if its regioselectivity improved after recycling.

Another type of encapsulation procedure was demonstrated in [45] for the synthesis of a Rh@Y catalyst. First, $RhCl_3$ was stirred for 1 h with 3-[2-(2-Aminoethylamino) ethylamino]propyltrimethoxysilane (AEPTS); after this, alkaline, $NaAlO_2$, and SiO_2 gel were added, and the mixture was subjected to crystallization, followed by calcination. According to XPS, XAS, and HAADF-STEM data, Rh was present in the catalyst in the form of atomically dispersed $Rh^{\delta+}$ species with 1 < δ+ < +3, coordinated mostly to light atoms, presumably oxygen with its state being mainly unchanged after use in hydroformylation. The catalyst was more active in hexene-1 hydroformylation compared to several impregnated Rh/zeolite catalysts, including Rh/Y. The TOF for Rh@Y reached 6567 h^{-1}, which is among the highest values for phosphorous-free hydroformylation catalysts. Rh@Y was tested in the hydroformylation of 22 unsaturated substrates, including aromatic and sterically hindered ones. It demonstrated moderate to quantitative yields of aldehydes (35–100%) in all the reactions, in addition to 1,3-butadiene and isoprene hydroformylation (0%). For terminal aliphatic olefins, a slight excess of linear product was formed (*n:iso* up to 1.38 at conversions >90%) [45]. Also, the catalyst was successfully used in five recycles of hexene-1 hydroformylation, with no detectable Rh leaching observed.

In their study, the authors of [42] treated zeolite H-Beta with nitric acid in order to remove Al. After this, they treated the material with solutions of rhodium and cobalt nitrates (the resulting catalyst is noted as 0.08Rh-0.36Co-DeAlBEA) or rhodium and zinc nitrates (0.08Rh-0.36Zn-DeAlBEA). Following the removal of Al, silanol nests were formed within the structure, which were subsequently occupied by the introduced metal ions (Figure 5a). The catalysts were subjected to hydrogen current treatment and subsequently evaluated in the hydroformylation of ethylene. The stability of the catalysts was ascertained to be satisfactory, as evidenced by the absence of a decline in the propanal yield over the course of 300 h experiments in a continuous flow regime. The rhodium–cobalt catalyst exhibited significantly higher activity than the rhodium–zinc catalyst (Table 5).

Table 5. Catalytic performance of 0.08Rh-0.36Co-DeAlBEA and 0.08Rh-0.36Zn-DeAlBEA [42].

Catalyst	CN (M–Rh):CN (Rh–Rh) [1]	Propanal Selectivity %	TOF, h^{-1}
0.08Rh-0.36Co-DeAlBEA	1.63	69	153
0.08Rh-0.36Zn-DeAlBEA	0.83	79	12

Reaction conditions: m (catalyst) = 20 mg; T = 130 °C; total feed flow rate = 90 mL/min; $p(C_2H_4)$ = 33.8 kPa; p(CO) = 33.8 kPa; $p(H_2)$ = 11.3 kPa, balanced with He up to atmospheric total pressure. [1] Coordination number (by X-ray absorption spectroscopy).

The XAS data indicate that no oxygen atoms remain in the rhodium near-coordination sphere after hydrogen treatment, and the occurrence of Rh-Rh and Rh-M (M = Zn or Co) bonds was recorded. In the case of the more active rhodium–cobalt catalyst, the ratio of Rh-M/Rh-Rh coordination numbers was higher than for the rhodium–zinc catalyst. Therefore, the rhodium–cobalt catalyst was characterized by a higher dispersion of rhodium centers (Figure 5b). It is noteworthy that, though cobalt complexes can be active in hydroformylation on their own, both the cobalt and zinc catalysts, when tested without the addition of rhodium, demonstrated no activity in the hydroformylation of ethylene under the conditions investigated. This indicates that the discrepancy in activity between the rhodium–cobalt and rhodium–zinc catalysts is attributable to the state of the rhodium rather than the direct involvement of the second metal in the reaction.

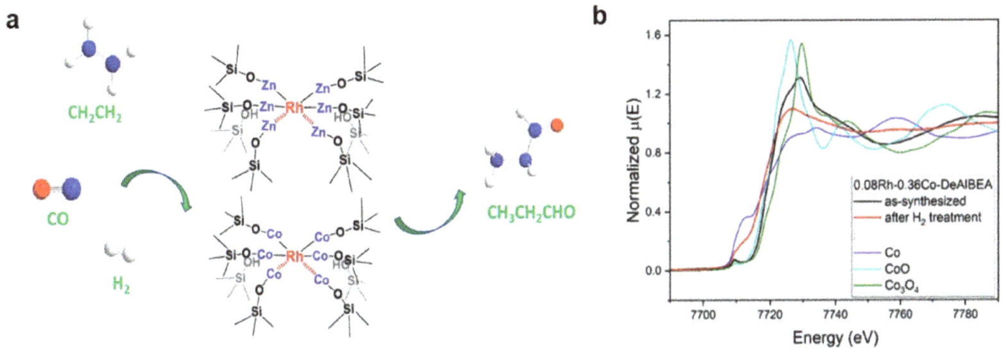

Figure 5. (**a**) Schematic image of Rh-Zn-DeAlBEA and Rh-Co-DeAlBEA catalysts; (**b**) Co K-edge near-edge spectra of 0.08Rh-0.36Co-DeAlBEA, as-prepared and after H_2 treatment at 200 °C and of cobalt foil, CoO, and Co_3O_4 standards [42].

In [41], the concept of the artificial creation of silanol nests within zeolite channels of siliceous MFI zeolite was also employed. This was achieved by replacing a portion of TEOS with diethoxydimethylsilane during synthesis. The mixture was then subjected to drying, resulting in the formation of an amorphous material comprising methyl groups. It was subsequently crystalized at an elevated temperature in the presence of TPAOH. Upon calcination, the methyl groups converted to hydroxyl groups (S1-OH). In NMR spectra (Figure 6a), in addition to the signal of the terminal silanol (~2.0 ppm) mainly located on the external surface of zeolite crystals, S1-OH exhibited an additional broad and obvious signal at 3.0–8.0 ppm, which is attributed to the silanols with hydrogen bond interactions, located within the channels. Rhodium was deposited on the resulting material through impregnation with $RhCl_3$ solution, followed by reduction with hydrogen. The catalyst contained rhodium in the form of nanoparticles with an average diameter of approximately 2.5 nm (Figure 6b). Despite containing only 0.06% rhodium by mass, the catalyst exhibited remarkably high activity in hydroformylation, with a TOF value of 50,000 h^{-1} (by calculation, taking into account the Rh dispersion) for the hydroformylation of styrene. This surpasses the activity of numerous homogeneous rhodium catalysts. It is also worth mentioning that the catalyst was considerably more active compared to its analogue obtained from the zeolite without the silanol nests inside the channels. The yields of the aldehyde decreased slightly with repeated use, but the authors of [41] highlight that there was a corresponding decrease in the mass of the catalyst. It is noteworthy that the hydroformylation of linear olefins predominantly yielded branched aldehydes, whereas in the hydroformylation of *p*-methylstyrene, the *n:iso* ratio was 1.1:1. This suggests that the catalyst was highly active in the isomerization of linear olefins.

Tsubaki et al. [38] added pre-formed rhodium nanoparticles to the mixture prior to the hydrothermal synthesis of the S-1 zeolite-based catalyst (Figure 7). The *n:iso* ratio in the hydroformylation of hexene-1 at an aldehyde yield of 85% was 2.5:1, which is slightly higher than the values typically observed in classical phosphorus-free systems. The catalyst was successfully employed in three consecutive experiments, and according to X-ray fluorescence analysis, no notable decline in rhodium content was discerned.

Despite Rh being the most active metal in hydroformylation, its high cost causes the interest in the cheaper alternatives. In 2022, the Au hydroformylation catalysts based on S-1 zeolite were reported [46]. The authors of [46] prepared encapsuled Au@S-1 catalysts by adding $HAuCl_4$/ethylenediamine solution to the mixture before the zeolite hydrothermal synthesis, as well as Au/S-1 catalysts via the impregnation of the pre-synthesized S-1 with $HAuCl_4$/ethylenediamine solution. Further, the catalysts of both types were calcinated. According to TEM and HAADF-STEM data, in Au/S-1 catalysts, gold nanoparticles and nanoclusters were observed on the catalyst surface, while for encapsuled Au@S-1 catalysts, surface Au was detected neither by TEM nor by XPS, although EDS mapping showed a uniform distribution of gold for these samples, indicating that Au species were encapsulated inside the material. Interestingly, in propylene hydroformylation, for both types of catalysts, a decrease in aldehyde yields was observed with an increase in Au loading in a catalyst (Figure 8a,b) (general Au amount was kept constant). Encapsulated catalysts had superior activity to the impregnated ones. The most active of the series was encapsulated catalyst Au(0.04%)@S-1 with a TOF of about 200 h^{-1}, which is comparable to some Rh catalysts. According to EXAFS data, encapsulated catalysts with lower Au loadings are characterized by the highest proportion of Au-O coordination compared to encapsulated catalysts with more Au as well as those impregnated with similar Au loading; for both, Au-Au coordination predominated (Figure 8c). This indicates that Au single atoms are the most active gold species in the catalysts of this series. The Au(0.2%)@S-1 catalyst was successfully used in five recycles with no loss in the yield of C_4 aldehydes.

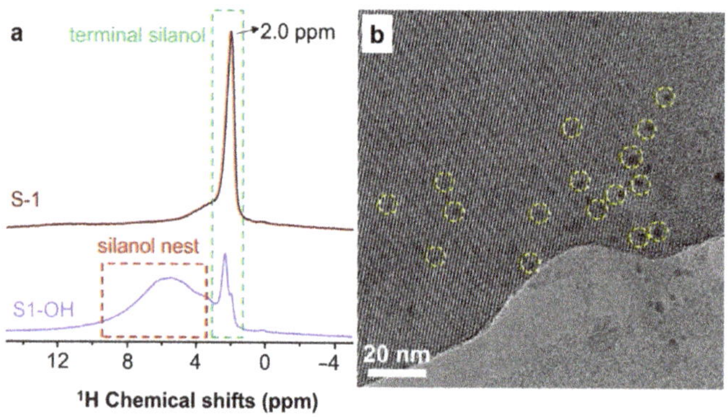

Figure 6. (**a**) ^{1}H MAS NMR spectra of S1-OH and S-1 zeolites; (**b**) TEM image of Rh/S1-OH [41].

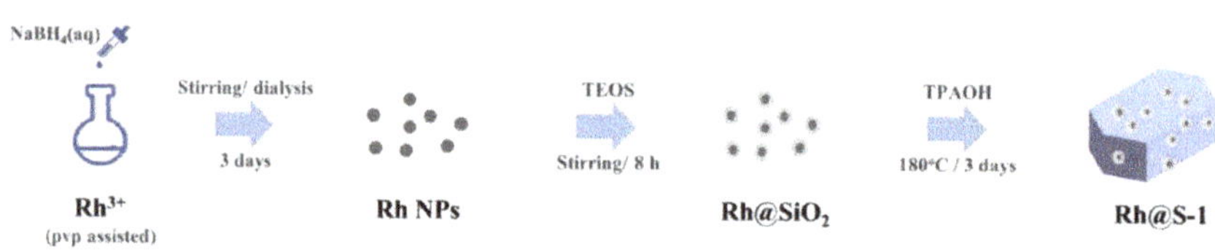

Figure 7. Schematic illustration of the steps for the preparation of the Rh@S-1 [38].

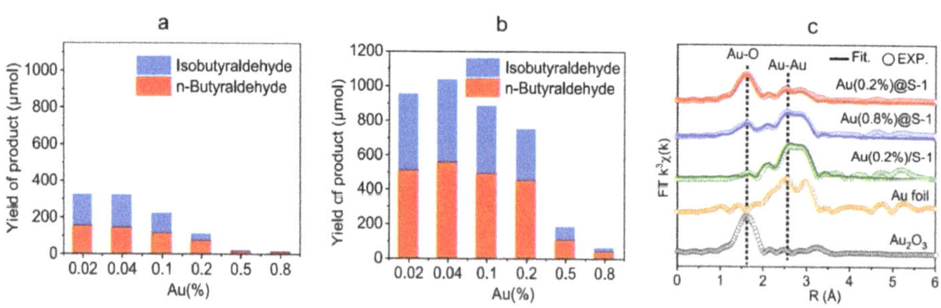

Figure 8. (**a**) Catalytic performances of different Au contents in Au/S-1. A certain amount of catalyst (corresponding to $2.54 \times 3 \times 10^{-4}$ mmol of Au), propylene (0.4 MPa), syngas (4 MPa; CO/H_2 = 1:1), 15 mL toluene; (**b**) catalytic performances of different Au contents in Au@S-1. A certain amount of catalyst (corresponding to $2.54 \times 3 \times 10^{-4}$ mmol of Au), propylene (0.4 MPa), syngas (4 MPa; CO/H_2 = 1:1), 15 mL toluene; (**c**) EXAFS spectra of the species at the Au L3-edge [46].

3. Zeolite-Based Catalysts for Tandem and One-Pot Processes Involving Hydroformylation

Recently, researchers have turned their attention to the topic of tandem reactions based on hydroformylation on zeolite catalysts. These allow for the synthesis of more complex products in a single stage, which would otherwise be obtained through two or more sequential reactions in separate units and on different catalysts in classical production [47]. In [48], the acidic function of zeolite was employed for the formation of acetals from aldehydes obtained by the hydroformylation reaction, thereby facilitating a tandem hydroformylation–acetalization reaction (Figure 9a). Acetals of the acyclic and cyclic structure are considered as promising solvents, detergents, emulsifiers, components of cosmetics, technical liquids, fuel additives, agrochemicals, and also as fragrances [49–53]. In [48], Rh was introduced into the ZSM-35(10) structure in the form of a carbonyl complex with sterically hindered diphosphorus ligands. This resulted in a high regioselectivity in the hydroformylation stage (*n:iso* > 30.5).

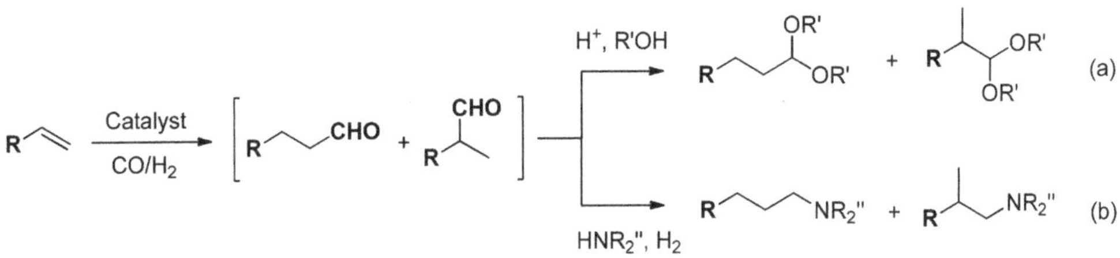

Figure 9. (**a**) Tandem hydroformylation–acetalization and (**b**) hydroaminomethylation.

Hydroaminomethylation is a sequence of hydroformylation, condensation of the formed aldehydes with secondary or primary amines, and hydrogenation of its products to tertiary or secondary amines, respectively (Figure 9b). It is an effective way to produce amines, which are used to give access to biologically active molecules, dyes, solvents, and functional materials [54,55]. In [56], this sequence was performed on a rhodium catalyst based on ZSM-5. The catalyst contained a diphosphine sterically hindered ligand, sulfoxantphos. This ligand has polar sulfogroups, which enhanced the interaction of the rhodium complex with the polar zeolite support. Furthermore, the introduction of the ligand resulted in the attainment of high regioselectivity (*n:iso* > 120).

As hydroformylation is a large-scale process using a C1 reagent, it is often regarded from the perspective of low-carbon economy [57,58]. This includes the studies of the use of CO_2 as a possible alternative C1 source for hydroformylation. For one-pot hydroformylation with CO_2, most often, homogeneous systems are proposed [59–63]; also, recently, polymeric POP catalysts for this process were reported [63,64]. In these systems, CO_2 is first converted to CO via a Reverse Water Gas Shift reaction (RWGS) with the excess of H_2, and the unreacted H_2 is further involved in hydroformylation along with the formed CO. De Vos et al. [65] recently reported the use of the optimized encapsulated Pt catalyst based on NaA zeolite for RWGS in such a system. RWGS is followed by hydroformylation catalyzed by classical Rh/Phosphine homogeneous systems in the same reactor (Figure 10). The main point of the zeolite introduction was to physically part RWGS from hydroformylation in a more economically effective one-pot system to prevent the hydrogenation of an unsaturated substrate with hydrogen, which is activated during RWGSR. This was achieved for the RWGS-hydroformylation of hexene-1 under comparatively mild conditions (ligand L, 1.5 MPa of CO_2, 0.5 MPa of H_2, 120 °C, 20 h, 70% yield of aldehydes, *n:iso* = 24), as well as with other high olefins, while ethylene and propylene were mainly hydrogenated to the corresponding alkanes [65]. Presumably, low-molecular substrates effectively diffused into the pores of zeolite, where they were hydrogenated on Pt centers, simultaneously restraining the access of CO_2 to them, thus inhibiting its transformation to

CO. Using common Pt-supported catalysts (e.g., Pt/Al_2O_3) or Pt on zeolites with larger pores resulted in the complete hydrogenation of octene-1.

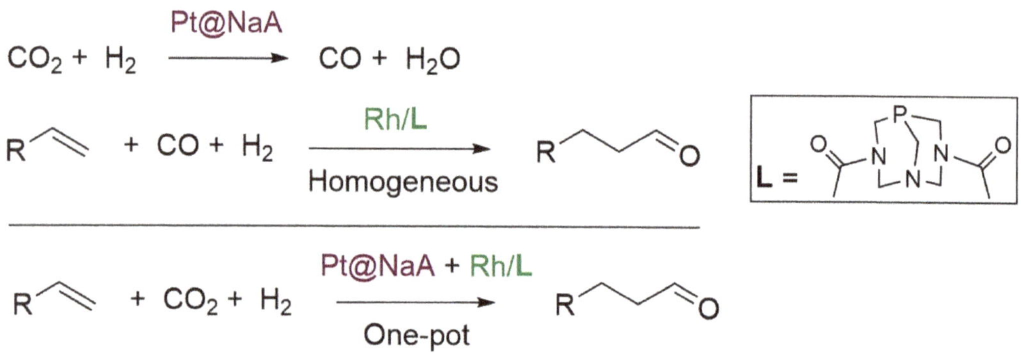

Figure 10. RWGS-hydroformylation over Pt@NaA and Rh/L [65].

Another interesting process allowing for the involvement of CO_2 in hydroformylation is CO_2-assisted dehydrogenation and reforming of ethane to produce ethylene, CO, and H_2 followed by the hydroformylation reaction between the obtained ethylene, CO, and H_2. Through this process, C3 oxygenates could be synthesized from two greenhouse gases [66]. For now, the two stages of this process are conducted in two separate fixed bed reactors. In [67], ethane was converted by the Fe-Ni/CeO_2 catalyst at 600–800 °C, and hydroformylation was catalyzed by Rh-Co/MCM-41 at 200 °C. Presumably, taking into account their high efficiency in gaseous-phase hydroformylation, zeolite-based catalysts are good candidates for the second stage of this promising sequence.

4. Conclusions and Future Prospects

The combination of shape selectivity and the possibility of fine tuning of the metal center electronic properties makes zeolites excellent candidates as carriers for heterogeneous hydroformylation catalysts. They can be regarded as solid analogues of classical organic ligands, characterized by more environmentally friendly synthetic procedures. The catalytic properties are defined by the characteristics of zeolite channels, the method of deposition of the active metal, the size of metal particles, and the microenvironment. Some advantages were achieved by such modifications of initial materials as the creation of a mesoporous network, silanol nests within zeolite channels, and the formation of a secondary hierarchic structure. The introduction of heteroatoms (metal or non-metal) inside the material regulates the electronic properties of the active metal center. Notably, the exceptionally high activity in the hydroformylation of styrene (TOF 50,000 h^{-1}) was achieved on the catalyst based on MFI-type material as well as the highest known *n*-regioselectivity in the hydroformylation of linear α-olefins among phosphorous-free catalytic systems (close to 100%). For propylene hydroformylation, the best results were obtained using an MEL zeolite-based Rh catalyst (TOF = 6500 h^{-1}, *n:iso* = 99). As a field of future work, the development of the selective zeolite-based catalysts for the hydroformylation of a broader scope of substrates, such as functionalized olefins, conjugated dienes, and bio-derived molecules, can be regarded. For a deep understanding of structure–activity relations, essential is high-throughput investigation of metal active centers and specific features of the zeolite framework, with a combination of up-to-date analytical techniques, such as high-resolution TEM, DRIFT, ex situ, and operando XAS. Methods of computational chemistry could also provide some significant insights into the optimization of the catalyst structure and reaction regioselectivity. Additionally, considering the achievements in this field during the last five years, pilot-scale mini-plant investigations are highly desirable.

Author Contributions: Conceptualization, writing—original draft preparation, M.V.N.; writing—review and editing, D.N.G. All authors have read and agreed to the published version of the manuscript.

Funding: This research was funded by a grant from the Ministry for Education and Science of the Russian Federation (agreement of 24 April 2024 no. 075-15-2024-547).

Data Availability Statement: No new data were created or analyzed in this study.

Conflicts of Interest: The authors declare no conflicts of interest.

References

1. Franke, R.; Selent, D.; Börner, A. Applied Hydroformylation. *Chem. Rev.* **2012**, *112*, 5675–5732. [CrossRef]
2. Börner, A.; Franke, R. (Eds.) *Hydroformylation*; Wiley-VCH Verlag GmbH & Co. KGaA: Weinheim, Germany, 2016; ISBN 9783527677931.
3. van Leeuwen, P.W.N.M.; Kamer, P.C.J. Featuring Xantphos. *Catal. Sci. Technol.* **2018**, *8*, 26–113. [CrossRef]
4. Yu, Z.; Zhang, S.; Zhang, L.; Liu, X.; Jia, Z.; Li, L.; Ta, N.; Wang, A.; Liu, W.; Wang, A.; et al. Suppressing Metal Leaching and Sintering in Hydroformylation Reaction by Modulating the Coordination of Rh Single Atoms with Reactants. *J. Am. Chem. Soc.* **2024**, *146*, 11955–11967. [CrossRef] [PubMed]
5. Zhao, K.; Wang, X.; He, D.; Wang, H.; Qian, B.; Shi, F. Recent Development towards Alkene Hydroformylation Catalysts Integrating Traditional Homo- and Heterogeneous Catalysis. *Catal. Sci. Technol.* **2022**, *12*, 4962–4982. [CrossRef]
6. Gorbunov, D.N.; Volkov, A.V.; Kardasheva, Y.S.; Maksimov, A.L.; Karakhanov, E.A. Hydroformylation in Petroleum Chemistry and Organic Synthesis: Implementation of the Process and Solving the Problem of Recycling Homogeneous Catalysts (Review). *Pet. Chem.* **2015**, *55*, 587–603. [CrossRef]
7. Liu, B.; Wang, Y.; Huang, N.; Lan, X.; Xie, Z.; Chen, J.G.; Wang, T. Heterogeneous Hydroformylation of Alkenes by Rh-Based Catalysts. *Chem* **2022**, *8*, 2630–2658. [CrossRef]
8. Gao, W.; Liu, S.; Wang, Z.; Peng, J.; Zhang, Y.; Yuan, X.; Zhang, X.; Li, Y.; Pan, Y. Outlook of Cobalt-Based Catalysts for Heterogeneous Hydroformylation of Olefins: From Nanostructures to Single Atoms. *Energy Fuels* **2024**, *38*, 2526–2547. [CrossRef]
9. Samanta, P.; Canivet, J. MOF-Supported Heterogeneous Catalysts for Hydroformylation Reactions: A Minireview. *ChemCatChem* **2024**, *16*, e202301435. [CrossRef]
10. Yu, X.; Ji, Y.; Jiang, Y.; Lang, R.; Fang, Y.; Qiao, B. Recent Development of Single-Atom Catalysis for the Functionalization of Alkenes. *Catalysts* **2023**, *13*, 730. [CrossRef]
11. Sandee, A.J.; van der Veen, L.A.; Reek, J.N.H.; Kamer, P.C.J.; Lutz, M.; Spek, A.L.; van Leeuwen, P.W.N.M. A Robust, Environmentally Benign Catalyst for Highly Selective Hydroformylation. *Angew. Chem. Int. Ed.* **1999**, *38*, 3231–3235. [CrossRef]
12. Gorbunov, D.; Safronova, D.; Kardasheva, Y.; Maximov, A.; Rosenberg, E.; Karakhanov, E. New Heterogeneous Rh-Containing Catalysts Immobilized on a Hybrid Organic–Inorganic Surface for Hydroformylation of Unsaturated Compounds. *ACS Appl. Mater. Interfaces* **2018**, *10*, 26566–26575. [CrossRef]
13. Gorbunov, D.; Nenasheva, M.; Naranov, E.; Maximov, A.; Rosenberg, E.; Karakhanov, E. Tandem Hydroformylation/Hydrogenation over Novel Immobilized Rh-Containing Catalysts Based on Tertiary Amine-Functionalized Hybrid Inorganic-Organic Materials. *Appl. Catal. A Gen.* **2021**, *623*, 118266. [CrossRef]
14. Gorbunov, D.N.; Nenasheva, M.V.; Kuvandykova, E.A.; Kardashev, S.V.; Karakhanov, E.A. Promising Applications of Polyethyleneimine as a Ligand in Rhodium-Catalyzed Tandem Hydroformylation/Hydrogenation of Olefins. *Pet. Chem.* **2023**, *63*, 594–606. [CrossRef]
15. Zhuchkov, D.P.; Nenasheva, M.V.; Terenina, M.V.; Kardasheva, Y.S.; Gorbunov, D.N.; Karakhanov, E.A. Polymeric Heterogeneous Catalysts in the Hydroformylation of Unsaturated Compounds. *Pet. Chem.* **2021**, *61*, 1–14. [CrossRef]
16. Wang, H.; Sun, J.; Cai, X.; Zhu, Y. Porous Organic Polymers Supported Heterogeneous Catalysts for Hydroformylation Reactions. *Eur. J. Inorg. Chem.* **2024**, *27*, e202400254. [CrossRef]
17. Sun, Q.; Dai, Z.; Liu, X.; Sheng, N.; Deng, F.; Meng, X.; Xiao, F.-S. Highly Efficient Heterogeneous Hydroformylation over Rh-Metalated Porous Organic Polymers: Synergistic Effect of High Ligand Concentration and Flexible Framework. *J. Am. Chem. Soc.* **2015**, *137*, 5204–5209. [CrossRef]
18. Feng, S.; Jiang, M.; Song, X.; Qiao, P.; Yan, L.; Cai, Y.; Li, B.; Li, C.; Ning, L.; Liu, S.; et al. Sulfur Poisoning and Self-Recovery of Single-Site Rh 1 /Porous Organic Polymer Catalysts for Olefin Hydroformylation. *Angew. Chem.* **2023**, *135*, e202304282. [CrossRef]
19. Gorbunov, D.N.; Nenasheva, M.V.; Terenina, M.V.; Kardasheva, Y.; Naranov, E.R.; Bugaev, A.L.; Soldatov, A.V.; Maximov, A.L.; Tilloy, S.; Monflier, E.; et al. Phosphorus-Free Nitrogen-Containing Catalytic Systems for Hydroformylation and Tandem Hydroformylation-Based Reactions. *Appl. Catal. A Gen.* **2022**, *647*, 118891. [CrossRef]
20. Martínez, C.; Corma, A. Zeolites. In *Comprehensive Inorganic Chemistry II*; Elsevier: Amsterdam, The Netherlands, 2013; pp. 103–131.
21. Yilmaz, B.; Müller, U. Catalytic Applications of Zeolites in Chemical Industry. *Top. Catal.* **2009**, *52*, 888–895. [CrossRef]
22. Zhang, H.; Samsudin, I.b.; Jaenicke, S.; Chuah, G.-K. Zeolites in Catalysis: Sustainable Synthesis and Its Impact on Properties and Applications. *Catal. Sci. Technol.* **2022**, *12*, 6024–6039. [CrossRef]

23. Ulmanu, M.; Anger, I. (Eds.) Zeolite Utilization in Solar Energy Storage and Heat Exchange. In *Handbook of Natural Zeolites*; Bentham Science Publishers: Sharjah, United Arab Emirates, 2012; pp. 369–398.
24. Sánchez-López, P.; Kotolevich, Y.; Yocupicio-Gaxiola, R.I.; Antúnez-García, J.; Chowdari, R.K.; Petranovskii, V.; Fuentes-Moyado, S. Recent Advances in Catalysis Based on Transition Metals Supported on Zeolites. *Front. Chem.* **2021**, *9*, 716745. [CrossRef]
25. Zhang, Y.; Shi, S.; Wang, Z.; Lan, H.; Liu, L.; Sun, Q.; Guo, G.; He, X.; Ji, H. Propane Dehydrogenation on Ir Single-atom Catalyst Modified by Atomically Dispersed Sn Promoters in Silicalite-1 Zeolite. *AIChE J.* **2024**, *70*, e18431. [CrossRef]
26. Zhang, Y.; Sun, Q.; Guo, G.; Cheng, Y.; Zhang, X.; Ji, H.; He, X. Trace Pt Atoms as Electronic Promoters in Pd Clusters for Direct Synthesis of Hydrogen Peroxide. *Chem. Eng. J.* **2023**, *451*, 138867. [CrossRef]
27. Ma, M.; Li, G.; Yan, W.; Wu, Z.; Zheng, Z.; Zhang, X.; Wang, Q.; Du, G.; Liu, D.; Xie, Z.; et al. Single-Atom Molybdenum Engineered Platinum Nanocatalyst for Boosted Alkaline Hydrogen Oxidation. *Adv. Energy Mater.* **2022**, *12*, 2103336. [CrossRef]
28. Han, M.; He, Y.; Yu, T.; Peng, P.; Shi, J.; Liu, H.; Liu, L.; Ye, C.; Chen, Q. Atomically Precise Control of Silver Species Encaged in Zeolite Catalysts with Minimal Loading for Maximal Performance. *ACS Catal.* **2024**, *14*, 8856–8864. [CrossRef]
29. Mantovani, E.; Palladino, N.; Zanobi, A. Zeolite Entrapped Rhodium Carbonyl Clusters as Catalyst for Liquid Phase Hydroformylation of Olefins. *J. Mol. Catal.* **1978**, *3*, 285–291. [CrossRef]
30. Arai, H. Hydroformylation and Hydrogenation of Olefins over Rhodium Zeolite Catalyst. *J. Catal.* **1982**, *75*, 188–189. [CrossRef]
31. Takahashi, N. Comparison of Ethylene with Propylene Hydroformylation over a Rh-Y Zeolite Catalyst under Atmospheric Pressure. *J. Catal.* **1984**, *85*, 89–97. [CrossRef]
32. Rode, E.; Davis, M.E.; Hanson, B.E. Hydroformylation of Propene with Zeolite-Supported Rhodium Phosphine Complexes. *J. Chem. Soc. Chem. Commun.* **1985**, 1477–1478. [CrossRef]
33. Mukhopadhyay, K.; Nair, V.S.; Chaudhari, R.V. *Hydroformylation of Olefins Using Encapsulated HRh (CO)((PPh3)3 in NaY Zeolite as a Catalyst*; Elsevier: Amsterdam, The Netherlands, 2000; pp. 2999–3003.
34. Mukhopadhyay, K.; Chaudhari, R.V. Heterogenized $HRh(CO)(PPh_3)_3$ on Zeolite Y Using Phosphotungstic Acid as Tethering Agent: A Novel Hydroformylation Catalyst. *J. Catal.* **2003**, *213*, 73–77. [CrossRef]
35. Zhao, M.; Li, C.; Gómez, D.; Gonell, F.; Diaconescu, V.M.; Simonelli, L.; Haro, M.L.; Calvino, J.J.; Meira, D.M.; Concepción, P.; et al. Low-Temperature Hydroformylation of Ethylene by Phosphorous Stabilized Rh Sites in a One-Pot Synthesized Rh-(O)-P-MFI Zeolite. *Nat. Commun.* **2023**, *14*, 7174. [CrossRef]
36. Zhang, J.; Sun, P.; Gao, G.; Wang, J.; Zhao, Z.; Muhammad, Y.; Li, F. Enhancing Regioselectivity via Tuning the Microenvironment in Heterogeneous Hydroformylation of Olefins. *J. Catal.* **2020**, *387*, 196–206. [CrossRef]
37. Sun, G.; Fan, S.; Lv, J.; Zhao, T. Effective Hollow Rh@H-S-1 Catalyst for Hydroformylation of 1-Hexene. *J. Catal.* **2024**, *439*, 115770. [CrossRef]
38. Fang, Y.; Zhao, H.; Guo, X.; He, Y.; Yang, G.; Tsubaki, N. Silicalite-1 Encapsulated Rhodium Nanoparticles for Hydroformylation of 1-Hexene. *Catal. Today* **2023**, *410*, 150–156. [CrossRef]
39. Liu, C.; Zhang, J.; Liu, H.; Qiu, J.; Zhang, X. Heterogeneous Ligand-Free Rhodium Oxide Catalyst Embedded within Zeolitic Microchannel to Enhance Regioselectivity in Hydroformylation. *Ind. Eng. Chem. Res.* **2019**, *58*, 21285–21295. [CrossRef]
40. Dou, X.; Yan, T.; Qian, L.; Hou, H.; Lopez-Haro, M.; Marini, C.; Agostini, G.; Meira, D.M.; Zhang, X.; Zhang, L.; et al. Regioselective Hydroformylation with Subnanometre Rh Clusters in MFI Zeolite. *Nat. Catal.* **2024**, *7*, 666–677. [CrossRef]
41. Liu, Y.; Liu, Z.; Hui, Y.; Wang, L.; Zhang, J.; Yi, X.; Chen, W.; Wang, C.; Wang, H.; Qin, Y.; et al. Rhodium Nanoparticles Supported on Silanol-Rich Zeolites beyond the Homogeneous Wilkinson's Catalyst for Hydroformylation of Olefins. *Nat. Commun.* **2023**, *14*, 2531. [CrossRef] [PubMed]
42. Qi, L.; Das, S.; Zhang, Y.; Nozik, D.; Gates, B.C.; Bell, A.T. Ethene Hydroformylation Catalyzed by Rhodium Dispersed with Zinc or Cobalt in Silanol Nests of Dealuminated Zeolite Beta. *J. Am. Chem. Soc.* **2023**, *145*, 2911–2929. [CrossRef] [PubMed]
43. Zhang, X.; Yan, T.; Hou, H.; Yin, J.; Wan, H.; Sun, X.; Zhang, Q.; Sun, F.; Wei, Y.; Dong, M.; et al. Regioselective Hydroformylation of Propene Catalysed by Rhodium-Zeolite. *Nature* **2024**, *629*, 597–602. [CrossRef]
44. Sun, G.; Chen, W.; Li, Y.; Fan, S.; Lv, J.; Zhao, T. In Situ Synthesis of Rh@NaX Catalyst for 1-Hexene Hydroformylation. *Fuel* **2024**, *373*, 132327. [CrossRef]
45. Shang, W.; Qin, B.; Gao, M.; Qin, X.; Chai, Y.; Wu, G.; Guan, N.; Ma, D.; Li, L. Efficient Heterogeneous Hydroformylation over Zeolite-Encaged Isolated Rhodium Ions. *CCS Chem.* **2023**, *5*, 1526–1539. [CrossRef]
46. Wei, B.; Liu, X.; Chang, Q.; Li, S.; Luo, H.; Hua, K.; Zhang, S.; Chen, J.; Shao, Z.; Huang, C.; et al. Single-Atom Gold Species within Zeolite for Efficient Hydroformylation. *Chem. Catal.* **2022**, *2*, 2066–2076. [CrossRef]
47. Petricci, E.; Cini, E. Domino Reactions Triggered by Hydroformylation. In *Hydroformylation for Organic Synthesis*; Springer: Berlin/Heidelberg, Germany, 2013; pp. 117–149.
48. Shu, Z.; Zhao, X.-X.; Zheng, Z.; Zhang, X.; Zhang, Y.; Sun, S.; Chen, J.; Xie, C.; Yuan, B.; Jia, X. Co-Catalysis of Rhodium/Phosphoramidite Catalyst and ZSM-35(10) for the Tandem Hydroformylation–Acetalization of Olefins. *Chem. Commun.* **2023**, *59*, 5237–5240. [CrossRef] [PubMed]
49. Gorbunov, D.N.; Nenasheva, M.V.; Sinikova, N.A.; Kardasheva, Y.S.; Maksimov, A.L.; Karakhanov, E.A. Tandem Hydroformylation–Acetalization Using a Water-Soluble Catalytic System: A Promising Procedure for Preparing Valuable Oxygen-Containing Compounds from Olefins and Polyols. *Russ. J. Appl. Chem.* **2018**, *91*, 990–995. [CrossRef]
50. Frusteri, F.; Spadaro, L.; Beatrice, C.; Guido, C. Oxygenated Additives Production for Diesel Engine Emission Improvement. *Chem. Eng. J.* **2007**, *134*, 239–245. [CrossRef]

51. De Torres, M.; Jiménez-osés, G.; Mayoral, J.A.; Pires, E.; de los Santos, M. Glycerol Ketals: Synthesis and Profits in Biodiesel Blends. *Fuel* **2012**, *94*, 614–616. [CrossRef]
52. García, E.; Laca, M.; Pérez, E.; Garrido, A.; Peinado, J. New Class of Acetal Derived from Glycerin as a Biodiesel Fuel Component. *Energy Fuels* **2008**, *22*, 4274–4280. [CrossRef]
53. Hölscher, B. Acetals and Ketals as Fragrances and Flavors. European Patent EP2757905A1, 19 August 2015.
54. Chen, C.; Dong, X.-Q.; Zhang, X. Recent Progress in Rhodium-Catalyzed Hydroaminomethylation. *Org. Chem. Front.* **2016**, *3*, 1359–1370. [CrossRef]
55. Roose, P.; Eller, K.; Henkes, E.; Rossbacher, R.; Höke, H. Amines, Aliphatic. In *Ullmann's Encyclopedia of Industrial Chemistry*; Wiley: Hoboken, NJ, USA, 2015; pp. 1–55.
56. Yuan, X.; Zhang, X.; Zheng, Z.; Sun, S.; Jia, X.; Dong, S. Highly Active and Regioselective Hydroaminomethylation of Olefins Catalyzed by Rh/Sulfoxantphos with ZSM-5. *Chem. Commun.* **2024**, *60*, 4667–4670. [CrossRef]
57. Skrydstrup, T.; Bonde, A.; Jakobsen, J.; Ahlers, A.; Huang, W.; Jackstell, R.; Beller, M. Integrating Hydroformylations into a Methanol Economy. *Res. Sq.* **2024**. [CrossRef]
58. Jolly, B.J.; Pung, M.J.; Liu, C. Integrated Electrochemical CO_2 Reduction and Hydroformylation. *Dalton Trans.* **2024**, *53*, 18834–18838. [CrossRef]
59. Gorbunov, D.N.; Nenasheva, M.V.; Terenina, M.V.; Kardasheva, Y.S.; Kardashev, S.V.; Naranov, E.R.; Bugaev, A.L.; Soldatov, A.V.; Maximov, A.L.; Karakhanov, E.A. Transformations of Carbon Dioxide under Homogeneous Catalysis Conditions (A Review). *Pet. Chem.* **2022**, *62*, 1–39. [CrossRef]
60. Wu, A.; Ding, J.; Zhao, L.; Li, H.; He, L. Hydroformylation of Olefins with CO_2/H_2 and Hydrosilane by Copper/Cobalt Tandem Catalysis. *ChemSusChem* **2024**, *17*, e202400608. [CrossRef] [PubMed]
61. Qian, C.; Zheng, Q.; Ji, J.; Liang, G.; Tu, B.; Tu, T. Imidazolium Carboxylates Boost Ru-Catalyzed Hydroaminomethylation of Olefins with CO_2. *Chem Catal.* **2023**, *3*, 100787. [CrossRef]
62. Tominaga, K.; Sasaki, Y. Ruthenium-Catalyzed One-Pot Hydroformylation of Alkenes Using Carbon Dioxide as a Reactant. *J. Mol. Catal. A Chem.* **2004**, *220*, 159–165. [CrossRef]
63. Sun, J.; Zhao, K.; Wang, H.; Shi, F. Heterogeneous Bisphosphine Ligand-Encapsulated Rh Catalysts for Heterogeneous Hydroformylation of Alkenes with CO_2. *ACS Sustain. Chem. Eng.* **2024**, *12*, 12736–12743. [CrossRef]
64. Wang, G.; Jiang, M.; Sun, Z.; Qian, L.; Ji, G.; Ma, L.; Li, C.; Wang, Z.; Yang, Y.; Lin, X.; et al. Synergistic Effect between Monophosphine Species for Regioselective Hydroformylation of Olefin with CO_2. *Chem. Eng. J.* **2023**, *476*, 146332. [CrossRef]
65. Van Dessel, H.; Van Minnebruggen, S.; Dedapper, J.; Paciok, P.; Usoltsev, O.; Krajnc, A.; Bugaev, A.; De Vos, D.E. Shape-Selective Zeolites for Tandem CO_2 Hydrogenation-Carbonylation Reactions. *Angew. Chem. Int. Ed.* **2024**, e202418670. [CrossRef]
66. Biswas, A.N.; Winter, L.R.; Xie, Z.; Chen, J.G. Utilizing CO_2 as a Reactant for C_3 Oxygenate Production via Tandem Reactions. *JACS Au* **2023**, *3*, 293–305. [CrossRef]
67. Xie, Z.; Xu, Y.; Xie, M.; Chen, X.; Lee, J.H.; Stavitski, E.; Kattel, S.; Chen, J.G. Reactions of CO_2 and Ethane Enable CO Bond Insertion for Production of C_3 Oxygenates. *Nat. Commun.* **2020**, *11*, 1887. [CrossRef]

Disclaimer/Publisher's Note: The statements, opinions and data contained in all publications are solely those of the individual author(s) and contributor(s) and not of MDPI and/or the editor(s). MDPI and/or the editor(s) disclaim responsibility for any injury to people or property resulting from any ideas, methods, instructions or products referred to in the content.

catalysts

MDPI

Communication

SiO_2@Fe(III)-Based Metal–Organic Framework Core–Shell Microspheres for Water-Purification-Based Photo-Fenton Processes

Kaihong Liu [†], Yuanli Zhu [†], Tanyu Cheng, Guohua Liu * and Chunxia Tan *

Shanghai Frontiers Science Center of Biomimetic Catalysis, Joint Laboratory of International Cooperation of Resource Chemistry of Ministry of Education, Shanghai Normal University, Shanghai 200234, China; 1000479980@smail.shnu.edu.cn (K.L.)

* Correspondence: ghliu@shnu.edu.cn (G.L.); tanchx@shnu.edu.cn (C.T.)

† These authors contributed equally to this work.

Abstract: In this study, SiO_2@MIL-88A(Fe) core–shell microspheres were successfully synthesized through a simple immobilization method for dye degradation via an MIL-88A(Fe)-mediated Fenton-like process. These microspheres were fabricated by in situ immobilizing MIL-88A(Fe) onto mesoporous organosilane spheres functionalized with -COOH groups. Structural analyses and characterizations confirmed the formation of well-defined MOF particles anchored on the silicate microspheres, with electron microscopy verifying their porous core–shell structure. The newly developed core–shell materials achieved a high degree of dye degradation, reaching up to 96% for 10 mg/L dye solutions in neutral aqueous conditions within 30 min at room temperature through the Fenton-like process. Furthermore, SiO_2@MIL-88A(Fe) exhibited excellent stability and recyclability, maintaining its performance over at least seven reuse cycles with minimal loss of activity. This material is easy to synthesize as well as cost-effective and demonstrates significant potential for wastewater purification involving a range of four different dyes.

Keywords: MOFs; core–shell microspheres; photo-Fenton process

1. Introduction

The urgent need for efficient technologies to address water pollution has become particularly pressing due to severe dye contamination caused by rapidly developing industries and economies [1–5]. As a result, various methods, including biological treatment [6,7], adsorption [8,9], membrane separation [10–12], column separation [13], and advanced oxidation [14,15], have been proposed to alleviate water pollution. Among these methods, the advanced oxidation process (AOP) [16] is widely regarded as an effective approach for tackling toxic, recalcitrant organic pollutants [17]. In particular, Fenton-type processes have garnered significant attention [18] due to their ability to degrade organic pollutants via highly reactive hydroxyl radicals, converting them into environmentally friendly products, such as carbon dioxide and water, using activated H_2O_2 via ferrous ion catalysis [19,20]. The Fenton process is recognized for its high efficiency, simplicity, and suitability for wastewater treatment. However, the key to its successful application in water pollution treatment lies in the use of appropriate, efficient, and highly stable materials [21–23].

Metal–organic frameworks (MOFs) are 3D porous materials consisting of periodically connected organic linkers and inorganic metal ions or clusters [24–28]. They have attracted significant attention as a new type of photocatalyst due to their remarkable ligand-metal charge transfer (LMCT) properties under light irradiation [29,30]. In particular, Fe-based MOFs show great potential as photocatalysts owing to their Fe_3–μ_3-oxo clusters and low toxicity, which enable a heterogeneous photo-Fenton-like reaction for wastewater treatment under light irradiation [31–33]. However, the large, tightly packed 3D structure of MOF particles can impede substrate diffusion, sometimes blocking it in the nanosized MOF channels [34,35]. Growing MOFs on larger matrixes can reduce crystal growth due to the matrix's nucleation effect [36], thereby enhancing mass transport speed.

Here, we directly grew MIL-88A(Fe) [37] on silica to produce an MIL-88A(Fe)@SiO_2 core–shell composite material. The MIL-88A(Fe)@SiO_2 was first prepared analogously to MIL-88A(Fe)@SiO_2, except with the addition of a 0.25 weight equivalent of SiO_2. The surface silanol groups on SiO_2 served as nucleation sites for the formation of fusiform-shaped MIL-88A(Fe) particles with an average dimension of ~0.06 × 0.2 μm. Due to the enhanced nucleation on the silica surface, the MOF particles in the composite material were smaller than pure MOF particles (0.3 × 2 μm) [37,38] and were homogeneously packed on the silica. The photo-Fenton catalytic activity of MIL-88A(Fe)@SiO_2 in the dye degradation reaction was investigated. In this technique, water was used as a solvent in mild conditions, and the catalyst showed excellent catalytic capacity. Furthermore, MIL-88A(Fe)@SiO_2 showed excellent stability and recyclability and could be reused at least seven times without significant loss of activity. The easy and cost-effective construction of this heterogeneous catalyst, MIL-88A(Fe)@SiO_2, makes it a candidate for applications in high-level environmental engineering.

2. Results and Discussion

2.1. Synthesis and Structural Characterization of the Core–Shell Catalyst

Core–shell-structured MIL-88A@SiO_2 microspheres were synthesized using a simple two-step postgrafting complexation procedure, as illustrated in Scheme 1. In the first step, the co-condensation of tetraethoxysilane (TEOS) and 3-aminopropyltriethoxysilane (APTES), followed by modification with succinic anhydride, resulted in carboxylate-terminated core–shell nanoparticles (COOH@SiO_2), as described in the reported method. In the second step, a postgrafting process was carried out by modifying MIL-88A onto the outer surface of the nanoparticles. The MIL-88A@SiO_2 composite was obtained through a complexation process in water and stirred for 12 h.

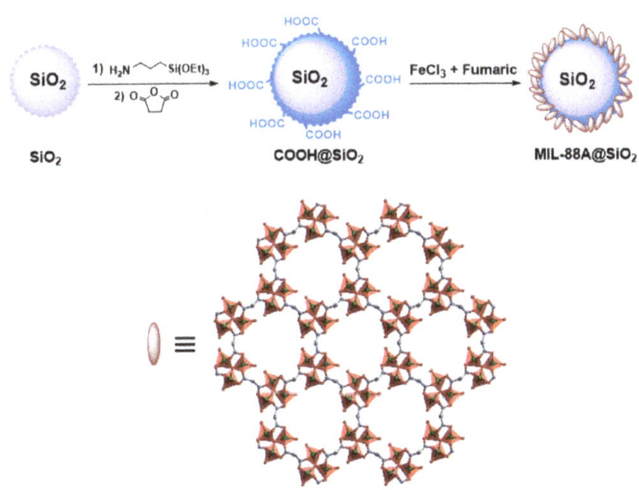

Scheme 1. Preparation of the core–shell-structured MIL-88A@SiO_2 microsphere particles.

The microstructure of the core–shell microspheres was observed by SEM (Figure 1). The core–shell microspheres exhibited a rough surface, densely covered with MIL-88A particles. Moreover, the morphology of the MIL-88A particles on the surface was consistent with their original polyhedral structure, confirming the successful preparation of the MIL-88A@SiO_2 core–shell material. The phase purity of MIL-88A@SiO_2 was verified by comparing the observed and simulated powder X-ray diffraction (PXRD) patterns (Figure 2). The thermal gravimetric analysis (TGA) indicated that the MIL-88A loading in the core–shell material was 11.953 mg (0.214 mmol) per gram of catalyst. This result aligns with the Fe loading of 0.198 mmol (11.06 mg) per gram of catalyst, as detected by inductively coupled plasma optical emission spectrometer (ICP-OES) analysis.

Figure 1. The SEM images of MIL-88A@SiO_2.

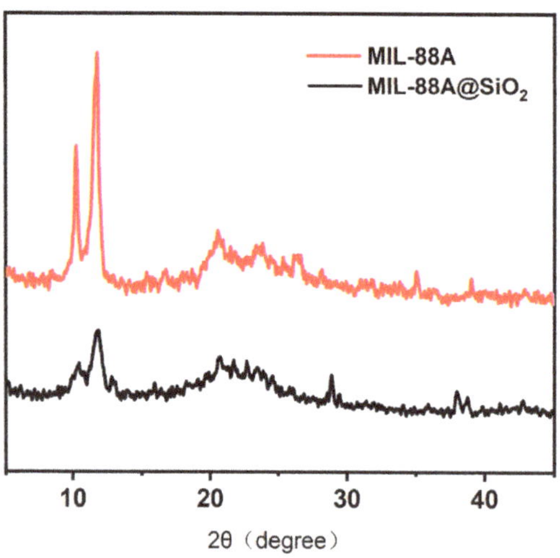

Figure 2. XRD patterns for MIL-88(A) and MIL-88(A)@SiO_2.

The FT-IR spectrum was collected to analyze the molecular structure of the as-synthesized core–shell material (Figure 3). The strong absorption bands corresponding to carboxylate groups were separated into asymmetric (ν_asCOO) and symmetric (ν_sCOO) stretching vibrations, appearing at 1607 and 1396 cm^{-1}, respectively. This indicates that the carboxyl groups of fumaric acid linkers were coordinated to the metal centers. The strong peak at 1093 cm^{-1} was attributed to the stretching vibration of the CO-NH bond in the carboxylate-terminated SiO_2 microspheres. XPS spectroscopy of the as-prepared MIL-88A@SiO_2 was conducted to examine the chemical composition and surface electronic states (Figure 4). The XPS survey spectrum confirmed the presence of C, O, and Fe in MIL-88A@SiO_2 (Figure S4). The high-resolution XPS spectrum of Fe 2p (Figure 4) shows two peaks at binding energies of 725.47 eV and 712.12 eV, corresponding to Fe $2p_1/_2$ and Fe $2p_3/_2$, respectively. These binding energies are consistent with the Fe^{3+} in α-Fe_2O_3 [20]. Additionally, the satellite peak is a characteristic feature of Fe^{3+}. The nitrogen adsorption–desorption isotherm (Figure S3) for MIL-88A@SiO_2 demonstrates its porous structure, exhibiting a typical type IV isotherm with an H1 hysteresis loop and a distinct step at P/P_0 = 0.82–0.95. The pore size distribution reveals uniform mesopores of approximately 3.8 nm (see Figure S3 in the SI), which is comparable to the pore structure of pure MIL-88A [39,40].

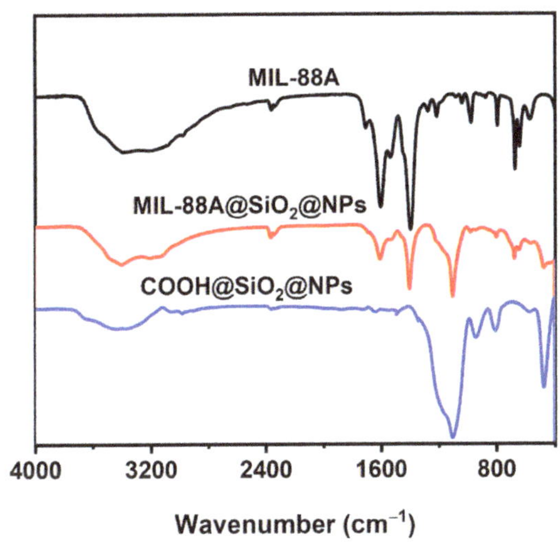

Figure 3. The FT-IR spectrum.

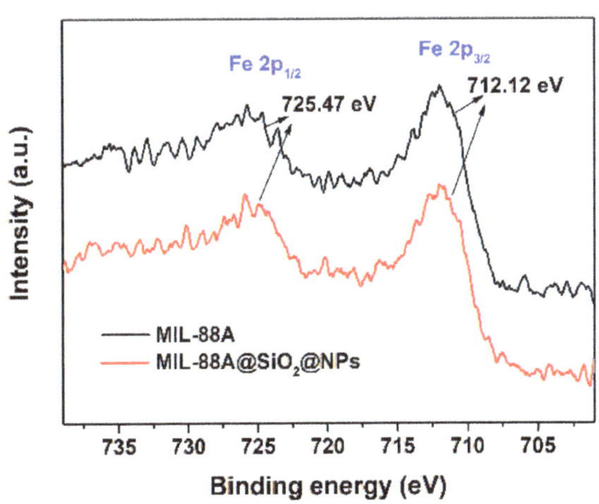

Figure 4. The XPS spectrum.

2.2. Catalytic Performance of the Heterogeneous Catalyst

Using the well-defined SiO_2@Fe(III) core–shell microspheres, acid red 18 (AR-18) was selected as a target pollutant due to being a common organic contaminant used in the assessment of various advanced oxidation processes (AOPs). A series of control experiments were conducted to evaluate the degradation of AR-18 under varying amounts of H_2O_2 and catalyst loadings, as shown in Figure 5a. It was observed that approximately 83% of the AR-18 was degraded after 20 min with the addition of 0.5 mL of H_2O_2. Excessive amounts of H_2O_2 were found to slow the degradation, while no degradation occurred in the absence of H_2O_2. Catalyst loading was also optimized for AR-18 degradation; without MIL-88A@SiO_2, degradation did not proceed, and the optimal degradation (up to 84%) was achieved with 20 mg of catalyst loading. These results confirm that MIL-88A@SiO_2 exhibits catalytic activity for the Fenton-like reaction. Similar findings have been reported by other researchers using different Fe-based MOFs as heterogeneous Fenton catalysts for the degradation of organic contaminants [39,40].

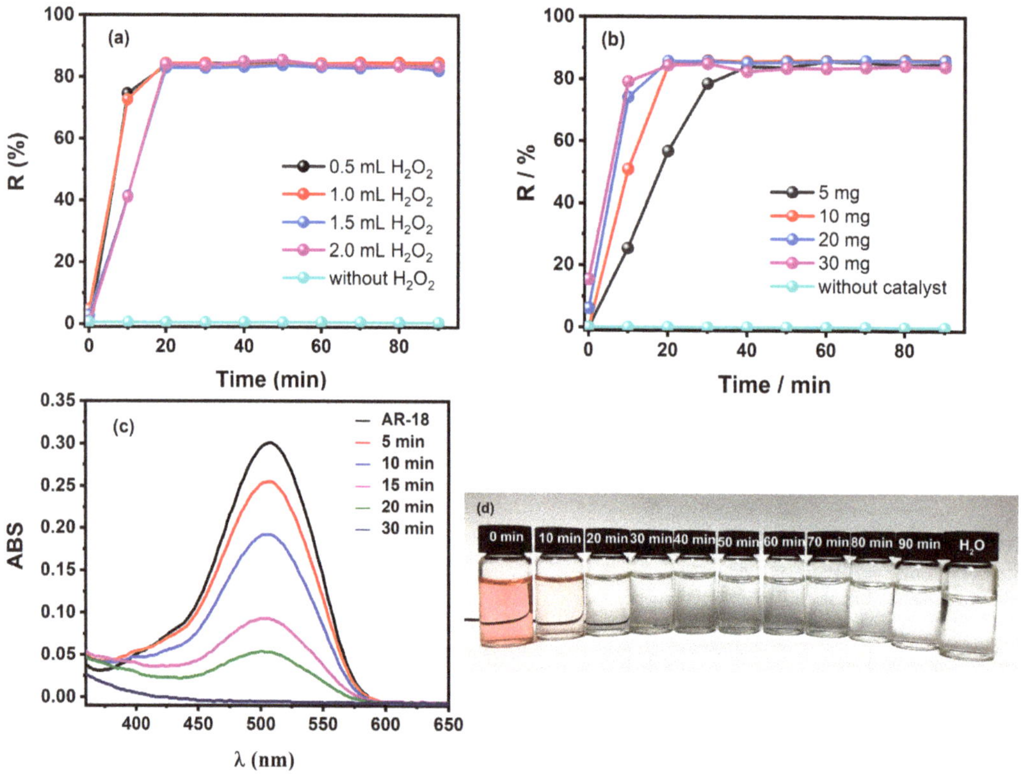

Figure 5. The degree of degradation for AR-18 under different (**a**) H_2O_2 amounts and (**b**) catalyst loadings. (**c**) The UV–visible spectrum and (**d**) the color change during the degradation of AR-18.

The color changes observed during the degradation of AR-18 in Figure 5d show that the solution became nearly colorless, similar to pure water, after 30 min. Correspondingly, the absorption band at 512 nm in the UV–visible spectrum (Figure 5c) nearly disappeared. Having established the catalytic system described above, the general applicability of the dye degradation process was further investigated using a series of dyes, including methyl orange (MO), methylene blue (MB), and rhodamine B (RHB). As expected, MIL-88A@SiO_2

effectively degraded these different dyes with degradation rates ranging from 81% to 96%, regardless of whether the dyes were neutral, anionic, or cationic (Figure 6d), consistent with results reported in the literature [14,20].

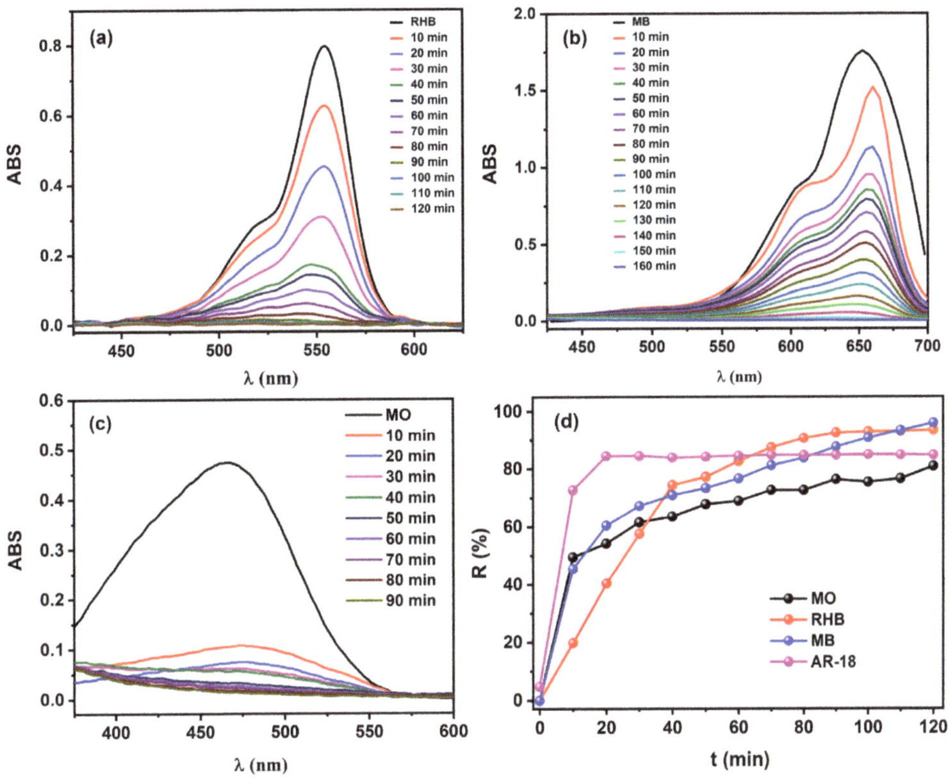

Figure 6. (**a**–**c**) The UV–visible spectrum during the degradation. (**d**) Changes in the degradation amount of four dyes over time.

The photocatalytic mechanism for the Fenton degradation process over SiO_2@MIL-88A(Fe) is proposed in Figure 7 based on references [17,18]. Under LED irradiation, the electrons in MIL-88A@SiO_2 are excited from the valence band (VB) to the conduction band (CB), leaving behind holes in the VB. These photogenerated electrons quickly transfer to the conduction band of MIL-88A, where they react with hydrogen peroxide (H_2O_2) to produce hydroxyl radicals (·OH). Simultaneously, the electrons react with dissolved oxygen (O_2), generating superoxide radicals ($\cdot O_2^-$). Additionally, the coordinated Fe(III) within the MIL-88A framework is reduced by H_2O_2 to Fe(II), which further decomposes H_2O_2 to generate more hydroxyl radicals (·OH). These reactive species, including hydroxyl radicals (·OH) and superoxide radicals ($\cdot O_2^-$), possess strong oxidative properties, enabling the effective degradation of organic pollutants. Finally, the dyes are oxidized into CO_2 and H_2O, as confirmed by NMR analysis (Figure S8), which show no other organic compounds in the residual solution.

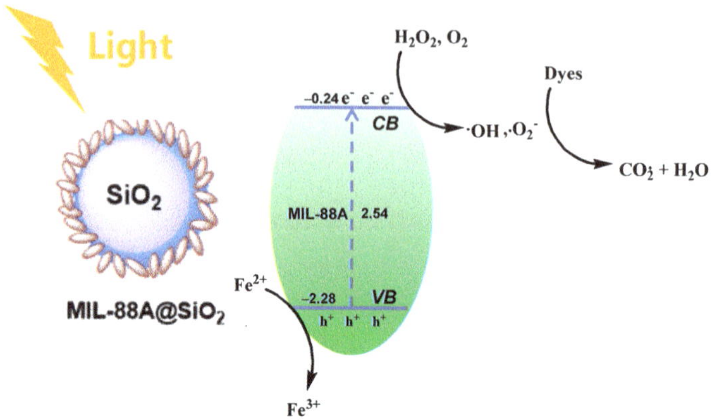

Figure 7. Schematic illustration of the proposed mechanism for photo-Fenton degradation of dyes over SiO_2@MIL-88A(Fe).

In addition to the goal of developing a modified hollow-shell Fenton-like catalyst for dye degradation, another key consideration in the design of SiO_2-based MIL-88A materials is the ease of separation via simple centrifugation and the ability to maintain catalytic activity after multiple recycling cycles. It was found that MIL-88A@SiO_2 could be easily recovered by centrifugation at 10,000 rpm for 5 min. The MIL-88A@SiO_2 was then Soxhlet-extracted with ethanol and water until no organic compounds were detected in the eluent. The recovered MIL-88A@SiO_2 was reactivated at 80 °C under a vacuum overnight and reused for subsequent reactions. Notably, after seven consecutive degradation cycles, the recycled MIL-88A@SiO_2 still achieved at least 79.3% degradation (Figure 8). Moreover, the recovered MIL-88A@SiO_2 remained structurally intact, as confirmed by XRD analysis (Figure S2).

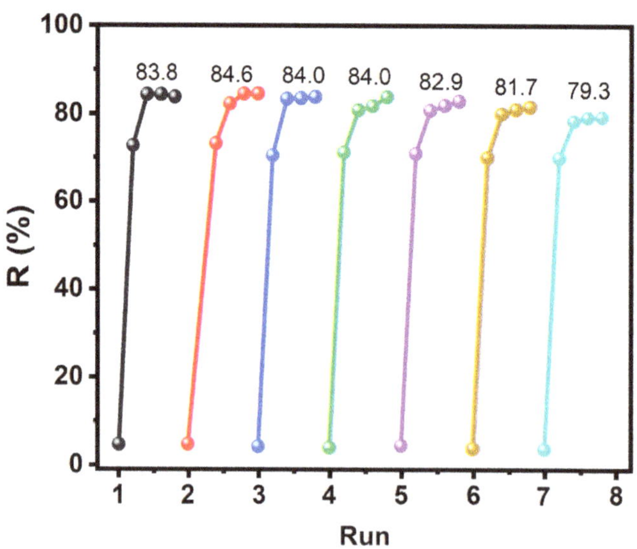

Figure 8. The degradation recycles for MO.

3. Experimental Section

3.1. Characterization

Fourier-transform infrared (FT-IR) spectra were collected using a Nicolet Magna 550 (Madison, WI, USA) spectrometer by the KBr pellet method. Scanning electron microscopy (SEM) images were acquired on a JEOL JSM-6380LV microscope (Akishima, Japan) operating at 29 kV. Nitrogen adsorption isotherms were measured at 77 K with a Quantachrome Nova 4000 analyzer (Graz, Austria), with the samples degassed overnight at 423 K. Pore size distributions were calculated using the BJH model, while specific surface areas (S_{BET}s) were determined from the linear portions of BET plots (P/P_0 = 0.05–0.3). Powder X-ray diffraction (XRD) patterns were recorded on a RINT2000 vertical goniometer (Rigaku, Akishima, Japan) over a 2θ range of 5–50°. The concentrations of MO, AR18, RhB, and MB were determined using the external standard method on a UV–Vis spectrophotometer (Hitachi U-3900, Tokyo, Japan) based on their absorbance at peak wavelengths of 465, 507, 554, and 664 nm, respectively. Fe loading in the catalysts was analyzed with an inductively coupled plasma optical emission spectrometer (ICP-OES, Varian VISTA-MPX, Palo Alto, CA, USA).

3.2. Synthesis of Carboxylate-Terminated SiO_2 Microspheres

The first step was the synthesis of the silicate yolk according to the literature [41]; 10 mL of 1-propanol, 1.04 mL of deionized water, and 0.95 mL of ammonium hydroxide (25%) were added into a round-bottomed flask. Then, the mixture was stirred at 10 °C, and a propanol solution of tetraethoxysilane (2.6 mL, 0.23 mol/L) was dropped into the above mixture under vigorous stirring. After 30 min, 75 mL of 1-propanol, 17.1 mL of deionized water, and 12 mL of ammonium hydroxide (25%) were added, and the mixture was stirred under 450 rpm at 25 °C for 10 min. Subsequently, 48 mL of ammonium hydroxide (25%) and propanol solution of tetraethoxysilane (177 mL, 0.23 mol/L) was dropped into the mixture over a period of 15 min under vigorous stirring. The mixture continued to be stirred for 8 h, and the milky silicate yolk (SiO_2) was obtained after being filtered and washed with excess ethanol and dried at 60 °C under vacuum for 8 h.

The second step was for the coating above the yolk. In a 100 mL round-bottomed flask, 500 mg of silicate yolk was dispersed into 50 mL of ethanol. The mixture was stirred at 80 °C under 800 rpm, and 3-aminopropyltriethoxysilane (APTES, 2 mL) was added dropwise. The mixture was stirred for 6 h, and, after cooling the mixture down to room temperature, the solid was filtered and washed with ethanol and dried at 60 °C for 8 h to produce NH_2@SiO_2). Subsequently, the NH_2@SiO_2 (500 mg) was dispersed into 30 mL *N*, *N*-dimethylformamide (DMF) under stirring at 500 rpm. Then, succinic anhydride (1.2 g) was added, and the mixture was refluxed at 85 °C for 8 h. After cooling the reactant to room temperature, he COOH@SiO_2 was obtained as white powder by filtering and washing with deionized water and drying at 60 °C under a vacuum for 8 h. IR (KBr, cm^{-1}): 3742–3115 (w), 3070 (s), 2975(s), 2348 (s), 1648(s), 1494(s), 1331 (s), 1099(m), 947 (m), 800 (m), 579 (m), 474 (m).

3.3. Preparation of the Core–Shell MIL-88A@ SiO_2

First, the $FeCl_3·6H_2O$ (0.338 g) was completely dissolved in 10.0 mL of deionized water in a 25 mL sealed tube; then, 10 mg COOH@SiO_2 was added to the mixture and stirred (150 rpm) at 65 °C for 12 h. Third, 0.0725 g fumaric acid was added, and the mixture stirred slowly for 12 h at 65 °C after being subject to ultrasonication for 30 min. After cooling the above mixture to room temperature, the target MIL-88A@SiO_2 (Catalyst **1**) was filtered and washed with excess water and ethanol, then dried overnight under a vacuum. The ICP analysis indicated that the content of Fe was 11.953 mg (0.214 mmol) per gram of heterogeneous catalyst. IR (KBr, cm^{-1}): 3680–2895 (w), 2358 (s), 1675 (s), 1607 (m), 1529 (s), 1396 (m), 1093 (m), 975 (s), 795 (s) 670 (m), 642 (s), 557 (s), 467 (s).

3.4. General Procedure for the Photo-Fenton Processes

The photo-Fenton processes were conducted under a 65 W LED lamp. Catalyst **1** (20.0 mg, 0.358 mmol of Fe based on the ICP analysis) was added into 50 mL of dye solution (10 mg/L) containing H_2O_2 (10 mmol/L). Then, we turned on the LED irradiation, and 2 mL of solution was collected at regular time intervals (10 min). During the degradation process, the catalyst was uniformly dispersed within the solution via vigorous magnetic stirring at about 500 rpm to avoid concentration gradients, ensuring the catalyst contacted the reactants, thereby enhancing the reaction efficiency.

The concentration of contaminants was measured by UV–Vis spectra. The separation or degree of degradation was calculated via the following equation:

$$R = \left(1 - \frac{C_f}{C_0}\right) \times 100\%$$

where C_f and C_0 are the contaminant concentration of the original and reaction solutions, respectively.

After completion of the reaction determined by the UV–Vis spectra, the mixture was centrifuged at 10,000 rpm for 5 min, and the precipitate was Soxhlet extracted with ethanol and H_2O until no organic compounds were detected in the eluent. The recovered solid was reactivated at 80 °C under vacuum overnight and then reused for the next runs directly.

4. Conclusions

In conclusion, SiO_2@MIL-88A(Fe) core–shell-structured microspheres were successfully synthesized. The structure of SiO_2@MIL-88A(Fe) was characterized using XRD, BET, FT-IR, ICP, XPS, and electron microscopy. As demonstrated in this study, SiO_2@MIL-88A(Fe) enables an efficient Fenton-like degradation of dyes, achieving up to 96% degradation. Moreover, SiO_2@MIL-88A(Fe) can be easily recovered by filtration and washing, maintaining its high catalytic activity over six consecutive cycles in the Fenton process. This work provides a promising approach for designing heterogeneous catalysts with both high catalytic efficiency and cost-effectiveness.

Supplementary Materials: The following supporting information can be downloaded at: https://www.mdpi.com/article/10.3390/catal15010023/s1, Figure S1: FT-IR spectra; Figure S2: XRD patterns; Figure S3: The N2 adsorption-desorption isotherm; Figure S4: XPS spectra; Figure S5: SEM image; Figure S6: TGA curve; Figure S7: UV-vis spectra; Table S1: The heterojunctions constructed with Fe-MOFs for the degradation of organic pollutants in the photo-Fenton system; Figure S8: NMR.

Author Contributions: Conceptualization and methodology, C.T.; formal analysis, investigation, and data curation, K.L. and Y.Z. writing—original draft preparation, K.L.; writing—review and editing, T.C. and G.L.; visualization, supervision, and project administration, C.T.; funding acquisition, C.T. and G.L. All authors have read and agreed to the published version of the manuscript.

Funding: We are grateful to the China National Natural Science Foundation (22071154, 22001171), the Shanghai Sciences and Technologies Development Fund (20070502600), the Shanghai Rising-Star Program (23QA1407200), and Shanghai Sailing Program (2020YF1435200) for financial support.

Data Availability Statement: Data is contained within the article.

Conflicts of Interest: The authors declare no conflicts of interest.

References

1. Ghosh, A.; Nayak, A.K.; Pal, A. Nano-Particle-Mediated Wastewater Treatment: A Review. *Curr. Pollut. Rep.* **2017**, *3*, 17–30. [CrossRef]
2. Che, H.; Che, G.; Dong, H.; Hu, W.; Hu, H.; Liu, C.; Li, C. Fabrication of Z-scheme $Bi_3O_4Cl/g\text{-}C_3N_4$ 2D/2D heterojunctions with enhanced interfacial charge separation and photocatalytic degradation various organic pollutants activity. *Appl. Surf. Sci.* **2018**, *455*, 705–716. [CrossRef]
3. Xie, A.; Dai, J.; Cui, J.; Lang, J.; Wei, M.; Dai, X.-H.; Li, C.; Yan, Y. Novel graphene oxide–confined nanospace directed synthesis of glucose-based porous carbon nanosheets with enhanced adsorption performance. *ACS Sustain. Chem. Eng.* **2017**, *5*, 11566–11576. [CrossRef]
4. Xie, A.; Dai, J.; Chen, X.; Ma, P.; He, J.; Li, C.; Zhou, Z.; Yan, Y. Ultrahigh adsorption of typical antibiotics onto novel hierarchical porous carbons derived from renewable lignin via halloysite nanotubes-template and in-situ activation. *Chem. Eng. J.* **2016**, *304*, 609–620. [CrossRef]
5. Che, H.; Liu, C.; Hu, W.; Hu, H.; Li, J.; Dou, J.; Shi, W.; Li, C.; Dong, H. NGQD active sites as effective collectors of charge carriers for improving the photocatalytic performance of Z-scheme $g\text{-}C_3N_4/Bi_2WO_6$heterojunctions. *Catal. Sci. Technol.* **2018**, *8*, 622–631. [CrossRef]
6. Li, X.; Chen, S.; Angelidaki, I.; Zhang, Y. Bio-electro-Fenton processes for wastewater treatment: Advances and prospects. *Chem. Eng. J.* **2018**, *354*, 492–506. [CrossRef]
7. Sudmalis, D.; Da Silva, P.; Temmink, H.; Bijmans, M.; Pereira, M. Biological treatment of produced water coupled with recovery of neutral lipids. *Water Res.* **2018**, *147*, 33–42. [CrossRef] [PubMed]
8. Cervantes-Avilés, P.; Keller, A.A. Incidence of metal-based nanoparticles in the conventional wastewater treatment process. *Water Res.* **2021**, *189*, 116603. [CrossRef]
9. Xie, A.; Cui, J.; Chen, Y.; Lang, J.; Li, C.; Yan, Y.; Dai, J. Simultaneous activation and magnetization toward facile preparation of auricularia-based magnetic porous carbon for efficient removal of tetracycline. *J. Alloy. Compd.* **2019**, *784*, 76–87. [CrossRef]
10. Ezugbe, E.O.; Rathilal, S. Membrane Technologies in Wastewater Treatment: A Review. *Membranes* **2020**, *10*, 89. [CrossRef] [PubMed]
11. Goh, P.S.; Ismail, A.F.; Ng, B.C.; Abdullah, M.S. Recent Progresses of Forward Osmosis Membranes Formulation and Design for Wastewater Treatment. *Water* **2019**, *11*, 2043. [CrossRef]
12. Li, J.; Wang, H.; Yuan, X.; Zhang, J.; Chew, J.W. Metal-organic framework membranes for wastewater treatment and water regeneration. *Coordination Chem. Rev.* **2020**, *404*, 213116. [CrossRef]
13. Reshmy, R.; Thomas, D.; Philip, E.; Paul, S.A.; Madhavan, A.; Sindhu, R.; Binod, P.; Pugazhendhi, A.; Sirohi, R.; Tarafdar, A.; et al. Potential of nanocellulose for wastewater treatment. *Chemosphere* **2021**, *281*, 130738. [CrossRef]
14. Wang, Z.; Yan, S.; Sun, Y.; Xiong, T.; Dong, F.; Zhang, W. Bi metal sphere/graphene oxide nanohybrids with enhanced direct plasmonic photocatalysis. *Appl. Catal. B Environ.* **2017**, *214*, 148–157. [CrossRef]
15. Zhang, X.W.; Wang, F.; Wang, C.C.; Wang, P.; Fu, H.; Zhao, C. Photocatalysis activation of peroxodisulfate over the supported Fe_3O_4 catalyst derived from MIL-88A(Fe) for efficient tetracycline hydrochloride degradation. *Chem. Eng. J.* **2021**, *426*, 131927. [CrossRef]
16. Dang, T.T.; Do, V.M.; Trinh, V.T. Nano-Catalysts in Ozone-Based Advanced Oxidation Processes for Wastewater Treatment. *Curr. Pollut. Rep.* **2020**, *6*, 217–229. [CrossRef]
17. Rajput, H.; Kwon, E.E.; Younis, S.A.; Weon, S.; Jeon, T.H.; Choi, W.; Kim, K.H. Photoelectrocatalysis as a high-de platform for pulping wastewater treatment and energy production. *Chem. Eng. J.* **2021**, *412*, 128612. [CrossRef]
18. Jain, B.; Singh, A.K.; Kim, H.; Lichtfouse, E.; Sharma, V.K. Treatment of organic pollutants by homogeneous and heterogeneous Fenton reaction processes. *Environ. Chem. Lett.* **2018**, *16*, 947–967. [CrossRef]
19. Meyerstein, D. Re-examining Fenton and Fenton-like reactions. *Nat. Rev. Chem.* **2021**, *5*, 595–597. [CrossRef]
20. Lu, Z.; Cao, X.; Wei, H.; Huo, W.; Wang, Q.; Li, K. Strong enhancement effect of bisulfite on MIL-68(Fe)-catalyzed Fenton-like reaction for organic pollutants degradation. *Appl. Surf. Sci.* **2021**, *542*, 148631. [CrossRef]
21. Sohrabi, M.R.; Khavaran, A.; Shariati, S. Removal of Carmoisine edible dye by Fenton and photo Fenton processes using Taguchi orthogonal array design. *Arab. J. Chem.* **2017**, *10*, S3523–S3531. [CrossRef]
22. Belloa, M.M.; Ramana, A.A.A.; Asghar, A. A review on approaches for addressing the limitations of Fentonoxidation for recalcitrant wastewater treatment. *Process Saf. Environ. Prot.* **2019**, *126*, 119–140. [CrossRef]
23. Zhu, Y.; Zhu, R.; Xi, Y.; Zhua, J.; Zhue, G.; He, H. Strategies for enhancing the heterogeneous Fenton catalytic reactivity: A Review. *Appl. Catal. B Environ.* **2019**, *255*, 117739. [CrossRef]
24. Wang, Q.; Astruc, D. State of the Art and Prospects in Metal–Organic Framework (MOF)-Based and MOF-Derived Nanocatalysis. *Chem. Rev.* **2020**, *120*, 1438–1511. [CrossRef] [PubMed]
25. Li, H.; Eddaoudi, M.; O'Keeffe, M.; Yaghi, O.M. Design and synthesis of an exceptionally stable and highly porous metal-organic framework. *Nature* **1999**, *402*, 276. [CrossRef]

26. Tranchemontagne, D.J.; Mendoza-Cortes, J.L.; O'Keeffe, M.; Yaghi, O.M. Secondary building units, nets and bonding in the chemistry of metal–organic frameworks. *Chem. Soc. Rev.* **2009**, *38*, 1257. [CrossRef]
27. Kitagawa, S.; Kitaura, R.; Noro, S.-I. Functional porous coordination polymers. *Angew. Chem. Int. Ed.* **2004**, *43*, 2334. [CrossRef] [PubMed]
28. Ferey, G. Hybrid porous solids: Past, present, future. *Chem. Soc. Rev.* **2008**, *37*, 191. [CrossRef] [PubMed]
29. Gao, S.; Cen, W.; Li, Q.; Li, J.; Lu, Y.; Wang, H.; Wu, Z. A mild one-step method for enhancing optical absorption of amine-functionalized metal-organic frameworks. *Appl. Catal. B Environ.* **2018**, *227*, 190–197. [CrossRef]
30. Yang, H.; He, X.-W.; Wang, F.; Kang, Y.; Zhang, J. Doping copper into ZIF-67 for enhancing gas uptake capacity and visible-light-driven photocatalytic degradation of organic dye. *J. Mater. Chem.* **2012**, *22*, 21849–21851. [CrossRef]
31. Roy, D.; Neogi, S.; De, S. Mechanistic investigation of photocatalytic degradation of Bisphenol-A using MIL-88A(Fe)/MoS2 Z-scheme heterojunction composite assisted peroxymonosulfate activation. *Chem. Eng. J.* **2022**, *428*, 131028. [CrossRef]
32. Liu, N.; Huang, W.; Zhang, X.; Tang, L.; Wang, L.; Wang, Y.; Wu, M. Ultrathin graphene oxide encapsulated in uniform MIL-88A(Fe) for enhanced visible light-driven photodegradation of RhB. *Appl. Catal. B Environ.* **2018**, *221*, 119–128. [CrossRef]
33. Xue, B.; Du, L.; Jin, J.; Meng, H.; Mi, J. In situ growth of MIL-88A into polyacrylate and its application in highly efficient photocatalytic degradation of organic pollutants in water. *Appl. Surf. Sci.* **2021**, *564*, 150404. [CrossRef]
34. Dhakshinamoorthy, A.; Li, Z.; Garcia, H. Catalysis and photocatalysis by metal organic frameworks. *Chem. Soc. Rev.* **2018**, *47*, 8134–8172. [CrossRef] [PubMed]
35. Pascanu, V.; Miera, G.G.; Inge, A.K.; Martín-Matute, B. Metal–Organic Frameworks as Catalysts for Organic Synthesis: A Critical Perspective. *J. Am. Chem. Soc.* **2019**, *141*, 7223–7234. [CrossRef]
36. Ji, P.; Feng, X.; Oliveres, P.; Li, Z.; Murakami, A.; Wang, C.; Lin, W. Strongly Lewis Acidic Metal–Organic Frameworks for Continuous Flow Catalysis. *J. Am. Chem. Soc.* **2019**, *141*, 14878–14888. [CrossRef] [PubMed]
37. Serre, C.; Mellot-Draznieks, C.; Surblé, S.; Audebrand, N.; Filinchuk, Y.; Férey, G. Role of Solvent-Host Interactions That Lead to Very Large Swelling of Hybrid Frameworks. *Science* **2007**, *315*, 1828–1831. [CrossRef] [PubMed]
38. Wang, L.; Zhang, Y.; Li, X.; Xie, Y.; He, J.; Yu, J.; Song, Y. The MIL-88A-Derived Fe_3O_4-Carbon Hierarchical Nanocomposites for Electrochemical Sensing. *Sci. Rep.* **2015**, *5*, 14341. [CrossRef] [PubMed]
39. Xie, A.; Cui, J.; Yang, J.; Chen, Y.; Lang, J.; Li, C.; Yan, Y.; Dai, J. Graphene oxide/Fe(III)-based metal-organic framework membrane for enhanced water purification based on synergistic separation and photo-Fenton processes. *Appl. Catal. B Environ.* **2020**, *264*, 118548. [CrossRef]
40. Fu, H.; Song, X.-X.; Wu, L.; Zhao, C.; Wang, P.; Wang, C.-C. Room-temperature preparation of MIL-88A as a heterogeneous photo-Fenton catalyst for degradation of rhodamine B and bisphenol a under visible light. *Mater. Res. Bull.* **2020**, *125*, 110806. [CrossRef]
41. Zhang, W.; Yan, Z.; Gao, J.; Tong, P.; Liu, W.; Zhang, L. Metal–organic framework UiO-66 modified magnetite@silicacore–shell magnetic microspheres for magnetic solid-phaseextraction of domoic acid from shellfish samples. *J. Chromatogr. A* **2015**, *1400*, 10–18. [CrossRef] [PubMed]

Disclaimer/Publisher's Note: The statements, opinions and data contained in all publications are solely those of the individual author(s) and contributor(s) and not of MDPI and/or the editor(s). MDPI and/or the editor(s) disclaim responsibility for any injury to people or property resulting from any ideas, methods, instructions or products referred to in the content.

catalysts

Review

A Mini-Review of Recent Progress in Zeolite-Based Catalysts for Photocatalytic or Photothermal Environmental Pollutant Treatment

Shenhao Zhang [1,†], Le Xu [1,†], Jie Xu [2,*] and Boxiong Shen [1,2,*]

[1] School of Energy and Environmental Engineering, Hebei University of Technology, Tianjin 300401, China; 202431304031@stu.hebut.edu.cn (S.Z.); 202211301010@stu.hebut.edu.cn (L.X.)
[2] School of Chemical Engineering and Technology, Hebei University of Technology, Tianjin 300401, China
* Correspondence: j_xu@hebut.edu.cn (J.X.); shenbx@hebut.edu.cn (B.S.)
† These authors contributed equally to this work and should be considered as co-first authors.

Abstract: Atmospheric and water pollution has led to serious harm to the global environment and human health. Photocatalysis and photothermal catalysis technologies have been considered as promising methods to handle pollutants in the atmosphere and water due to their energy savings and environmental friendliness. Zeolite catalysts have been widely used in the field of photocatalytic and photothermal catalytic removal of environmental pollutants due to their well-developed pore structure, high stability, and tunable surface chemistry. In this review, we have elaborated the photocatalytic and photothermal catalytic mechanisms and summarized the recent progress in zeolite-based catalysts for photocatalytic or photothermal catalytic environmental pollutant treatment. In summary, it is found that the strategies of elemental doping and surface structure modification directly affect the adsorption performance of zeolite for target pollutants, and the construction of a bifunctional structure promotes the generation of intrinsic active species and photogenerated charge separation. Finally, the paper presents current challenges and perspectives on zeolite-based catalysts for photocatalytic and photothermal catalytic treatment of environmental pollutants.

Keywords: zeolite; photocatalysis; photothermal catalysis; VOCs; water pollutants

1. Introduction

With the advancement and development of industrialization, the issue of global environmental pollution has intensified, with excessive emissions of pollutants into the atmosphere and water environments, leading to a severe threat to the global ecosystem and human health. Consequently, developing an appropriate and effective treatment strategy has emerged as an imperative priority [1,2]. Water pollutants, including organic dyes, heavy metals, and other pollutants, mainly come from industrial and agricultural wastewater. Volatile organic compounds (VOCs), which have been regarded as some of the main atmospheric pollutants, mainly come from automobile exhaust, industrial waste gases, and synthetic materials [3]. Therefore, the treatment of water pollutants and VOCs is urgently needed [4]. The current technologies for treating water pollutants and VOCs include traditional adsorption, photocatalytic oxidation, plasma technology, catalytic combustion, advanced oxidation, membrane separation technology, and biological methods [5]. Therein, the utilization of solar energy to realize pollutant degradation and purification, which has been regarded as a promising treatment strategy, has the advantages of cleanliness, energy savings, and environmental friendliness. Therefore, photocatalytic technology has attracted much attention for pollutant degradation under mild conditions with low operating costs [6]. In recent years, in order to strengthen the degradation effect of photocatalytic technology, photothermal catalytic technology had been proposed. Combining the characteristics of photocatalysis and thermal catalysis, photothermal synergistic catalytic technology could effectively avoid the problems of needing a large external energy supply and low degradation efficiency, and it recently has become a research hotspot [7,8].

However, since pollutant molecules in the atmosphere and water are often complex and difficult to deeply oxidize, it is important to develop novel and efficient catalysts for photocatalysis or photothermal catalysis [9]. Currently, common catalysts include metal oxides, precious metal catalysts, semiconductor materials, carbon-based materials, metal-organic frameworks, and zeolite [10]. Non-zeolite carrier catalysts have good light absorption ability for photothermal catalysis, effectively absorbing light energy and using UV or visible light to excite electron–hole pairs to promote catalytic reactions.

Therein, zeolites are a class of microporous materials with a regular pore structure that are usually composed of silicon, aluminum, oxygen, and metal elements. The pore size and shape of zeolite can be precisely controlled, which allows it to selectively adsorb pollutant molecules [11]. Zeolite catalysts have significant advantages in the field of photocatalytic and photothermal catalytic degradation of environmental pollutants due to their well-developed pore structure, high stability, and tunable surface chemistry [12,13]. Based on the differences in the pore structure of zeolites, the common catalysts can be categorized as A-type [14], X-type [15], Y-type [16], β-type [17], mordenite zeolite [18], ZSM-5 zeolite [19], and USY zeolite [20]. As shown in Table 1, the well-developed pore structure always endows zeolite with a high specific surface area (200–1000 m^2/g), and the adsorption performance is not hindered when the pore size of zeolite is larger than that of the adsorbed molecules. Due to the different combinations of the basic unit in zeolite, the significant difference in the pore structure and size greatly influences the adsorption performance [21,22], and the selection of a suitable zeolite is very important for different catalytic reactions [8,23].

Table 1. Properties and characteristics of different zeolites.

Types of Zeolite	Structural Characteristics	Crystal Structure	Pore Size (Å)	Silicon to Aluminum Ratio	Type of Pore Channel	Exchange Capacity (meq/g)	Specific Surface Area (m^2/g)	Acid Strength
A-Type Zeolite	Cubic Crystal Structure	LTA Type	4	1:1–3:1	Cubic Diagonal Channel	0.3–0.5	300–600	Weakly Acidic
X-Type Zeolite	Cubic Crystal Structure	FAU Type	8–9	1:1–2:1	Octahedral Pore	3.0–4.5	500–900	Strong Acid
Y-Type Zeolite	Cubic Crystal Structure	FAU Type	7–9	>2:1	Octahedral Pore	0.7–1.2	600–1000	Strong Acid
β-Type Zeolite	Hexagonal Crystal Structure	BEA Type	5.6–6.6	>10:1	Double Channel	1.0–1.5	500–800	Moderate Acidity
Mordenite	Monoclinic Crystal Structure	MOR Type	6.5–7.0	5:1–20:1	Straight and Curved Ducts	0.8–1.2	350–500	Moderate Acidity
ZSM-5	Monoclinic Crystal Structure	MFI Type	5.3–5.7	>10:1	Straight Channel	2.5–3.5	300–500	Strong Acid
USY	Cubic Crystal Structure	FAU Type	7.4	>3:1	Octahedral Pore	0.5–3.5	400–700	Strong Acid

At present, although considerable research about the application of zeolite for photocatalytic degradation of environmental pollutants has been reported, reviews focused on zeolite in the photothermal catalytic degradation of pollutants are still scarce, and a comprehensive review to summarize the recent progress in zeolite-based catalysts in the field of photocatalysis and photothermal catalysis is lacking. Accordingly, this review briefly outlines photocatalytic and photothermal catalytic theory, the modification strategies of zeolite for catalytic reactions, and research progress in the photocatalytic and photothermal catalytic degradation of pollutants in the atmosphere and water environments. This work provides a summary for the application of zeolite-based catalysts in photocatalysis and photothermal catalysis, which can guide catalyst design in the future.

2. Principles of Photocatalysis and Photothermal Catalysis

2.1. Principle of Photocatalysis

The basic principle of photocatalysis (PC) is that the generated electron–hole pairs after excitation of the catalyst by light energy triggers a series of redox reactions to achieve chemical transformation (Figure 1). The photocatalytic pollutant degradation reaction on semiconductors mainly includes the following steps [24]: (1) Semiconductor light absorption. The semiconductor catalyst harvests the photon energy after light irradiation; (2) Formation of photogenerated electron–hole pairs. When the photon energy harvested by the semiconductor is larger than the forbidden band width (E_g) between the valence band (VB) and conduction band (CB), the electrons on the VB will absorb the energy to jump into the CB to generate photogenerated electron–hole pairs; (3) Migration of photogenerated electron–hole pairs. Photogenerated electron–hole pairs can migrate to the surface of the photocatalyst and undergo redox reactions with the pollutants adsorbed on the surface. Photogenerated electrons can be captured by O_2 to generate superoxide radicals ($\cdot O_2^-$) and photogenerated holes can be captured by H_2O to generate hydroxyl radicals ($\cdot OH$). These radicals can help the pollutants degrade to produce CO_2 and H_2O molecules [25].

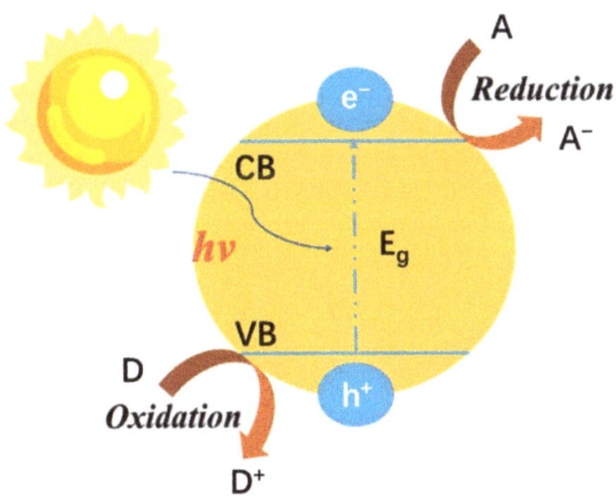

Figure 1. Schematic diagram of photocatalytic principle [26].

2.2. Principle of Photothermal Catalysis

Photothermal catalysis (PTC) is an advanced technology that utilizes both light and thermal energy to augment the catalytic reaction, which integrates the advantages of photocatalysis (PC) and thermal catalysis (TC) to enhance catalytic performance through the photothermal synergetic effect. Photothermal catalysis can simultaneously mitigate the substantial energy consumption associated with thermal catalysis and amplify the degradation efficiency of photocatalysis. Based on a previous report [27], photothermal catalysis has been proven to be an efficient and promising technology for pollutant degradation due to its performance always greatly outperforming the linear superposition of photocatalysis and thermal catalysis. In photothermal catalysis, in addition to the photocatalytic effect, the addition of thermal energy changes the activation energy of the reaction, strengthens the oxidation processes of reactants, and promotes the desorption and release processes of the final products. Photothermal catalysis is always divided into four reaction modes: thermal-assisted photocatalysis (TAPC), photo-assisted thermal catalysis (PATC), photo-driven thermal catalysis (PDTC), and photothermal co-catalysis (PTCC) [3]. In thermal-assisted photocatalysis, light is the main driving force for catalytic reactions, while heat plays an auxiliary role to enhance the photocatalytic effect. In photo-assisted thermal catalysis, heat is the main driving force for catalytic reactions, while light is used to facilitate the thermal catalytic reaction. In current research, photo-driven thermal catalysis and photothermal co-catalysis are the main reaction modes for pollutant degradation (Figure 2) and, therefore, the principles of these two reaction modes are introduced in detail here.

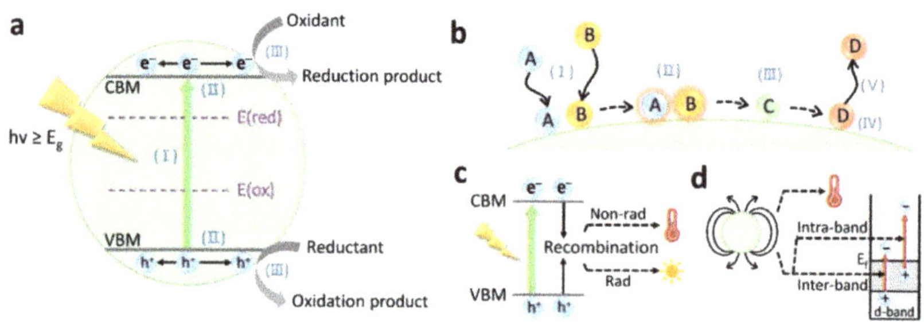

Figure 2. Catalytic principle and pathway of photothermal catalysis: (**a**) Thermal-assisted photocatalysis; (**b**) Photo-assisted thermal catalysis; (**c**) Photothermal effect induced by recombination of charge carriers over semiconductor catalysts; (**d**) Photothermal effect and hot carrier generation induced by the decay of plasmons over plasmonic metal catalysts [27].

Photo-induced heat is the main driving force for catalytic reactions in photo-driven thermal catalysis. Light indirectly drives thermal catalysis through the photothermal effect by raising the temperature to meet the demand of thermal catalysis rather than directly participating in the photocatalytic reaction. In this reaction mode, the required temperature for thermal catalysis is exclusively derived from the conversion of light energy on the catalyst surface. To achieve the synergistic effect in photo-driven thermal catalysis, the catalyst should meet the following requirements: (1) the catalyst should show strong absorption in the entire solar spectral region and can efficiently convert the absorbed solar energy into heat, thereby raising the temperature above the required temperature for thermal catalysis; and (2) the catalyst needs to have good thermal catalytic activity and a relatively low reaction temperature. Under suitable conditions, the activity of photo-driven thermal catalysis is always higher than that of single thermal catalysis, which can achieve an energy-saving effect and high solar utilization efficiency [27].

Light and heat are both driving forces for catalytic reactions in photothermal co-catalysis. Light directly drives photocatalysis and heat directly drives thermal catalysis, while the synergistic effect of light and heat can achieve higher activity than the sum of photocatalysis and thermal catalysis. In this reaction mode, the catalyst should simultaneously possess photocatalytic and thermal catalytic activities, which is the key point for successful application of photothermal co-catalytic technology. By reasonably designing a catalyst with optimal activity and stability under photothermal co-catalytic conditions, this technology can not only increase the reaction rate but also enhance product selectivity, which allows the catalyst to exhibit superhigh performance beyond that of single photocatalysis or thermal catalysis in specific reactions.

3. Modification Strategies for Zeolite Catalysts

3.1. Construction of Bifunctional Catalysts

Construction of bifunctional catalysts is an efficient strategy to improve the catalytic activity of zeolite, which is aimed at integrating multiple active sites into a monolithic catalyst to achieve directional acceleration of different reaction steps. The key to this strategy lies in methods such as metal or metal oxide loading of zeolite catalysts and enhancement of acidic sites, so that multiple catalytic functions can be achieved simultaneously on the overall catalyst.

Guo et al. first prepared the specific precursor by activating hydroxyl radicals under UV irradiation, as in the aging process, and then a bifunctional composite of UV-CDs/Zeolite-4A/TiO_2 with adsorption-photocatalytic functions was prepared by a hydrothermal method (Figure 3), which showed a removal efficiency of 90.63% for methylene blue within 90 min. After five consecutive tests (7.5 h), the methylene blue decolorization rate of UV-CZT still reached 82.94%, demonstrating a long service life [28]. Tang et al. combined hydrophilic {001} TiO_2 nanosheets with hydrophobic NaY zeolite to prepare a TiO_2@HYZ bifunctional photocatalyst, which achieved spatial separation of the catalytic center and adsorption center and could achieve high photocatalytic degradation activity toward gaseous toluene under UV irradiation (Figure 4), with a removal rate of 96.6% and a mineralization efficiency of 87.4% within 120 min. The characterization results showed that the combination of HYZ and TiO_2 into a bifunctional catalyst enhanced the adsorption of toluene by HYZ and facilitated the efficient separation of photoexcited carriers on TiO_2. The deposition of degradation intermediates over active sites was also avoided and the mineralization performance was improved. Significant decreases in toluene removal and CO_2 yield were observed for pristine TiO_2 after five photodegradation cycles (12 h), indicating that it incurred severe deactivation. Meanwhile, the color of pristine TiO_2 generally changed from white to dark brown, implying that a certain amount of intermediates from toluene mineralization accumulated on its surface. Unlike for pristine TiO_2, the toluene removal efficiency and CO_2 yield of TiO_2@HYZ were constant at roughly 96.8% and 211.5 μmol/L, respectively, after five cycles, and its surface remained white. This indicated that TiO_2@HYZ showed better photocatalytic durability than pristine TiO_2 owing to the incorporation of HYZ [29]. Zakaria et al. used zeolite nanoclay (ZNC) and constructed Cu/WO_{3-x}@ZNC bifunctional composites by hydrothermal and ball milling methods for efficient photocatalytic degradation of ciprofloxacin (CIP) and photothermal desalination of water (Figure 5). Under IR and visible irradiation, the CIP degradation efficiencies reached 88.3% and 81.7%, respectively, while the evaporation efficiencies of water reached 97.5% and 72.8%, respectively. After three full cycles (3 h), the composite showed good photocatalytic reusability [30]. Sacco et al. developed an adsorption-photocatalytic method to remove caffeine by loading ZnO on the surface of commercial zeolite particles (ZnO/ZEO), which exhibited high catalytic activity in adsorption/photocatalytic processes. The characterization results showed that ZnO nanoparticles loaded on the surface of ZEO were in the fibrillated zincite phase, which was mainly present in the mesoporous structure of ZEO. The experimental results showed that the ZnO/ZEO particles of the constructed bifunctional catalyst removed 60% of caffeine in 120 min under dark conditions and up to 100% of caffeine under UV irradiation [31].

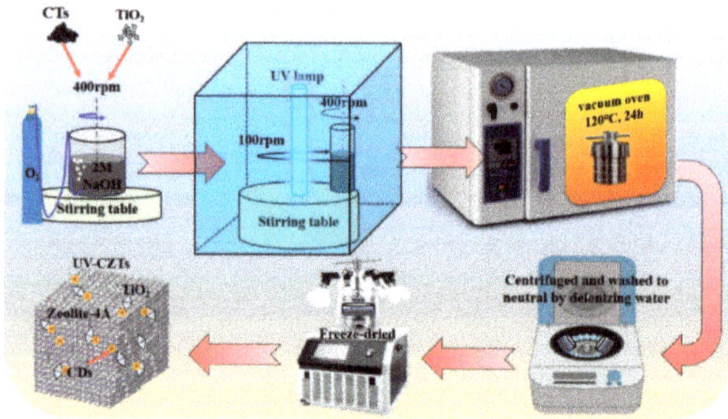

Figure 3. Schematic diagram of synthesis process for UV-CDs/Zeolite-4A/TiO_2 [28].

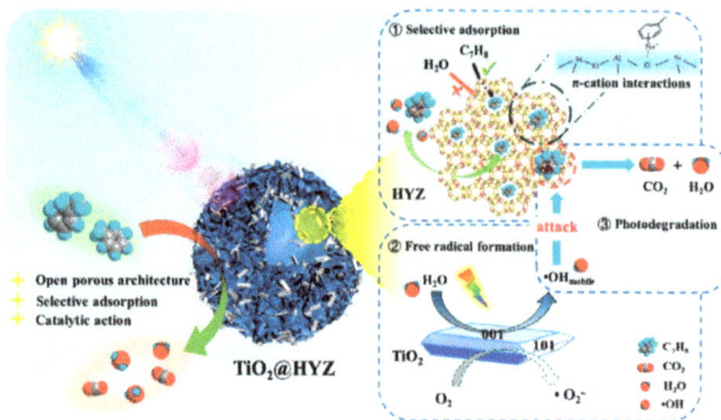

Figure 4. Possible mechanism of photocatalytic toluene degradation over TiO_2@HYZ under UV irradiation [29].

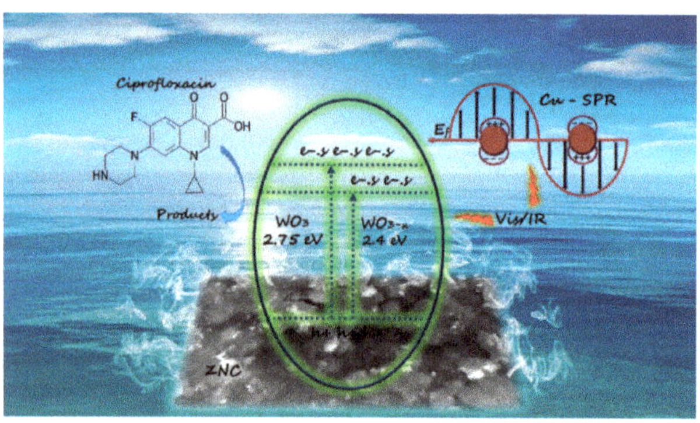

Figure 5. Schematic diagram of Cu/WO_{3-x}@ZNC for catalytic degradation of CIP and photothermal evaporation of water [30].

Based on the above studies, the strategy of constructing bifunctional catalysts using zeolite can simultaneously facilitate multistage reaction steps and significantly diminish the conversion barrier of intermediates, thereby enhancing the holistic reaction efficiency. Meanwhile, the precise regulation of reaction pathways can be achieved to selectively obtain the desired transformed products.

3.2. Elemental Doping

Elemental doping is an efficient strategy to improve the physical, chemical, or electrical properties of zeolite through the addition of a quantitative elemental precursor during catalyst synthesis. In the field of photocatalysis and photothermal catalysis, metal doping is particularly important for improving the activity of zeolite because it provides emerging active sites for surface reactions. By accurately controlling the type, amount, and distribution of doped metal, the catalytic performance can be optimized to obtain excellent activity in catalytic reactions.

Hu et al. used titanium-containing blast furnace slag as the raw material and promoted transfer of the Ti element to the NaZSM-5 molecular sieve by a hydrothermal method to realize the successful doping of metal Ti into zeolite. The synthesized Ti-NaZSM-5 composite catalysts contained a backbone of tetrahedrally coordinated Ti species and amorphous exo-backbone Ti species and exhibited a high degradation efficiency of 100% for methyl orange under UV irradiation [32]. Ke et al. loaded Ag-doped Y-type zeolite (Ag@Y) on the surface of an alumina nanofiber membrane by a wet chemical method for visible light-driven photocatalysis (Figure 6). The integrated membrane could maintain a large flux of 200 Lm^{-2} h^{-1} bar^{-1} and a stable transmittance selectivity of 85% during the photocatalytic degradation of dyes. Meanwhile, the degradation efficiency of methylene blue could reach about 40% within 60 min [33]. Elimian et al. introduced noble metal Pt species into the USY molecular sieve to achieve metal doping by a sol-gel method. The enhanced oxygen vacancy and Ti^{3+} species could improve the light absorption of the catalyst, and the prepared Pt-mTiO_2/USY nanocomposites were used for photo-driven thermal catalytic toluene oxidation (Figure 7). The composite catalysts could obtain a maximum toluene conversion of 86.6% and a CO_2 formation of 74.5% when the mass fraction of Pt was 0.9 wt.% [34].

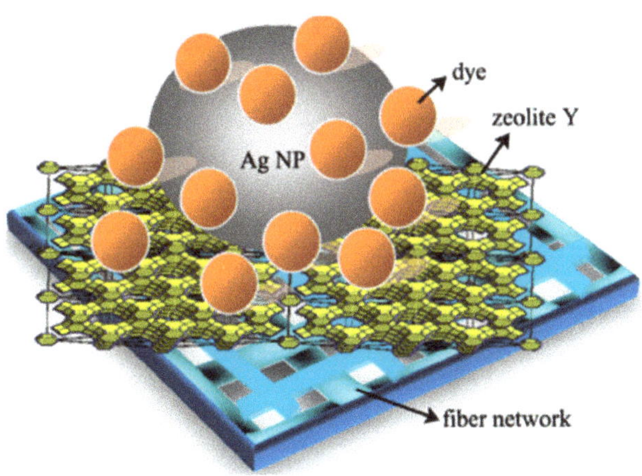

Figure 6. Schematic diagram of the structure of a nanofiber membrane functionalized with silver nanoparticles [33].

Based on the above studies, element doping can significantly enhance the catalytic activity of zeolite in photocatalysis and photothermal catalysis. The introduced metal species can expand the light absorption range of the catalyst and improve the charge separation efficiency. Meanwhile, element doping can provide suitable active sites for catalytic reactions and strengthen the adsorption performance of pollutant molecules, thereby improving the catalytic reaction rate and selectivity. The introduced metal species can intensify the photothermal effect on catalytic reactions, promote the generation of various intermediates, and reduce the reaction energy barrier.

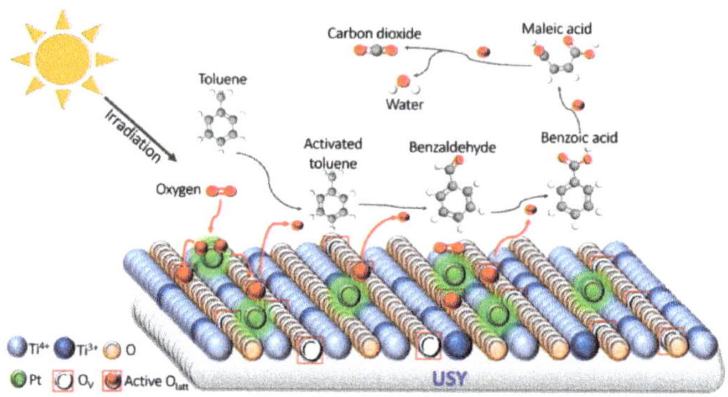

Figure 7. Schematic diagram of toluene oxidation reaction over 0.9Pt-mTiO_2/USY [34].

3.3. Structural Modification of Catalysts

Structural modification of zeolite includes the adjustment and optimization of atomic arrangement, crystal structure, pore structure, and active sites on the catalyst. By modifying the structure of zeolite catalysts by physical, chemical, or physicochemical methods, the catalytic reaction activity, selectivity, and stability can be significantly improved for efficient pollutant removal.

Alakhras et al. used zeolite and prepared Zeo-TiO_2 and Zeo-ZnO composite photocatalysts by a co-precipitation method for photocatalytic degradation of RhB dye under UV light. The TiO_2 nanoparticles were fully encapsulated on natural zeolite with a uniform spherical structure, while the ZnO nanoparticles were not fully encapsulated on the zeolite with the irregular block shape. In photocatalysis, superoxide radicals ($\cdot O_2^-$) and hydroxyl radicals ($\cdot OH$) played dominant roles in the degradation of RhB dye, while the contribution of holes (h^+) was negligible (Figure 8). The results showed that Zeo-TiO_2 exhibited a degradation efficiency of 100% for RhB dye with 80 min, while Zeo-ZnO only exhibited a degradation efficiency of 81% for RhB dye. The Zeo-TiO_2 catalyst retained almost the same performance with a very small activity loss, and the degradation efficiency after five cycles (400 min) was 93.8%, which indicated good recycling stability. Meanwhile, the degradation rate of RhB by the Zeo-ZnO catalyst was decreased significantly with increasing number of reuse times, and the degradation efficiency obtained after five successive cycles of degradation tests was 50.15%. The large reduction in photocatalytic efficiency may have been caused by the unavoidable desilication of the zinc oxide mass into the surface of zeolite tuff during the washing process [35]. Vaiano et al. introduced MoO_x as a functionalized component into magnesia-alkali zeolite by an ion exchange method. The formed polymetallic species on the catalyst surface could improve reaction selectivity by changing the surface structure and the distribution of active sites, which obtained a high selectivity of 80% in the degradation reaction of benzene [36]. Liu et al. used fly ash as the raw material to prepare honeycomb ZnO nanospheres on porous zeolite material to obtain a high specific surface area and photocatalytic performance. The composite catalysts showed high photocatalytic activity, with a degradation efficiency of 90% for methylene blue dye in water under UV irradiation within 30 min. The characterization results showed that the methylene blue molecules were first adsorbed on the surface of the ZnO spheres and outside or inside the zeolite, while the photocatalytic degradation process occurred on the surface of the ZnO spheres. When the concentration of methylene blue molecules on the surface of the ZnO spheres gradually decreased, the methylene blue molecules on the zeolite were transferred to the surface of the ZnO spheres, which played an important role in the degradation process [37]. Znad et al. prepared TiO_2/ZSM-5 photocatalysts by a direct template technology for MO dye mineralization in the wastewater. The multilayer structure of TiO_2/ZSM-5 with a high specific surface area (S_{BET}) of 1151 $m^2\ g^{-1}$ was beneficial to light utilization in multiple internal spaces. The catalyst showed a high degradation efficiency of 99.55% and mineralization efficiency of 99% for MO dye within 180 min under solar irradiation [38].

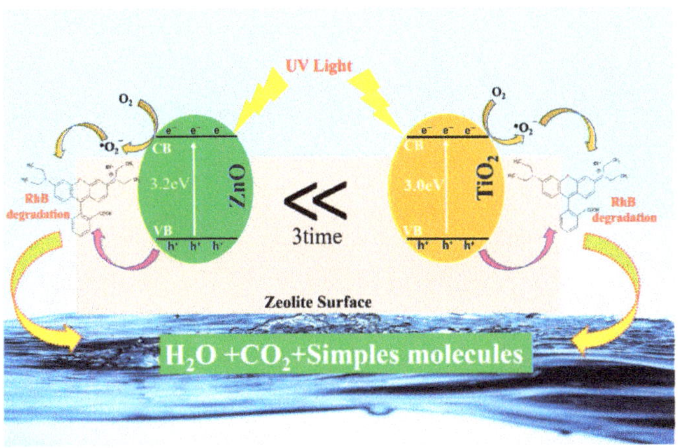

Figure 8. Photocatalytic oxidation mechanism of RhB on Zeo-TiO_2 and Zeo-ZnO catalysts [35].

Based on the above studies, structural modifications can tailor the intrinsic architecture of catalysts to provide more reactive sites and expedite the pollutant degradation rate. The reaction pathway can also be orchestrated to enhance product selectivity when handling a specific contaminant. The catalytic stability and structural framework can be improved to avoid impurity adsorption and enhance resistance to poisoning conditions. These properties can preclude the activity decline of zeolite in complex environments as much as possible and promote the deep eradication of pollutants.

4. Advances in Zeolite-Based Catalysts for Photocatalytic/Photothermal Pollutant Treatment

4.1. Photocatalytic and Photothermal Catalytic Degradation of VOCs

Photocatalytic degradation of VOCs can efficiently and selectively convert VOCs into harmless substances under ambient conditions, as well as having low energy consumption, ease of operation, and sustainability. The photocatalytic degradation efficiencies of several zeolite composite catalysts for VOCs removal are summarized in Table 2. The application of zeolite can greatly enhance the conversion efficiency of various VOCs under light illumination, exhibiting potential for the construction of composite catalysts for photocatalytic VOCs degradation. Photothermal catalytic VOC degradation can efficiently improve the reaction rate and selectivity by raising the reaction temperature via the photothermal synergistic effect, decreasing the activation energy barrier of the reaction, and overcoming thermodynamic limitations. Therefore, photothermal catalysis breaks the bottlenecks of low catalytic efficiency and limited mass transfer in photocatalysis and decreases energy consumption in thermal catalysis. The degradation efficiencies of several zeolite composites in photothermal catalysis for VOC removal ae summarized in Table 3. The zeolite composite catalysts exhibit high VOC conversion efficiency and CO_2 selectivity under photothermal conditions.

Table 2. The photocatalytic oxidation performance of different catalysts for VOCs.

Catalyst	VOC	VOC Concentration	Catalyst Mass (mg)	Light	Conversion (%)	S_{BET} (m^2/g)	Reference
TiO_2/ZSM-5	Propene	100 ppmv	110	8 W/365 nm	82	234	[39]
TiO_2/silicalite	Trichloroethylene	25 ppm	30	8 W	74	396	[40]
TiO_2/zeolite	Toluene	42.5 ppm	—	4 W/365 nm	89	—	[41]
FeZSM-5-HT	Ethylene	1000 ppm	200	4 W/254 nm	52	—	[42]
HZSM-5	Isopropyl alcohol	150 ppm	200	8 W/254 nm	94	308	[43]
TiO_2/HZSM-5	Formaldehyde	10 ppmv	—	1.6 W/365 nm	80	78	[44]
TiO_2/HZSM-5	Acetaldehyde	10 ppmv	—	1.6 W/365 nm	50	78	[44]
TiO_2/HZSM-5	Acetic acid	10 ppmv	—	1.6 W/365 nm	98	78	[44]
TiO_2/HZSM-5	Toluene	10 ppmv	—	1.6 W/365 nm	70	78	[44]
TiO_2/zeolite	Ethanol	700 ppm	9	100 W/367 nm	100	335	[45]
TiO_2/zeolite	Diethyl sulfide	375 ppm	9	100 W/367 nm	100	335	[45]
5Fe-10W-NaY	Acetaldehyde	—	1000	>400 nm	80	18.4	[46]
5Fe-10W-NaY	O-xylene	—	1000	>400 nm	76	18.4	[46]
TiO_2/ZSM-5	Formaldehyde	15 ppm	30	8 W	95	422	[47]
TiO_2/Zeolite Y	trichloroethylene	25 ppm	30	8 W	80	776	[47]
TiO_2@HYZ	Toluene	760 ppm	100	300 W	96.6	86.6	[29]

Improving the adsorption performance and increasing the reactive species in the reaction are the key points in VOC degradation for zeolite catalysts. Surface modification, optimization of regeneration conditions, and regulation of pore size are the main methods to improve the adsorption performance of zeolite. Kim et al. constructed a practical-scale photocatalytic air purifier using TiO_2/H-ZSM-5 composite as the filter to remove indoor VOCs under UV illumination. The TiO_2/H-ZSM-5 composite exhibited stable photocatalytic performance and could significantly remove various VOCs, including formaldehyde, acetaldehyde, and toluene. H-ZSM-5 zeolite provided the adsorption sites to strengthen the adsorption of VOCs, which accelerated the photocatalytic conversion of VOCs to CO_2 over the TiO_2 photocatalyst. In addition, due to its strong adsorption capacity, the composite filter completely prevented the formation of formaldehyde from the oxidation of toluene. The regeneration and durability tests also proved the sustainability of this composite filter for VOC removal [44]. Javier et al. prepared TiO_2/nano-zeolite composites through two different synthesis methods: adding pre-synthesized zeolite into a TiO_2 precursor and adding pre-synthesized TiO_2 into a ZSM-5 zeolite precursor. The experimental results indicated that introducing zeolite into a TiO_2 precursor to prepare TiO_2/nano-zeolite composite could obtain high photocatalytic activity for propylene degradation, which was attributed to the excellent adsorption property for propylene molecules [39]. Kovalevskiy et al. loaded TiO_2 on the surface of zeolite to construct a composite catalyst for the photocatalytic degradation of ethanol. The presence of zeolite greatly increased the adsorption capacity of the composite catalyst and prevented the desorption and release of intermediates during the photocatalytic reaction. Therefore, the TiO_2/zeolite catalyst significantly suppressed the secondary contamination of harmful intermediates [45]. Ma et al. used MFI zeolite as the support for the loading of active FeO_x nanoparticles to prepare FeO_x/ZSM-5 and FeO_x/S-1 (Silicate-I) catalysts (Figure 9), which achieved n-butane conversion efficiencies of 70.5% and 70.8%, respectively. The FeO_x/ZSM-5 catalyst also achieved efficient acetophenone oxidation, with a yield of 13.0 mmol $molFe^{-1}$ h^{-1}. In the reaction process, FeO_x species acted as the active components and zeolite efficiently trapped N-hydroxyl molecules in the pore channel, which accelerated the generation of N-oxygen radicals for the oxidation reaction [48]. Improving the adsorption performance of zeolite can bring multiple advantages for catalytic reactions, including the precise enrichment of reactants and exclusion of impurities by selective adsorption and the enhancement of light utilization efficiency and thermal catalytic activity.

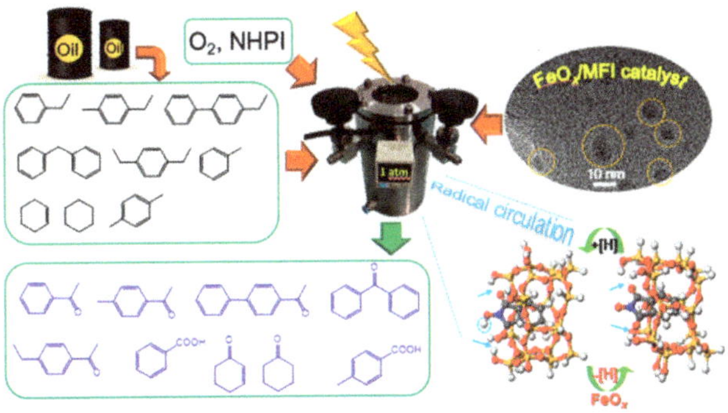

Figure 9. Schematic diagram of photocatalytic n-butane conversion over FeOx/MFI catalyst [48].

Table 3. The photothermal catalytic oxidation performance of different catalysts for VOCs.

Catalyst	VOC	VOC Concentration	Catalyst Mass (mg)	Temperature (°C)	Optical Density (mW/cm^2)	Water Vapor (vol.%)	Conversion (%)	CO_2 Selectivity (%)	S_{BET} (m^2/g)	Average Pore Size (nm)	Reference
0.93%Pt-mTiO_2/USY	Toluene	200	30	243	490	5	86.6	74.5	487.6	2.3	[34]
CuO_x-CeO_{2-x}-STO/USY	Toluene	200	30	227	700	10	86.2	76.6	404.6	5.4	[49]
20%CuO_x-WO_x/mTiO_{2-x}-USY	Toluene	200	30	235	500	5	90.4	82	380.8	7.99	[50]

Reactive species, intermediates that are highly active in photocatalytic and photothermal reactions, can rapidly participate in redox reaction processes. Wei et al. prepared a N-TiO_2/zeolite composite catalyst with a porous structure by loading TiO_2 doped with N element on the surface of the Ca-5A molecular sieve, which exhibited high photocatalytic activity for toluene degradation, with a removal efficiency of 96.7%. The strong absorption of UV–visible light and the abundant surface hydroxyl groups greatly promoted the photocatalytic degradation process [51]. Sastre et al. prepared β-zeolites containing internal silanol groups for photocatalytic methane conversion to C1 products (methanol, formaldehyde, and formic acid) at room temperature, and the selectivity for these products exceeded 95%. The surface property of zeolite greatly affected the conversion efficiency, and the all-silica β-zeolite prepared in OH^- water medium exhibited the highest photocatalytic activity and selectivity due to the existence of abundant internal silanol groups [52]. Hao et al. prepared Cu-MOR/g-C_3N_4 composite catalysts by four methods (liquid-phase ion exchange, equivalent-volume impregnation, solid-state ion exchange, and hydrothermal method), and compared their photothermal catalytic activity for methane oxidation. The catalyst prepared by liquid-phase ion exchange showed the highest activity, with a methanol yield of 3.09 μmol h^{-1} $gcat^{-1}$ at 200 °C under visible light irradiation. The strong interaction between Cu_xO_y and g-C_3N_4 decreased the interfacial charge transfer resistance, and the interface provided a large number of active sites to improve the photocatalytic activity [53]. Elimian et al. designed CuO-CeO_2 bimetallic oxides on USY zeolite by an impregnation method for photothermal catalytic oxidation of toluene. The optimized composite catalyst achieved a toluene conversion efficiency of 86.2% and CO_2 yield of 76.6% within 90 min. The interaction between CuO_x and CeO_{2-x} on the surface of USY zeolite increased the concentration of oxygen species, which was responsible for the improvement in catalytic oxidation activity [49]. Whang et al. prepared a Cu/Ti-ZSM-5 composite catalyst by hydrothermal and ion exchange methods for photothermal catalytic conversion of methane to methanol. The characterization results showed that Ti species were introduced into the skeleton of ZSM-5 in the form of TiO_4, which acted as a photo absorber and effectively prevented pore clogging (Figure 10). Under optimal conditions, the catalyst could achieve a methanol yield of 67.41 μmol h^{-1} $gcat^{-1}$ with a selectivity of 92%. Cu species were introduced into the pore channel of ZSM-5 to form Cu_xO_y and acted as the active sites to obtain excellent catalytic activity and stability in the reaction process [54]. Elimian et al. prepared yCuO_x-WO_x/mTiO_{2-x}-USY composite catalyst by an impregnation method (Figure 11) and improved the photothermal catalytic toluene degradation efficiency to 90.4% by changing the Cu loading content. The synergistic effect of CuO_x and WO_x promoted the activation of molecular oxygen and the migration of oxygen species, which accelerated the oxidation of toluene and intermediates and improved the photothermal catalytic activity. There were no significant decreases in the toluene conversion and CO_2 yield after 10 cycles, with each cycle lasting for 90 min, although the toluene conversion decreased slightly in the presence of water vapor. The above results suggested that 20CuO_x-WO_x/mTiO_{2-x}-USY exhibited good catalytic stability [50]. The increase in active reactive species can accelerate the redox process, improve reaction selectivity, and facilitate the energy transformation, thereby resulting in a prominent enhancement in catalytic activity.

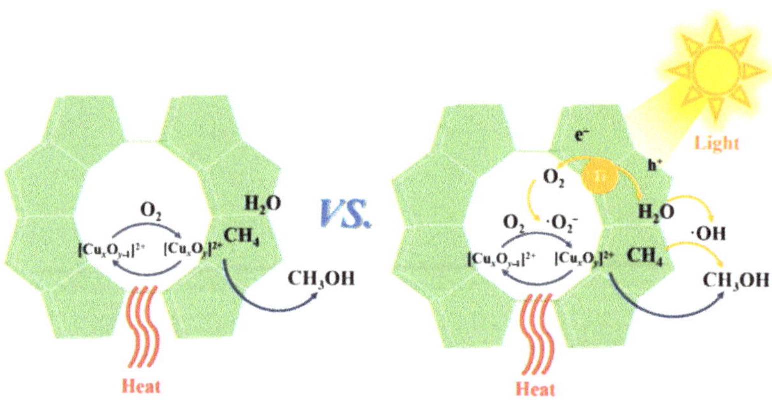

Figure 10. Schematic diagram of methane oxidation to methanol over Cu/Ti-ZSM-5 through thermal catalysis and photothermal catalysis [54].

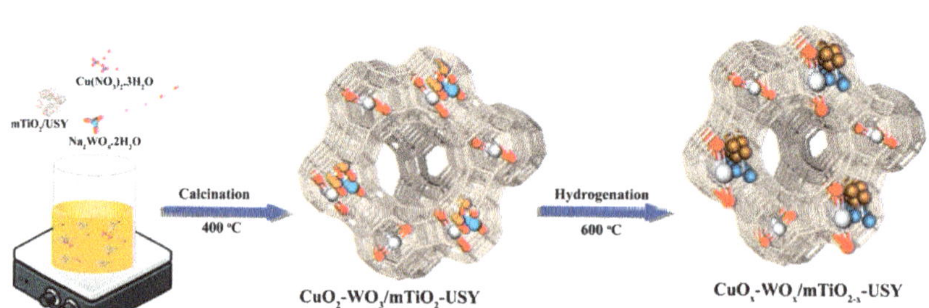

Figure 11. Schematic illustration of two-step synthesis process for the CuO_x-WO_x/mTiO_{2-x}-USY catalyst [50].

4.2. Application of Photocatalysis and Photothermal Catalysis in Water Treatment

With the assistance of zeolite, photocatalytic and photothermal catalytic technologies can effectively decompose pollutants in water, such as chlorinated hydrocarbons, benzene compounds, and organic dyes. Typically, deposition of the active phase leads to a decrease in the specific surface area and pore volume. However, when the catalytic material is loaded onto a suitable zeolite, the specific surface area and pore volume are not significantly increased, but the dispersion of active components can be improved through its high specific surface area and regular microporous structure, thus increasing the catalytic activity. The degradation performance of several zeolite composite catalysts for water pollutants are summarized in Table 4. The introduction of zeolite can enhance the conversion efficiency of water pollutants (organic dyes or heavy metals) under light irradiation, exhibiting the potential for wastewater treatment.

Table 4. The catalytic oxidation performance of different catalysts for water pollutants.

Catalyst	Water Pollutant	Pollutant Concentration (mg/L)	Catalyst Concentration (g/L)	Light	Time (min)	Degradation Rate (%)	S_{BET} (m^2/g)	Reference
CuO/NaX	Methylene blue	9.6	1	—	180	94	412	[15]
NiO/zeolite	Malachite green	30	0.25	12 W/460 nm	250	83	720	[55]
ZnO-natural zeolite	Procion red	50	4	15 W/254 nm	120	75.54	134.35	[56]
TiO_2/Fe-ZSM-5	COD	602	2	8 W	240	80	304.6	[57]
TiO_2/HSZ-385	Sulfamethazine	10	0.2	365 nm	360	66.7	424.22	[58]
TiO_2/5A	Oxytetracycline	50	0.1	16 W/254 nm	210	100	539.51	[59]
Chabazite-TiO_2	Rhodamine 6G	14.37	0.25	Sunlight	90	97.9	—	[60]
10%$TiO_{2-x}N_x$/Beta	Methylene blue	15	0.3	>460 nm	175	85	315.6	[61]
TiO_2-HX	Acetaminophen	1	1	15 W/245 nm	120	95.45	—	[62]
TiO_2-zeolite	Pentafluoropropionic acid	10	0.5	16 W/185 nm	480	58.7	270	[63]
UV-CDs/zeolite-4A/TiO_2	Methylene blue	10	0.1	500 W/365 nm	60	90.63	237.55	[28]
ZnO/$ZnFe_2O_4$/zeolite	Rhodamine B	50	0.2	30 W/395 nm	60	98.55	62.97	[64]
ZSM-5/graphene	Methyl orange	20	0.5	450 W	180	92	—	[65]
20%TiO_2/ET4	Methyl orange	10	0.5	150 W	240	95	—	[66]
Cu/TiO_2/NaY	Reactive blue dye	10	1	7 W/254 nm	240	53	13.29	[67]
NaP1	Methylene blue	100	1	—	720	96	24.86	[68]
TiO_2/zeolite	Cr(VI)	25	2	500 W	180	100	53.59	[69]
10.4%CuO/X zeolite	O-phenylenediamine	25	0.3	Sunlight	240	90	—	[70]
TiO_2/HY	2,4-D	200	2	8 W/254 nm	300	100	—	[71]
ZnO/zeolite	Methylene blue	10	1	30 W/365 nm	180	90	395	[37]
NH_4ZSM-5	Methylene blue	1	0.5	125 W/365 nm	180	77.5	—	[72]
NH_4BETA	Rhodamine B	1	0.5	125 W/365 nm	180	83.3	—	[72]
RGO@1%Pt/Ti-MFI-NSs	Methylene blue	220	2.5	300 W/420 nm	90	99	—	[73]
15%TiO_2/5A	Oxytetracycline	50	0.5	16 W/254 nm	210	100	—	[74]
10%TiO_2/13X	Oxytetracycline	50	0.5	16 W/254 nm	210	100	—	[74]
10%ZnO/NaX	Reactive blue 5G	10	1	250 W/310–350 nm	30	100	239	[75]
Ti-NaZSM-5	Methyl orange	10	0.33	300 W	90	100	325.2	[32]
15%Ag_3PO_4@Zeolite-A	Methylene blue	10	1	300 W	150	100	—	[76]
4A/WO_3/CuO	Methyl orange	10	0.15	15 W	30	99.12	10.76	[77]
4A/WO_3/CuO	Indigo carmine	10	0.15	15 W	30	97.24	10.76	[77]
Zeo-TiO_2	Rhodamine B	5	1	35 W	80	100	172	[35]
Zeo-ZnO	Rhodamine B	5	1	35 W	80	81	158	[35]
5%ZnO/FeY	Dichlorophenoxyacetic acid	18	1	250 W	300	57	523.41	[78]
5%ZnO/FeY	Dichlorophenoxyacetic acid	18	1	Sunlight	300	85	523.41	[78]
SnO_2-hierarchical zeolite	Methylene blue	40	1	15 W/254 nm	120	97	229	[79]
6%Co_3O_4/ABW	Bordeaux dye	2	2	—	30	90	29.45	[80]
TiO_2/ZSM-5	Methyl orange	20	2	550 W	180	99.55	1151	[38]
MT-ZLSH-Li+	Methylene blue	15	0.5	300–800 nm	180	77	—	[81]
TiO_2@Zeolite-Y	Polyphenols	111	1	15 W/254 nm	480	77	216	[82]

Inhibition of photogenerated charge recombination is a key point for wastewater treatment, which can be achieved by designing a suitable energy band structure, constructing a heterojunction, and adding co-catalysts into the system. Ghribi et al. prepared a NiO/zeolite catalyst for the photocatalytic degradation of malachite green dye in aqueous solution. The catalyst showed a high degradation efficiency of 83% and photogenerated electrons and holes could be effectively separated after the combination of NiO and zeolite [55]. Foroughi et al. used 4A molecular sieves loaded with nano-WO_3 and CuO to construct Z-type heterojunctions of 4A/WO_3/CuO by a hydrothermal method, and the charge recombination was obviously inhibited after formation of the heterojunctions. The catalyst showed high photocatalytic degradation efficiencies of 99.12% and 97.24% for methyl orange and indigo carmine dyes, respectively. The performance of the 4A/WO_3/CuO catalyst remained above 90% for up to 6 cycles of reuse. This indicated the stability of the catalyst structure and its effective performance in degrading the targeted dyes [77]. Haq et al. combined Ag_3PO_4 with linde A-type zeolite by hydrothermal and wet chemical methods to form the Ag_3PO_4@Al_2O_3-A composite catalyst for photocatalytic degradation of methylene blue dye. The zeolite provided a large surface area for the adsorption of MB dye and Ag_3PO_4 acted as a visible light absorber to achieve the degradation of MB dye (Figure 12). The results showed that the catalyst achieved an MB degradation efficiency of nearly 100% within 120 min. The characterization results indicated that Ag_3PO_4 was responsive to visible light, while the zeolite was responsive to UV light, and the combination of the two components improved the light response range and effectively inhibited charge recombination, which was responsible for the high catalytic activity. After 5 cycles, the composite batch activity only marginally decreased. This finding demonstrated the outstanding stability and recyclability of Ag_3PO_4@Al_2O_3-A [76]. Sodha et al. prepared the graphene/ZSM-5 composite catalyst by a hydrothermal method for photocatalytic dye removal in wastewater. When the loading content of graphene was 1%, labeled with sample code GZ 1, the GZ 1 catalyst showed high degradation efficiencies for methyl orange and methylene blue due to the efficient photogenerated charge separation (Figure 13). The holes played the most important role in the photocatalytic degradation process, followed by hydroxyl radicals and electrons [65]. Gallegos et al. prepared BiOI/mordenite zeolite composites by co-precipitation and solvothermal methods under different time and temperature conditions (Figure 14) for photocatalytic caffeic acid oxidation. The optimal catalyst was obtained under 187 °C for 9 h and showed the highest degradation efficiency for caffeic acid. The presence of BiOI on the surface of mordenite zeolite promoted the generation of electron and hole pairs under light irradiation, which interacted with water and dissolved oxygen to form hydroxyl and superoxide radicals. The active oxygen species reacted with caffeic acid molecules to achieve high removal efficiency. In addition, the presence of zeolite strengthened the structural stability and inhibited the recombination of electron–hole pairs [83]. Inhibiting the recombination of photogenerated charge can significantly improve catalytic activity, promote the conversion of light energy to chemical energy, and improve catalytic stability.

Improving the adsorption performance of zeolite can also enhance photocatalytic and photothermal performance in wastewater treatment. Sun et al. prepared a TiO_2/natural zeolite (Ti-ZE) catalyst from the simple hydrolysis and calcination of the $TiCl_4$ precursor. The catalyst calcined at 500 °C maintained the original structure of zeolite with high adsorption capacity and good TiO_2 crystallinity, which showed the highest photocatalytic reduction efficiency of nearly 100% for Cr(VI) within 240 min [69]. Zhao et al. designed and constructed a zeolite-chitosan-TiO_2@PPy (polypyrrole) aerogel (ZCTP) solar evaporator with excellent light absorption performance. The long-running ZCTP aerogel solar evaporator could make full use of solar energy during the daytime to improve water purification efficiency. The outstanding salt resistance of the ZCTP aerogel made it perfectly stable and reusable, and it exhibited a huge potential to be used as a high-efficiency photothermal material for large-scale wastewater treatment and seawater desalination [84]. Zhao et al. constructed a TiO_2/5A zeolite composite catalyst with a porous structure and large specific surface area. The catalyst showed a high degradation efficiency of 100% for hygromycin under light irradiation. The 5A molecular sieve showed a high adsorption capacity for hygromycin and prevented the particle agglomeration of TiO_2, which were important for high photocatalytic activity [59]. Chen et al. used waste incineration fly ash as the raw material to synthesize an NaP1 zeolite catalyst by a modified microwave-assisted hydrothermal method for photocatalytic degradation of methylene blue (Figure 15). The methylene blue removal efficiency could reach 96% within 12 h, which was much higher than that of original fly ash (38%). This excellent photocatalytic performance was attributed to the enhanced adsorption performance due to the 10-fold increase in surface area (24.864 m^2 g^{-1}), as well as the active metal elements embedded in the zeolite structure [68]. Based on the above studies, improving the surface area and strengthening pollutant adsorption on zeolite catalysts can greatly enhance photocatalytic and photothermal catalytic activity.

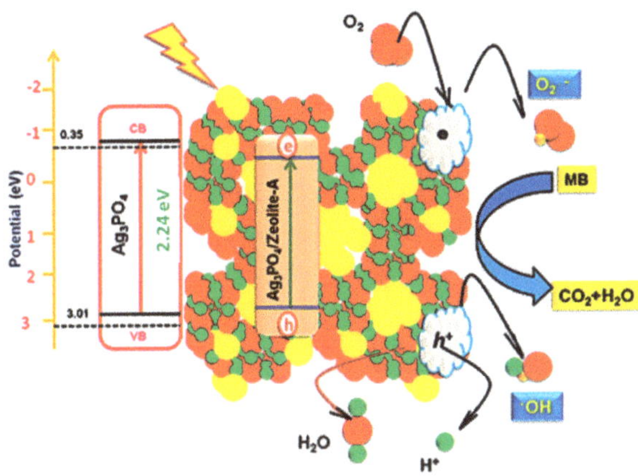

Figure 12. Photocatalytic mechanism of the degradation of MB dye over Ag_3PO_4@zeolite [76].

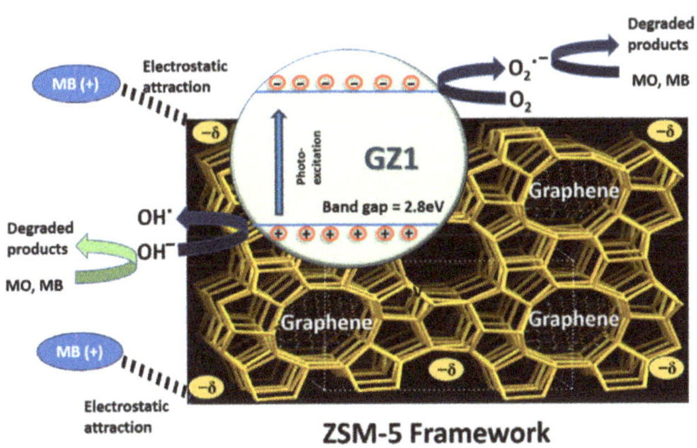

Figure 13. Degradation mechanism of methylene blue and methyl orange dye over GZ1 catalyst [65].

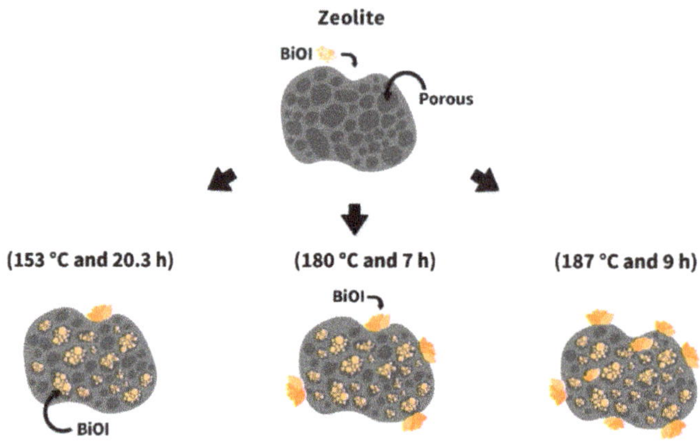

Figure 14. Schematic illustration of the formation of BiOI/mordenite composite catalyst [83].

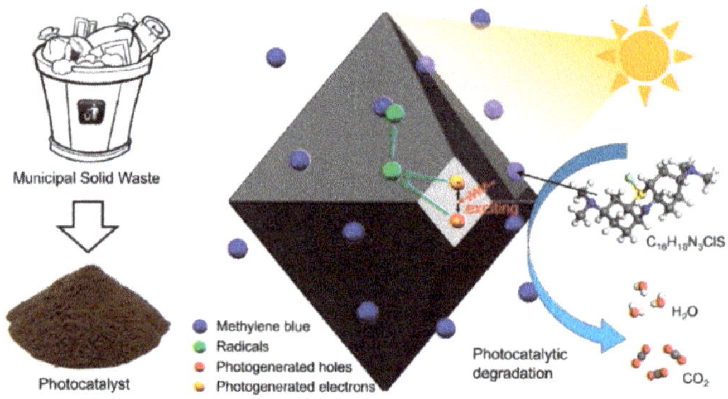

Figure 15. Schematic illustration of the transformation from municipal solid waste to a photocatalyst and the photocatalytic degradation of methylene blue under solar radiation [68].

4.3. Secondary Pollution

Zeolite itself has a good pore structure and strong adsorption capacity, which can effectively adsorb pollutants in water as well as VOCs and undergo photocatalytic and photothermal catalytic reactions on its surface to decompose them into harmless substances such as carbon dioxide and water [11]. Compared with conventional chemical treatments, photocatalytic and photothermal catalytic processes are usually carried out under mild conditions, avoiding side reactions that may be triggered by extreme conditions such as high temperatures and pressures, thus reducing the generation of harmful by-products [27].

Although photocatalytic and photothermal catalytic technologies have the advantages of high efficiency and environmental protection when treating pollutants and VOCs in water, they may still produce some secondary pollution problems [45]. During photocatalysis and photothermal catalysis, pollutants in water, such as organic dyes, may not be completely degraded to carbon dioxide and water, but instead intermediate products, such as small molecule organic compounds, are generated. These intermediate products may be toxic or bioaccumulative. Some intermediate products, such as aldehydes, ketones, and acids, may be produced during the photocatalytic and photothermal catalytic degradation of VOCs [29]. These intermediate products may be toxic or irritating and may further react to generate other harmful substances under certain conditions.

Tang et al. combined hydrophilic {001} TiO_2 nanosheets with hydrophobic NaY zeolite to prepare a TiO_2@HYZ bifunctional photocatalyst, which could achieve high photocatalytic degradation activity of gaseous toluene under UV irradiation. The carbonaceous intermediates generated from the oxidation of toluene were more strongly adsorbed on the surface of TiO_2 than the substrate molecules, and so could cover the active sites. Excessive accumulation of carbon-containing intermediates can lead to reduced photocatalytic efficiency and secondary pollution. Adsorbed carbonaceous intermediates can also hinder the transmission of light and reduce the utilization of light energy by TiO_2. But the incorporation of HYZ effectively inhibited the accumulation of carbonaceous intermediates and avoided catalyst deactivation [29]. Kovalevskiy et al. loaded TiO_2 on the surface of zeolite to construct a composite catalyst for the photocatalytic degradation of ethanol. Zeolite, as a carrier of TiO_2, greatly improved the adsorption capacity of the composite photocatalyst, thus reducing the concentration of intermediates desorbed from the surface of the photocatalyst and released into the gas phase during the PCO process. Titanium dioxide/zeolite composite photocatalysts can greatly inhibited secondary pollution by harmful intermediates [45]. Fan et al. synthesised tungsten-iron oxide molecular sieve composites with different Fe and W loadings by a wet impregnation method using NaY molecular sieves as carriers. The activity of photocatalytic degradation toward acetaldehyde and o-xylene under sunlight irradiation was investigated. It was found that some of the mineralization was to carbon dioxide and oxygenated intermediates, such as acetone and acetic acid, were formed. The accumulation of these intermediates led to a decrease in photocatalytic efficiency and secondary pollution. The zeolite acid sites adsorbed the VOCs and stabilized the intermediates, promoting their oxidation into final products [46]. Sun et al. prepared a TiO_2/natural zeolite (Ti-ZE) catalyst by simple hydrolysis and calcination of the $TiCl_4$ precursor. A portion of the electrons and holes migrated to the TiO_2 surface, the electrons reduced adsorbed Cr(VI)to Cr(III), and the holes oxidized H_2O to O_2 and H^+. Cr(III), as a reaction product, is less toxic than Cr(VI), but high concentrations are still harmful to the environment and the human body and are prone to secondary pollution [69].

In order to avoid secondary pollution, zeolite catalysts with high activity, selectivity, and stability can be developed. These catalysts are designed to efficiently degrade pollutants at lower temperatures and light intensities, while also minimizing the formation of intermediate products, thus reducing the overall environmental impact. Regulating the appropriate temperature and humidity during catalytic reactions can significantly improve catalyst activity and reaction efficiency. By optimizing these factors, the catalyst performs more effectively, leading to better reaction outcomes and higher efficiency. Too high or too low temperatures and humidity may affect the performance of the catalyst and lead to the accumulation of intermediate products. Fixing the catalyst on a suitable carrier can reduce leaching and migration of the catalyst and avoid its pollution in the environment [46]. Catalytic equipment should be regularly inspected and maintained and deactivated catalysts should be replaced in time. The treated wastewater should be further tested and treated, e.g., by filtration, adsorption, etc., to remove possible residual catalyst particles or incomplete degradation products [69].

5. Conclusions and Future Prospects

When zeolite catalysts are applied in photocatalytic and photothermal catalytic, there will be differences in energy efficiency and catalytic efficiency due to the different mechanisms of action of light and heat energy. The energy efficiency of photocatalysis is usually low because not all of the light energy is absorbed by the catalyst, and part of it may be reflected or dissipated. The reaction efficiency of photocatalysis is usually limited by the intensity of the light source and the available spectral range. Thermally catalyzed reactions are generally more energy efficient because the heat source can directly provide continuous heat, ensuring that the catalyst remains activated. Thermal catalysis usually exhibits high reaction efficiency, especially when the reaction requires higher temperature to overcome the energy barrier. The acidic sites and pore structure of zeolite catalysts are well represented in thermal catalysis and can promote the reaction efficiently. Photothermal catalysis combines the advantages of photocatalysis and thermal catalysis, using both a light source to excite the catalyst and an applied heat source to increase the energy of the reactants; this approach can increase the activity and efficiency of the reaction.

Zeolite catalysts exhibit long lifetimes in photocatalysis and photothermal catalysis due to their highly ordered microporous structure and good thermal stability. Different solvents also have effects on deactivation of the catalyst. The nature of the solvent affects the pore structure of the zeolite catalyst and the diffusion efficiency of the reactants. The polarity of the solvent may also affect the activity of acidic sites on the catalyst, thus affecting the lifetime of the catalyst. Impurities or reaction by-products in the solvent may also form deposits on the surface of the catalyst, leading to poisoning or deactivation of the catalyst.

Zeolite composite catalysts have been regarded as excellent materials for photocatalytic and photothermal catalytic degradation of VOCs and water pollutants. Although zeolite-based catalysts have attracted much attention due to their unique structure and excellent performance, the promotion in practical applications of zeolite-based catalysts still faces some challenges:

(1) More reliable zeolite-based catalysts with high catalytic performance need to be developed. In order to promote the application of zeolite-based catalysts, catalysts with high catalytic activity and stability are still scarce. By optimizing the structure of zeolite, element doping, or constructing composite materials, the light absorption, electron transfer, and structural stability of catalysts can be further improved in their practical application for pollutant treatment and energy conversion. In addition, the regeneration and recycling of zeolite is also important to realize the sustainability of environmental treatment.
(2) The role of zeolite in enhancing adsorption capacity needs to be further investigated to improve the catalytic performance, stability, and product selectivity. Improving the surface properties and pore structure of zeolites can help to enhance the adsorption capacity, which is beneficial to improving catalytic stability and selectivity for specific pollutants, thereby realizing environmental purification goals.
(3) The mechanisms and pathways of pollutant degradation over zeolite-based catalysts should be further investigated. Although the reaction systems show good catalytic activity, understanding the degradation process, electron transfer mechanism, and detailed reaction pathway is beneficial to catalyst structure design in photocatalysis and photothermal catalysis.
(4) Zeolite catalysts always show a narrow light absorption range and low utilization efficiency of sunlight, which limits their catalytic activity in practical applications. The regulation of band structure and enhancement of light absorption ability are beneficial to improving the generation of photogenerated charge and light-to-heat conversion ability.
(5) The fast recombination of photogenerated charge is a serious problem that limits the photocatalytic and photothermal catalytic activity. The improvement of charge separation efficiency is important for practical applications of zeolite-based catalysts.
(6) Studies of in situ characterizations during photocatalytic or photothermal reactions are still scarce, which restricts the in-depth understanding of the complex mechanisms in the reactions. Designing a suitable multifunctional in situ reaction cell is

also important to reveal the key active species in catalytic reactions. This will help to promote the practical application of zeolite-based catalysts in photocatalytic or photothermal technologies.

Author Contributions: Conceptualization, S.Z., L.X. and J.X.; methodology, S.Z. and L.X.; validation, L.X. and J.X.; formal analysis, L.X. and J.X.; investigation, S.Z. and L.X.; data curation, S.Z.; writing—original draft preparation, S.Z. and L.X.; writing—review and editing, S.Z., L.X. and J.X.; supervision, J.X. and B.S. All authors have read and agreed to the published version of the manuscript.

Funding: This research was funded by the Major National Science and Technology Project for Comprehensive Environmental Management in the Beijing-Tianjin-Hebei Region (2024ZD1200302), the National Natural Science Foundation of China (52376104, 52306133), the Natural Science Foundation of Hebei Province (E2023202105), the China Postdoctoral Science Foundation (2024T170209), the National Government Guides Local Science and Technology Development Fund Project (246Z3701G), the Project of Science and Technology at the Universities of Hebei Province (JZX2023006), and the Tianjin Science and Technology Plan Project (23YDPYSN00260).

Data Availability Statement: No data were reported in the study.

Conflicts of Interest: The authors declare no conflicts of interest.

References

1. Hu, G.; Yang, J.; Duan, X.; Farnood, R.; Yang, C.; Yang, J.; Liu, W.; Liu, Q. Recent Developments and Challenges in Zeolite-Based Composite Photocatalysts for Environmental Applications. *Chem. Eng. J.* **2021**, *417*, 129209. [CrossRef]
2. Dai, L.; Lu, Q.; Zhou, H.; Shen, F.; Liu, Z.; Zhu, W.; Huang, H. Tuning Oxygenated Functional Groups on Biochar for Water Pollution Control: A Critical Review. *J. Hazard. Mater.* **2021**, *420*, 126547. [CrossRef]
3. Ma, R.; Sun, J.; Li, D.H.; Wei, J.J. Review of Synergistic Photo-Thermo-Catalysis: Mechanisms, Materials and Applications. *Int. J. Hydrogen Energy* **2020**, *45*, 30288–30324. [CrossRef]
4. Rashtbari, S.; Dehghan, G.; Marefat, A.; Khataee, S.; Khataee, A. Proficient Sonophotocatalytic Degradation of Organic Pollutants Using Co3O4/TiO2 Nanocomposite Immobilized on Zeolite: Optimization, and Artificial Neural Network Modeling. *Ultrason. Sonochem.* **2024**, *102*, 106740. [CrossRef] [PubMed]
5. Velarde, L.; Nabavi, M.S.; Escalera, E.; Antti, M.-L.; Akhtar, F. Adsorption of Heavy Metals on Natural Zeolites: A Review. *Chemosphere* **2023**, *328*, 138508. [CrossRef] [PubMed]
6. Liu, Y.; Zhong, Q.; Xu, P.; Huang, H.; Yang, F.; Cao, M.; He, L.; Zhang, Q.; Chen, J. Solar Thermal Catalysis for Sustainable and Efficient Polyester Upcycling. *Matter* **2022**, *5*, 1305–1317. [CrossRef]
7. Pu, Y.; Xie, X.; Jiang, W.; Yang, L.; Jiang, X.; Yao, L. Low-Temperature Selective Catalytic Reduction of NOx with NH_3 over Zeolite Catalysts: A Review. *Chin. Chem. Lett.* **2020**, *31*, 2549–2555. [CrossRef]
8. Zhang, W.; Fang, D.; Huang, G.; Li, D.; Zheng, Y. Research and Application Development of Catalytic Redox Technology for Zeolite-Based Catalysts. *Catalysts* **2023**, *13*, 1197. [CrossRef]
9. Yu, Q.; Cheng, H.; Tang, X.; Yi, H.; Ren, X.; Li, Z. Progress in the Synthesis of Small-Pore Zeolites for Purifying NOx from Motor Vehicle Exhaust. *J. Clean. Prod.* **2022**, *381*, 135119. [CrossRef]
10. Zhu, Z.-J.; He, Z.-H.; Tian, Y.; Wang, S.-W.; Sun, Y.-C.; Wang, K.; Wang, W.; Zhang, Z.-F.; Liu, J.; Liu, Z.-T. Mass-Transfer Enhancement in the CO_2 Oxidative Dehydrogenation of Propane over GaN Supported on Zeolite Nanosheets with a Short *b* -Axis and Hierarchical Pores. *ACS Catal.* **2024**, *14*, 10376–10391. [CrossRef]
11. Liaquat, I.; Munir, R.; Abbasi, N.A.; Sadia, B.; Muneer, A.; Younas, F.; Sardar, M.F.; Zahid, M.; Noreen, S. Exploring Zeolite-Based Composites in Adsorption and Photocatalysis for Toxic Wastewater Treatment: Preparation, Mechanisms, and Future Perspectives. *Environ. Pollut.* **2024**, *349*, 123922. [CrossRef] [PubMed]
12. Narayanan, S.; Tamizhdurai, P.; Mangesh, V.L.; Ragupathi, C.; Santhana Krishnan, P.; Ramesh, A. Recent Advances in the Synthesis and Applications of Mordenite Zeolite—Review. *RSC Adv.* **2021**, *11*, 250–267. [CrossRef] [PubMed]
13. Baur, R.; Krishna, R. The Effectiveness Factor for Zeolite Catalysed Reactions. *Catal. Today* **2005**, *105*, 173–179. [CrossRef]
14. Manna, M.; Sen, S. A Mechanistic Evaluation for Total Removal of Toxic Hexavalent and Trivalent Chromium from Water by Oxygen Vacancy-Engineered Nanocomposite. *J. Environ. Chem. Eng.* **2024**, *12*, 111755. [CrossRef]
15. Nezamzadeh-Ejhieh, A.; Hushmandrad, S. Solar Photodecolorization of Methylene Blue by CuO/X Zeolite as a Heterogeneous Catalyst. *Appl. Catal. A Gen.* **2010**, *388*, 149–159. [CrossRef]

16. Umejuru, E.C.; Mashifana, T.; Kandjou, V.; Amani-Beni, M.; Sadeghifar, H.; Fayazi, M.; Karimi-Maleh, H.; Sithole, N.T. Application of Zeolite Based Nanocomposites for Wastewater Remediation: Evaluating Newer and Environmentally Benign Approaches. *Environ. Res.* **2023**, *231*, 116073. [CrossRef] [PubMed]
17. Hanaoka, T.; Miyazawa, T.; Shimura, K.; Hirata, S. Jet Fuel Synthesis from Fischer–Tropsch Product under Mild Hydrocracking Conditions Using Pt-Loaded Catalysts. *Chem. Eng. J.* **2015**, *263*, 178–185. [CrossRef]
18. Jaime-Acuña, O.E.; Villavicencio, H.; Díaz-Hernández, J.A.; Petranovskii, V.; Herrera, M.; Raymond-Herrera, O. Atomic and Electronic Structure of Quaternary CdxZnySδOγ Nanoparticles Grown on Mordenite. *Chem. Mater.* **2014**, *26*, 6152–6159. [CrossRef]
19. Rutkowska, M.; Pacia, I.; Basąg, S.; Kowalczyk, A.; Piwowarska, Z.; Duda, M.; Tarach, K.A.; Góra-Marek, K.; Michalik, M.; Díaz, U.; et al. Catalytic Performance of Commercial Cu-ZSM-5 Zeolite Modified by Desilication in NH_3-SCR and NH_3-SCO Processes. *Microporous Mesoporous Mater.* **2017**, *246*, 193–206. [CrossRef]
20. Yadav, V.K.; Choudhary, N.; Tirth, V.; Kalasariya, H.; Gnanamoorthy, G.; Algahtani, A.; Yadav, K.K.; Soni, S.; Islam, S.; Yadav, S.; et al. A Short Review on the Utilization of Incense Sticks Ash as an Emerging and Overlooked Material for the Synthesis of Zeolites. *Crystals* **2021**, *11*, 1255. [CrossRef]
21. Frising, T.; Leflaive, P. Extraframework Cation Distributions in X and Y Faujasite Zeolites: A Review. *Microporous Mesoporous Mater.* **2008**, *114*, 27–63. [CrossRef]
22. Chen, L.-H.; Sun, M.-H.; Wang, Z.; Yang, W.; Xie, Z.; Su, B.-L. Hierarchically Structured Zeolites: From Design to Application. *Chem. Rev.* **2020**, *120*, 11194–11294. [CrossRef]
23. Gao, W.; Tang, X.; Yi, H.; Jiang, S.; Yu, Q.; Xie, X.; Zhuang, R. Mesoporous Molecular Sieve-Based Materials for Catalytic Oxidation of VOC: A Review. *J. Environ. Sci.* **2023**, *125*, 112–134. [CrossRef] [PubMed]
24. Low, J.; Yu, J.; Jaroniec, M.; Wageh, S.; Al-Ghamdi, A.A. Heterojunction Photocatalysts. *Adv. Mater.* **2017**, *29*, 1601694. [CrossRef]
25. Razavi, Z.; Mirghaffari, N.; Alemrajabi, A.A.; Davar, F.; Soleimani, M. Adsorption and Photocatalytic Removal of SO2 Using Natural and Synthetic Zeolites-Supported TiO_2 in a Solar Parabolic Trough Collector. *J. Clean. Prod.* **2021**, *310*, 127376. [CrossRef]
26. Shi, M.; Meng, X. Photothermal Catalysis: From Principles to Applications. *Int. J. Hydrogen Energy* **2023**, *48*, 34659–34676. [CrossRef]
27. Fang, S.; Hu, Y.H. Thermo-Photo Catalysis: A Whole Greater than the Sum of Its Parts. *Chem. Soc. Rev.* **2022**, *51*, 3609–3647. [CrossRef] [PubMed]
28. Guo, J.; Fan, Y.; Dong, X.; Zeng, H.; Ma, X.; Fu, Y. Study on Preparation of UV-CDs/Zeolite-4A/TiO_2 Composite Photocatalyst Coupled with Ultraviolet-Irradiation and Their Application of Photocatalytic Degradation of Dyes. *J. Environ. Manag.* **2024**, *354*, 120342. [CrossRef]
29. Tang, J.; Fu, M.; Mao, Y.; Yang, S.; Lin, W.; Yu, Y.; Song, S. Decorating {0 0 1} TiO_2 Nanosheets on Hydrophobic NaY Zeolite: An Efficient Deactivation-Resistant Photocatalyst for Gaseous Toluene Removal. *Chem. Eng. J.* **2023**, *472*, 144883. [CrossRef]
30. Zakaria, H.; Li, Y.; Wei, H.; Fathy, M.M.; Zhang, C. Noble Metal-Free Plasmonic Cu/WO_3-x@ZNC Bi-Functional Composite for Ciprofloxacin Photocatalytic Degradation and Photothermal Water Evaporation via Visible-Infrared Exposure. *Chemosphere* **2024**, *364*, 143309. [CrossRef] [PubMed]
31. Sacco, O.; Vaiano, V.; Matarangolo, M. ZnO Supported on Zeolite Pellets as Efficient Catalytic System for the Removal of Caffeine by Adsorption and Photocatalysis. *Sep. Purif. Technol.* **2018**, *193*, 303–310. [CrossRef]
32. Hu, G.; Duan, X.; Yang, J.; Yang, C.; Liu, Q.; Ren, S.; Li, J.; Teng, L.; Liu, W. A Novel Conversion of Ti-Bearing Blast Furnace Slag into Ti-Containing Zeolites: Comparison Study between FAU and MFI Type Zeolites. *Adv. Powder Technol.* **2022**, *33*, 103559. [CrossRef]
33. Ke, X.; Ribbens, S.; Fan, Y.; Liu, H.; Cool, P.; Yang, D.; Zhu, H. Integrating Efficient Filtration and Visible-Light Photocatalysis by Loading Ag-Doped Zeolite Y Particles on Filtration Membrane of Alumina Nanofibers. *J. Membr. Sci.* **2011**, *375*, 69–74. [CrossRef]
34. Elimian, E.A.; Zhang, M.; Chen, J.; Jia, H.; Sun, Y.; He, J. Construction of Pt-mTiO_2/USY Multifunctional Catalyst Enriched with Oxygen Vacancies for the Enhanced Light-Driven Photothermocatalytic Degradation of Toluene. *Appl. Catal. B Environ.* **2022**, *307*, 121203. [CrossRef]
35. Alakhras, F.; Alhajri, E.; Haounati, R.; Ouachtak, H.; Addi, A.A.; Saleh, T.A. A Comparative Study of Photocatalytic Degradation of Rhodamine B Using Natural-Based Zeolite Composites. *Surf. Interfaces* **2020**, *20*, 100611. [CrossRef]
36. Vaiano, V.; Sannino, D. UV Light Driven Selective Oxidation of Cyclohexane in Gaseous Phase Using Mo-Functionalized Zeolites. *Surfaces* **2019**, *2*, 546–559. [CrossRef]
37. Liu, Z.; Liu, Z.; Cui, T.; Li, J.; Zhang, J.; Chen, T.; Wang, X.; Liang, X. Photocatalysis of Two-Dimensional Honeycomb-like ZnO Nanowalls on Zeolite. *Chem. Eng. J.* **2014**, *235*, 257–263. [CrossRef]
38. Znad, H.; Abbas, K.; Hena, S.; Awual, M.R. Synthesis a Novel Multilamellar Mesoporous TiO2/ZSM-5 for Photo-Catalytic Degradation of Methyl Orange Dye in Aqueous Media. *J. Environ. Chem. Eng.* **2018**, *6*, 218–227. [CrossRef]

39. Fernández-Catalá, J.; Sánchez-Rubio, M.; Navlani-García, M.; Berenguer-Murcia, Á.; Cazorla-Amorós, D. Synthesis of TiO_2/Nanozeolite Composites for Highly Efficient Photocatalytic Oxidation of Propene in the Gas Phase. *ACS Omega* **2020**, *5*, 31323–31331. [CrossRef] [PubMed]
40. Suárez, S. Silicalite-1 Synthesized with Geothermal and Ludox Colloidal Silica and Corresponding TiO_2/Silicalite-1 Hybrid Photocatalysts for VOC Oxidation. *Microporous Mesoporous Mater.* **2020**, *302*, 110202. [CrossRef]
41. Ko, S.; Pekarovic, J.; Fleming, P.D.; Ari-Gur, P. High Performance Nano-Titania Photocatalytic Paper Composite. Part I: Experimental Design Study for TiO_2 Composite Sheet Using a Natural Zeolite Microparticle System and Its Photocatalytic Property. *Mater. Sci. Eng. B* **2010**, *166*, 127–131. [CrossRef]
42. Yan, G.; Long, J.; Wang, X.; Li, Z.; Fu, X. Photoactive Sites in Commercial HZSM-5 Zeolite with Iron Impurities: An UV Raman Study. *Comp. Rend. Chim.* **2007**, *11*, 114–119. [CrossRef]
43. Wang, L.; Wang, W.; Liu, M.; Ge, H.; Zha, W.; Wei, Y.; Fei, E.; Zhang, Z.; Long, J.; Sa, R.; et al. Understanding Structure-Function Relationships in HZSM-5 Zeolite Catalysts for Photocatalytic Oxidation of Isopropyl Alcohol. *J. Catal.* **2019**, *377*, 322–331. [CrossRef]
44. Kim, S.; Kim, S.; Park, H.-J.; Park, S.; Kim, J.Y.; Jeong, Y.W.; Yang, H.H.; Choi, Y.; Yeom, M.; Song, D.; et al. Practical Scale Evaluation of a Photocatalytic Air Purifier Equipped with a Titania-Zeolite Composite Bead Filter for VOC Removal and Viral Inactivation. *Environ. Res.* **2022**, *204*, 112036. [CrossRef]
45. Kovalevskiy, N.S.; Lyulyukin, M.N.; Selishchev, D.S.; Kozlov, D.V. Analysis of Air Photocatalytic Purification Using a Total Hazard Index: Effect of the Composite TiO_2/Zeolite Photocatalyst. *J. Hazard. Mater.* **2018**, *358*, 302–309. [CrossRef]
46. Fan, Y.; Lyu, C.; Song, W.; Hao, M. Synthesis of Tungsten Iron-Oxide Zeolite Composites and Their Catalytic Degradation in VOC from Rubber Powder Modified Asphalt. *Alex. Eng. J.* **2024**, *94*, 212–218. [CrossRef]
47. Jansson, I.; Suárez, S.; Garcia-Garcia, F.J.; Sánchez, B. Zeolite–TiO_2 Hybrid Composites for Pollutant Degradation in Gas Phase. *Appl. Catal. B Environ.* **2015**, *178*, 100–107. [CrossRef]
48. Ma, R.; Chen, W.; Wang, L.; Yi, X.; Xiao, Y.; Gao, X.; Zhang, J.; Tang, X.; Yang, C.; Meng, X.; et al. *N*-Oxyl Radicals Trapped on Zeolite Surface Accelerate Photocatalysis. *ACS Catal.* **2019**, *9*, 10448–10453. [CrossRef]
49. Agbovhimen Elimian, E.; Zhang, M.; Li, Q.; Sun, Y.; He, J.; Jia, H. Tailoring CuO-CeO_2 Active Sites on Strontium Titanate/Ultra-Stable Y Zeolite Catalyst for Photothermal Catalytic Oxidation of Toluene. *Sep. Purif. Technol.* **2024**, *340*, 126771. [CrossRef]
50. Elimian, E.A.; Zhang, M.; Li, Q.; Chen, J.; Sun, Y.; Jia, H.; He, J. Light-Driven Photothermal Catalytic Oxidation of Toluene over CuOx-WOx/mTiO_{2-x}-USY: Revealing CuOx-WOx Synergy. *Appl. Catal. B Environ.* **2023**, *331*, 122702. [CrossRef]
51. Wei, Z.; Sun, J.; Xie, Z.; Liang, M.; Chen, S. Removal of Gaseous Toluene by the Combination of Photocatalytic Oxidation under Complex Light Irradiation of UV and Visible Light and Biological Process. *J. Hazard. Mater.* **2010**, *177*, 814–821. [CrossRef] [PubMed]
52. Sastre, F.; Fornés, V.; Corma, A.; García, H. Selective, Room-Temperature Transformation of Methane to C1 Oxygenates by Deep UV Photolysis over Zeolites. *J. Am. Chem. Soc.* **2011**, *133*, 17257–17261. [CrossRef] [PubMed]
53. Hao, J.-C.; Zhang, R.-X.; Ren, M.; Zhao, J.-X.; Gao, Z.-H.; Liu, L.; Zhang, Z.-X.; Zuo, Z.-J. Effect of the Preparation Method on Cu-MOR/g-C_3N_4 for Direct Methanol Synthesis from Methane Oxidation by Photothermal Catalysis. *Catalysts* **2023**, *13*, 868. [CrossRef]
54. Wang, W.; Hao, J.; Liu, L.; Zhang, Z.; Huang, W.; Zuo, Z. Photothermal Catalytic Methane Oxidation to Methanol on Cu/Ti-ZSM-5. *Fuel* **2024**, *376*, 132720. [CrossRef]
55. Preparation and Enhanced Visible-Light Photo-Catalytic Dye Degradation Activity of NiZY Composites. *Biointerface Res. Appl. Chem* **2022**, *13*, 72. [CrossRef]
56. Agustina, T.E.; Prajawita, A.U.; Sinaga, S.A.S. The Effect of pH on Synthesis of ZnO-Natural Zeolite Nanocomposite by Co-Precipitation Method. *IOP Conf. Ser. Mater. Sci. Eng.* **2020**, *845*, 012049. [CrossRef]
57. Ghasemi, Z.; Younesi, H.; Zinatizadeh, A.A. Preparation, Characterization and Photocatalytic Application of TiO_2/Fe-ZSM-5 Nanocomposite for the Treatment of Petroleum Refinery Wastewater: Optimization of Process Parameters by Response Surface Methodology. *Chemosphere* **2016**, *159*, 552–564. [CrossRef] [PubMed]
58. Fukahori, S.; Fujiwara, T. Modeling of Sulfonamide Antibiotic Removal by TiO_2/High-Silica Zeolite HSZ-385 Composite. *J. Hazard. Mater.* **2014**, *272*, 1–9. [CrossRef]
59. Zhao, C.; Zhou, Y.; Ridder, D.J.D.; Zhai, J.; Wei, Y.; Deng, H. Advantages of TiO_2/5A Composite Catalyst for Photocatalytic Degradation of Antibiotic Oxytetracycline in Aqueous Solution: Comparison between TiO_2 and TiO_2/5A Composite System. *Chem. Eng. J.* **2014**, *248*, 280–289. [CrossRef]
60. Ramírez-Aparicio, J.; Samaniego-Benítez, J.E.; Ramírez-Bon, R. TiO_2-Chabazite Semiconductor Composites for Photocatalytic Degradation of Rhodamine under Sunlight Irradiation. *Sol. Energy* **2016**, *139*, 258–265. [CrossRef]

61. Shen, Q.; Zhang, W.; Hao, Z.; Zou, L. A Study on the Synergistic Adsorptive and Photocatalytic Activities of TiO_2-xNx/Beta Composite Catalysts under Visible Light Irradiation. *Chem. Eng. J.* **2010**, *165*, 301–309. [CrossRef]
62. Malakootian, M.; Pourshaban-Mazandarani, M.; Hossaini, H.; Ehrampoush, M.H. Preparation and Characterization of TiO_2 Incorporated 13X Molecular Sieves for Photocatalytic Removal of Acetaminophen from Aqueous Solutions. *Process Saf. Environ. Prot.* **2016**, *104*, 334–345. [CrossRef]
63. Halder, S.; Chou, F.-C.; Wan, T.-K.; Tsai, Y.-T.; Cheng, P.-C.; Lin, Y.-C. Enhanced Photodegradation of Pentafluoropropionic Acid (PFPA) via TiO_2-Synthetic Zeolite Composites: A Sustainable Approach for Effective Defluorination. *J. Water Process Eng.* **2024**, *62*, 105296. [CrossRef]
64. Hu, T.; Li, J.; Wang, L.; Wang, H.; Zhang, Z.; Jiang, W.; Xue, C. ZnO/$ZnFe_2O_4$/Zeolite Composite Catalyst for Peroxymonosulfate Oxidation and Photocatalysis. *Mater. Lett.* **2023**, *330*, 133310. [CrossRef]
65. Sodha, V.; Bandyopadhyay, M.; Gaur, R.; Bandyopadhyay, R.; Shahabuddin, S. Synthesis and Application of ZSM-5/Graphene Composite for Photocatalytic Degradation of Industrial Dyes. *Adv. Nat. Sci. Nanosci. Nanotechnol.* **2024**, *15*, 015006. [CrossRef]
66. Guesh, K.J.; López-Muñoz, M.; Márquez-Álvarez, C.; Chebude, Y.; Diaz, I. Ethiopian Natural Zeolites for Photocatalysis. *Bull. Chem. Soc. Eth.* **2015**, *29*, 431. [CrossRef]
67. Subtil, G.W.; Vicentini, J.C.M.; De Oliveira, D.M.; De Castro-Hoshino, L.V.; Yassue-Cordeiro, P.H.; Vicentino, R.C.; Scaliante, M.H.N.O. The Influence of Different Zeolitic Supports on Hydrogen Production and Waste Degradation. *Can. J. Chem. Eng.* **2023**, *101*, 1345–1357. [CrossRef]
68. Chen, Q.; Zhao, Y.; Qiu, Q.; Long, L.; Liu, X.; Lin, S.; Jiang, X. Zeolite NaP1 Synthesized from Municipal Solid Waste Incineration Fly Ash for Photocatalytic Degradation of Methylene Blue. *Environ. Res.* **2023**, *218*, 114873. [CrossRef]
69. Sun, Q.; Hu, X.; Zheng, S.; Sun, Z.; Liu, S.; Li, H. Influence of Calcination Temperature on the Structural, Adsorption and Photocatalytic Properties of TiO_2 Nanoparticles Supported on Natural Zeolite. *Powder Technol.* **2015**, *274*, 88–97. [CrossRef]
70. Nezamzadeh-Ejhieh, A.; Salimi, Z. Solar Photocatalytic Degradation of O-Phenylenediamine by Heterogeneous CuO/X Zeolite Catalyst. *Desalination* **2011**, *280*, 281–287. [CrossRef]
71. Shankar, M.V.; Anandan, S.; Venkatachalam, N.; Arabindoo, B.; Murugesan, V. Fine Route for an Efficient Removal of 2,4-Dichlorophenoxyacetic Acid (2,4-D) by Zeolite-Supported TiO_2. *Chemosphere* **2006**, *63*, 1014–1021. [CrossRef]
72. Rakanović, M.; Vukojević, A.; Savanović, M.M.; Armaković, S.; Pelemiš, S.; Živić, F.; Sladojević, S.; Armaković, S.J. Zeolites as Adsorbents and Photocatalysts for Removal of Dyes from the Aqueous Environment. *Molecules* **2022**, *27*, 6582. [CrossRef]
73. Gong, P.; Li, B.; Kong, X.; Shakeel, M.; Liu, J.; Zuo, S. Hybriding Hierarchical Zeolite with Pt Nanoparticles and Graphene: Ternary Nanocomposites for Efficient Visible-Light Photocatalytic Degradation of Methylene Blue. *Microporous Mesoporous Mater.* **2018**, *260*, 180–189. [CrossRef]
74. Zhao, C.; Deng, H.; Li, Y.; Liu, Z. Photodegradation of Oxytetracycline in Aqueous by 5A and 13X Loaded with TiO_2 under UV Irradiation. *J. Hazard. Mater.* **2010**, *176*, 884–892. [CrossRef]
75. De Brites-Nóbrega, F.F.; Polo, A.N.B.; Benedetti, A.M.; Leão, M.M.D.; Slusarski-Santana, V.; Fernandes-Machado, N.R.C. Evaluation of Photocatalytic Activities of Supported Catalysts on NaX Zeolite or Activated Charcoal. *J. Hazard. Mater.* **2013**, *263*, 61–66. [CrossRef] [PubMed]
76. Haq, Z.U.; Pandith, A.H.; Bashir, A.; Nazir, I.; Tomar, R. Enhanced Charge Transfer and Photocatalytic Performance of Cube-Shaped Ag_3PO_4@Zeolite-A Nanocomposite. *Mater. Chem. Phys.* **2023**, *302*, 127701. [CrossRef]
77. Foroughi, M.; Peighambardoust, S.J.; Ramavandi, B.; Foroutan, R.; Peighambardoust, N.S. Simultaneous Degradation of Methyl Orange and Indigo Carmine Dyes from an Aqueous Solution Using Nanostructured WO_3 and CuO Supported on Zeolite 4A. *Sep. Purif. Technol.* **2024**, *344*, 127265. [CrossRef]
78. Nippes, R.P.; Frederichi, D.; Olsen Scaliante, E.M.H.N. Enhanced Photocatalytic Performance under Solar Radiation of ZnO through Hetero-Junction with Iron Functionalized Zeolite. *J. Photochem. Photobiol. A Chem.* **2021**, *418*, 113373. [CrossRef]
79. Kurniawan, T.; Saepurahman; Azis, M.A. Jayanudin Simultaneous Impregnation-Dealumination to Produce SnO_2-Hierarchical Zeolite for Methylene Blue Elimination via Adsorption-Photodegradation. *Case Stud. Chem. Environ. Eng.* **2024**, *9*, 100613. [CrossRef]
80. Li, C.J.; Zhang, Y.J.; Chen, H.; He, P.Y.; Zhang, Y.; Meng, Q. Synthesis of Fly Ash Cenospheres-Based Hollow ABW Zeolite for Dye Removal via the Coupling of Adsorption and Photocatalysis. *Adv. Powder Technol.* **2021**, *32*, 3436–3446. [CrossRef]
81. Guaya, D.; Debut, A.; Campoverde, J. A Novel Approach to Waste Recycling and Dye Removal: Lithium-Functionalized Nanoparticle Zeolites. *Molecules* **2024**, *29*, 4643. [CrossRef]
82. Speltini, A.; Maraschi, F.; Sturini, M.; Caratto, V.; Ferretti, M.; Profumo, A. Sorbents Coupled to Solar Light TiO_2-Based Photocatalysts for Olive Mill Wastewater Treatment. *Int. J. Photoenergy* **2016**, *2016*, 1–7. [CrossRef]

83. Gallegos-Alcaíno, A.; Robles-Araya, N.; Avalos, C.; Alfonso-Alvarez, A.; Rodríguez, C.A.; Valdés, H.; Sánchez-Flores, N.A.; Durán-Alvarez, J.C.; Bizarro, M.; Romero-Salguero, F.J.; et al. Synthesis of BiOI/Mordenite Composites for Photocatalytic Treatment of Organic Pollutants Present in Agro-Industrial Wastewater. *Nanomaterials* **2022**, *12*, 1161. [CrossRef] [PubMed]
84. Zhao, S.; Zhang, X.; Wei, G.; Su, Z. All-Weather Photothermal-Electrothermal Integrated System for Efficient Solar Steam Generation. *Chem. Eng. J.* **2023**, *458*, 141520. [CrossRef]

Disclaimer/Publisher's Note: The statements, opinions and data contained in all publications are solely those of the individual author(s) and contributor(s) and not of MDPI and/or the editor(s). MDPI and/or the editor(s) disclaim responsibility for any injury to people or property resulting from any ideas, methods, instructions or products referred to in the content.

catalysts

MDPI

Article

Experimental and Kinetic Studies on the Conversion of Glucose to Levulinic Acid Catalyzed by Synergistic Cr/HZSM-5 in GVL/H_2O Biphasic System

Han Wu [1,†], Rui Zhang [1,*,†], Jiantao Li [1], Jing Chang [1], Zhihua Liu [1], Jiale Chen [1], Jian Xiong [2], Yina Qiao [3], Zhihao Yu [4] and Xuebin Lu [2]

[1] School of Environmental and Municipal Engineering, Tianjin Chengjian University, Tianjin 300384, China; hanwu1220@163.com (H.W.); lijiantao@tcu.edu.cn (J.L.); changjinghit@163.com (J.C.); sdwfliu@tcu.edu.cn (Z.L.)
[2] School of Ecology and Environment, Tibet University, Lhasa 850000, China; jianxiong@utibet.edu.cn (J.X.); xbltju@tju.edu.cn (X.L.)
[3] School of Environment and Safety Engineering, North University of China, Taiyuan 030051, China; qiaoyina203@126.com
[4] School of Environmental Science and Engineering, Tianjin University, Tianjin 300350, China; zhihaoyupublic@tju.edu.cn
* Correspondence: rzhang@tcu.edu.cn
† Han Wu and Rui Zhang are co-first authors of the article.

Abstract: In this paper, modified HZSM-5 catalysts with different ratios of chromium (Cr/HZSM-5) were synthesized and the solvent effect of gamma valerolactone (GVL) on the enhancement of levulinic acid (LA) yield was investigated. Characterization of the Cr/HZSM-5 catalyst revealed that the introduction of Cr did not change the structure of HZSM-5. The LA yield was increased from 42.5% (H_2O solvent system) to 51.4% (GVL/H_2O solvent system) under optimal conditions. The influence of GVL on the reaction mechanism was investigated through kinetic analysis, revealing that the incorporation of GVL reduces the activation energy barrier for the conversion of glucose to LA, thereby enhancing the glucose dehydration process. The effect of GVL on the product (LA) was studied, based on molecular dynamics. It was found that the addition of GVL squeezes the water in the solvent system into the second solvation shell layer, which causes GVL to distribute around the carbonyl, hydroxyl, and carboxyl groups of LA, and reduces the likelihood of LA side reactions, thus increasing the yield of LA.

Keywords: Cr/HZSM-5 catalyst; reaction energy barrier; kinetic simulation; solvent distribution

1. Introduction

In the background of global fossil energy scarcity and environmental degradation, the development of renewable energy is given priority in the current trajectory of social progress [1–3]. As the exclusive renewable carbon source, it has attracted significant scrutiny from researchers owing to its environmentally friendly and sustainable characteristics. Cellulose, a natural polymer composed of glucose units linked by β-1,4-glycosidic bonds, represents one of the most abundant biomass resources on Earth. Glucose is broadly acknowledged as a pivotal precursor for synthesizing a wide range of high-value chemicals, including 5-hydroxymethylfurfural (HMF) [4], levulinic acid (LA), and 5-ethoxymethylfurfural (EMF) [5]. Among these distinguished chemicals, LA has been categorized as one of the 12 biomass-based platform compounds acknowledged by the

https://doi.org/10.3390/catal15020162

Department of Energy, United States, and it is extensively employed in the pharmaceutical and agricultural fields [6,7]. It can undergo further synthesis, producing valuable compounds like γ-valerolactone, resin, 2-methyltetrahydrofuran, etc. [8,9]. Consequently, the conversion of glucose to LA holds profound significance in addressing the energy crisis and mitigating environmental pollution [10].

The conversion of glucose into LA involved a series of complex, multi-step cascade reactions, where the catalyst played a pivotal role in orchestrating key reactions and optimizing target products [11,12]. In the non-homogeneous catalytic systems extensively employed in the glucose synthesis of LA, the catalyst carriers were usually carbon, metal oxides, and zeolite molecular sieves [13,14]. These carriers were subsequently loaded with active centers tailored to meet specific requirements. In the realm of catalyst carriers, zeolite molecular sieve was a prevalent choice, characterized by its impressive specific surface area, ample pore structure, and effortless loading of active sites [15]. These characteristics eased the diffusion of the reaction substrate into the pore structure and increased the contact possibility with the active sites [16]. Thus, molecular sieve-based catalysts with strong catalytic activity are frequently utilized in materials science for the high-value transformation of biomass.

Research has indicated that metal chromium exhibits a favorable catalytic effect in the reaction of glucose to LA [17]. Consequently, solid acid catalysts constructed by loading metal chromium onto zeolite molecular sieves were of interest. Nevertheless, when chromium was loaded onto molecular sieves as active sites, the interaction between metal and molecular sieve skeleton led to increasing load and clogging pore structure, hindering the contact between reactants and active sites, and thereby diminishing the product yield [18–21].

Confronted with these challenges, the solvent can be tailored to the reaction process, which can provide tunability and control the stable and ordered generation of products [22]. Mellmer et al. discovered that polar nonprotonic solvents containing chloride ions can envelop the reaction substrate and this configuration facilitated the protonation reactions [23]. Polar nonprotonic solvents could establish hydrophilic domains surrounding the reaction substrate. Notably, γ-valerolactone (GVL) as a quintessential polar nonprotonic solvent was extensively used in biomass conversion and diverse organic reactions, distinguished by its green and renewable properties [24,25]. Song et al. found that when the reaction system changed from pure water into solution of 75% GVL with 25% H_2O, the reaction rate constant decreased from 0.0034 min^{-1} to 0.0020 min^{-1} and the energy barrier decreased from 117 kJ/mol to 96 kJ/mol [26]. Therefore, the addition of a suitable solvent, such as GVL, to the reaction system could enhance the reaction rate, reduce the energy barrier, and thereby facilitate the overall reaction process. Despite extensive discourse on solvent effect at the macroscopic level, there has been scant research addressing the interaction mechanism between solvent system and reactants, and the inhibition of by-products at the microscopic level [27].

In this study, Cr/HZSM-5 catalysts were synthesized by incorporating chromium metal into HZSM-5 molecular sieves as precursor materials. The catalytic performance of Cr/HZSM-5 catalysts was assessed using LA derived from glucose conversion as a model reaction. Additionally, the impact of various solvent systems on the catalytic efficiency of Cr/HZSM-5 catalysts was examined. In addition, reaction kinetics and molecular dynamics were employed to delve deeper into the mechanism by which solvent enhancement impacts the catalytic performance of Cr/HZSM-5 catalysts. This study aims to prepare catalysts for efficient conversion of glucose to LA, and elucidate the fundamental interactions between the solvent and product at the microscopic level to further explain the mechanism of solvent-improving catalytic effect.

2. Results and Discussion

2.1. Catalyst Characterization

Figure 1A shows the XRD patterns of Cr/HZSM-5 catalysts with different loading ratios. As depicted in Figure 1A, it was apparent that Cr/HZSM-5 retained several noticeable diffraction peaks at 7.9°, 8.8°, 23.0°, and 23.9°, which were the distinctive peaks associated with MFI-type molecular sieves [28]. The characteristic peaks of MFI-type molecular sieves weakened upon loading Cr metal into the HZSM-5 catalyst, likely due to the structural changes induced by chromium incorporation. To confirm the chemical valence state of Cr in HZSM-5, XPS analysis was performed (Figure 1B), revealing that Cr species on the Cr/HZSM-5 catalyst surface existed as Cr_2O_3 with a binding energy of 576.62 eV. These findings confirm that elemental Cr was effectively loaded onto the HZSM-5 surface via a straightforward wet impregnation method.

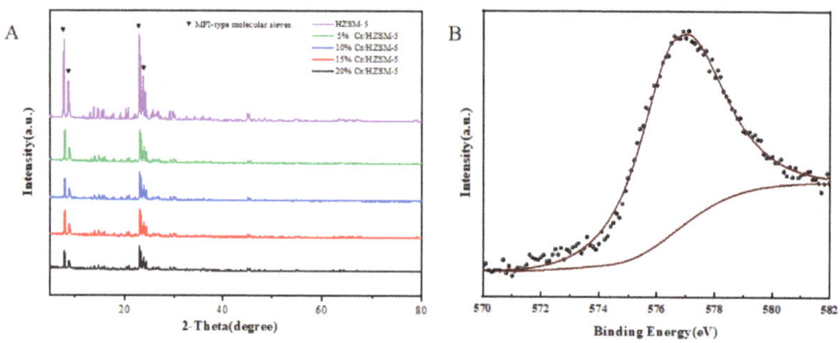

Figure 1. (**A**) XRD pattern of HZSM-5 and Cr/HZSM-5 catalyst. (**B**) XPS of 15% Cr/HZSM-5 catalyst.

Figure 2 shows the infrared spectra of HZSM-5 and Cr/HZSM-5. Figure 1 reveals characteristic peaks at 456 cm^{-1}, 546 cm^{-1}, 794 cm^{-1}, 1066 cm^{-1}, and 1644 cm^{-1} for both Cr/HZSM-5 and HZSM-5 across various loadings. These results show that the HZSM-5 framework remains constant with increasing chromium loading, which is consistent with the XRD findings. Vibrational peaks at 456 cm^{-1} and 1066 cm^{-1} were observed on both HZSM-5 and Cr/HZSM-5 catalysts, corresponding to the symmetric vibrations of SiO_4 and AlO_4 tetrahedra within the five-membered ring of HZSM-5 [29]. The characteristic peak at 546 cm^{-1} is attributed to the exo-bicyclic vibrations of the zeolite framework [30]. The vibrational peak at 794 cm^{-1} corresponds to the structure-sensitive band of the molecular sieve framework [31].

Figure 3 presents the TEM, SEM, and EDS analyses of HZSM-5 and 15% Cr/HZSM-5 catalysts. As observed in Figure 3, the HZSM-5 structure retained its prismatic crystalline form after Cr loading. To confirm the successful Cr loading, EDS-mapping images of Cr/HZSM-5 were obtained, demonstrating that Cr was effectively loaded and well-dispersed on the surface of HZSM-5.

The metal content and structural changes of HZSM-5 following the introduction of Cr were analyzed using ICP-OES and N_2 adsorption–desorption techniques, with the results presented in Table 1. It was observed that the Cr content in the Cr/HZSM-5 catalysts increased nearly proportionally with the increase in Cr loading. The increase in Cr loading led to varying reductions in the surface area and pore size of HZSM-5. As the Cr content in the Cr/HZSM-5 catalyst increased from 0.97 wt% to 3.09 wt%, the total specific surface area of the catalyst decreased from 262 m^2/g to 190 m^2/g, and the pore size decreased from 5.16 nm to 1.87 nm. This can be attributed to the Cr loading occupying the pore structure, reducing specific surface area and pore size.

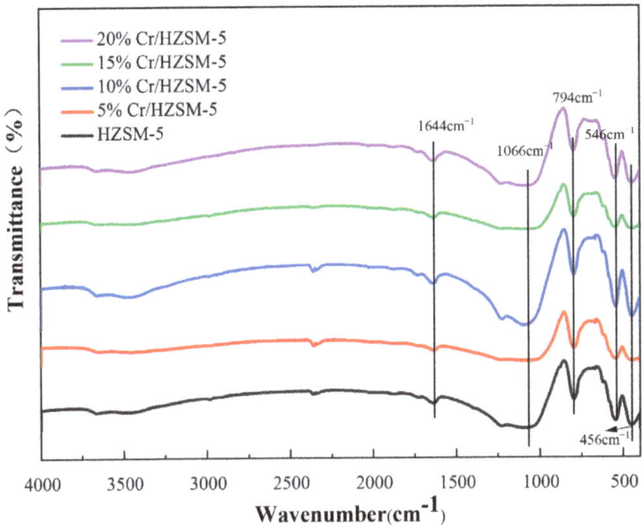

Figure 2. FT-IR spectra of HZSM-5 and Cr/HZSM-5 catalysts.

Figure 3. SEM and TEM images, and EDS spectra of HZSM-5 and 15% Cr/HZSM-5.

Table 1. Cr content, pore parameters, and acidic properties of HZSM-5 and Cr/HZSM-5 catalysts.

Catalysts	Cr (wt%)	Surface Area (m^2/g)	Pore Size (nm)	Weak Acidity Amount (mmol/g)	Moderate Acidity Amount (mmol/g)	Total Acidity (mmol/g)
HZSM-5	-	266	5.11	0.89	0.76	1.650
5% Cr/HZSM-5	0.97	262	5.16	2.813	1.154	3.967
10% Cr/HZSM-5	1.65	246	4.97	2.946	1.398	4.344
15% Cr/HZSM-5	2.48	231	3.50	2.994	1.581	4.575
20% Cr/HZSM-5	3.09	190	1.87	3.223	1.553	4.776

The acidity of the Cr/HZSM-5 catalysts was systematically evaluated via NH_3-TPD, with the resulting data presented in Figure S3. As depicted in Table 1, all catalysts exhibited a distinct peak at approximately 110 °C, corresponding to the weak acid sites, such as hydroxyl (OH) or carboxyl (COOH) groups. The strong acid sites observed in HZSM-5, characterized by a peak around 400 °C, are attributed to the Brønsted acid centers within the Si-OH-Al structure. The incorporation of Cr metal induces ion exchange between metal cations and the Brønsted acid sites, thereby enhancing the concentration of strong acid sites within the catalyst [32]. The aforementioned results demonstrate that, with increasing Cr metal loading, the total acidity of the Cr/HZSM-5 catalyst also rises proportionally. The 20% Cr/HZSM-5 catalyst exhibits the highest acidity, quantified at 4.776 mmol/g. This suggests that a greater amount of metal is incorporated into the pore structure of the HZSM-5 relative to other Cr/HZSM-5 catalysts, thereby providing a higher density of active acid sites.

2.2. Activity Comparison of HZSM-5 Catalyst with Different Cr Loading Ratios

The glucose conversion and downstream product distributions as a function of HZSM-5 catalyst with different Cr loading ratios are shown in Figure 4. The primary downstream products of glucose were FA and LA. The isomerization of glucose to fructose played a crucial role in the sequential conversion of glucose to LA. The interactions between the metal and oxygen atoms (Cr^{+3}-OH) actively promoted the isomerization of glucose [33,34]. In the blank control experiment, the glucose conversion and LA yield only reached 56.8% and 20.1%, respectively. Yet, with the progressive increase in metal loading (Cr) on the Cr/HZSM-5 catalyst, both the glucose conversion and LA yield first increased and then declined. The highest values of LA yield of 36.5% and glucose conversion of 95.6% were reached when the catalyst was 15% Cr/HZSM-5. It indicated that an elevated number of acidic sites did not consistently promote the reaction process from glucose to LA. Excessive Cr, functioning as Lewis acid sites, could potentially expedite concurrent side reactions, leading to the degradation of glucose and reaction intermediates into by-products, such as humins [35].

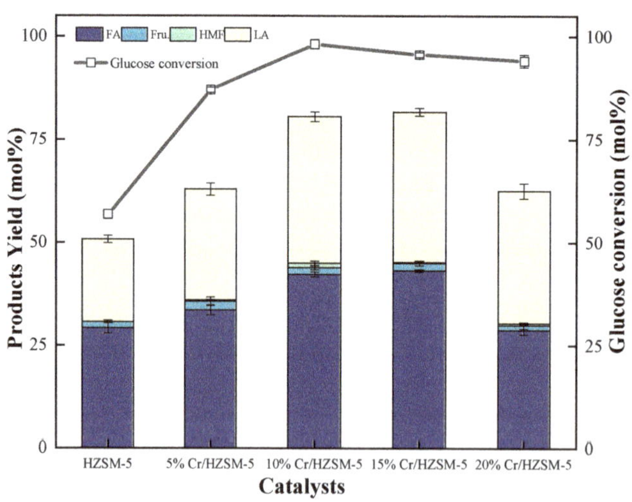

Figure 4. Effect of 0–20% Cr/HZSM-5 catalyst on glucose conversion and downstream product distribution. Other reaction conditions: Glucose concentration of 20 g/L, at 180 °C for 3 h, Cr/HZSM-5 dosage = 1 g.

2.3. Influence of Reaction Conditions on Catalytic Performance

The effects of 15% Cr/HZSM-5 catalyst on the conversion of glucose to LA under different reaction conditions were investigated. Research has demonstrated that temperature plays a crucial role in influencing the downstream products of glucose [36]. As shown in Figure 5A, both the desired product LA and the accompanying product FA exhibited a trend of initial increase followed by a subsequent decrease with the elevated temperature. Upon elevating the temperature from 140 °C to 180 °C, the yields of LA and FA exhibited an ascent from 4.8% and 19.2% to 38.7% and 58.3%, respectively. Simultaneously, glucose conversion rose from 24.7% to 97%, As the temperature increased further, the yields of LA and FA decreased. The maximal yields of LA and FA were obtained at 180 °C. This might because HMF generation was more favorable within the temperature range of 140–160 °C, and lower temperature was not conducive to the spontaneous advancement of this cascade reaction [37].

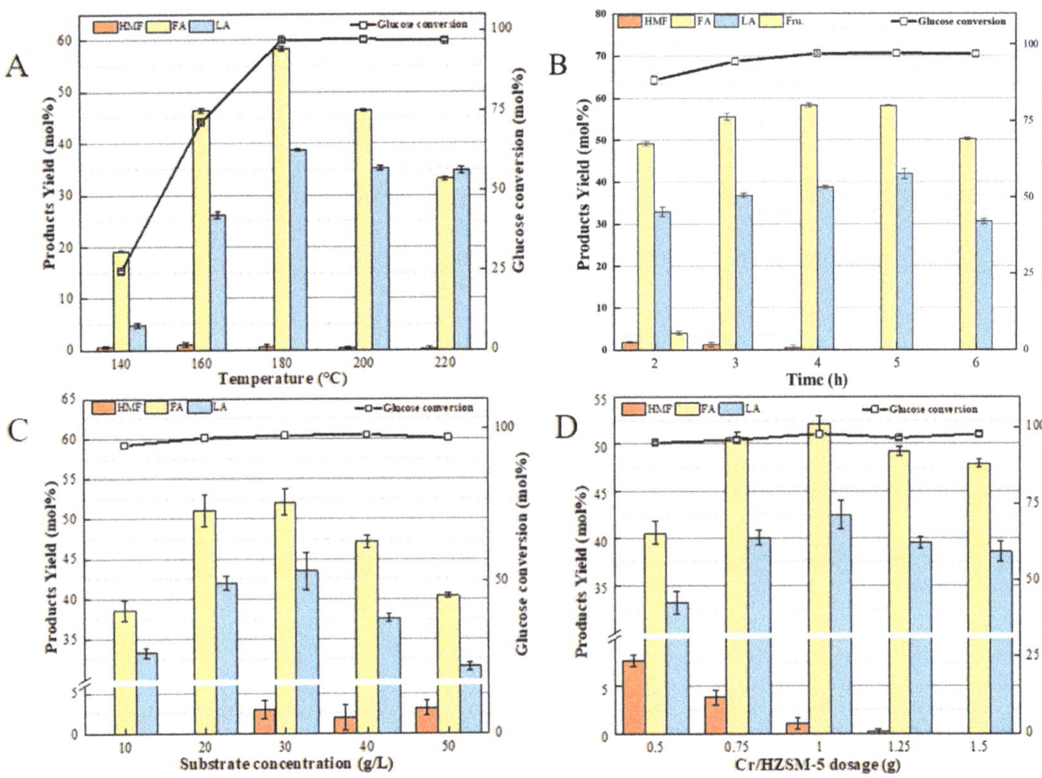

Figure 5. Effect of reaction conditions on glucose conversion and downstream product distribution. (**A**): temperature; (**B**): time; (**C**): substrate concentration; (**D**): catalyst dosage.

The influence of reaction time on the conversion of glucose to LA is illustrated in Figure 5B. It is evident that both fructose and HMF exhibited a gradual decrease when the reaction time was extended from 2 to 5 h, suggesting that these unstable intermediates undergo further rehydration to yield LA and FA (Figure 6). The yield of LA demonstrated a trend of initially increasing and subsequently decreasing with the prolongation of reaction time. This observation suggests that LA is more prone to form humic acid through aldol condensation, acetal cyclization, and dehydration processes under prolonged high-temperature conditions [38].

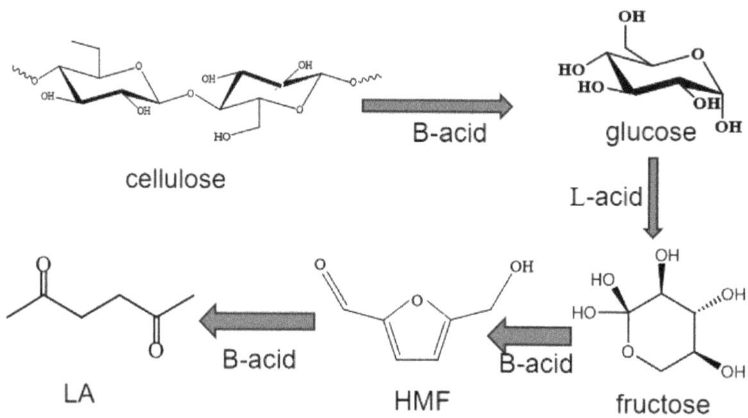

Figure 6. Reaction pathway diagram of cellulose to LA.

As shown in Figure 5C,D, the yields of both LA and FA exhibited a trend of initial increase followed by subsequent decrease as the glucose concentration was elevated along with the substrate concentration. This phenomenon could be attributed to the escalating concentration of glucose in the reaction system. As glucose concentration rose, the sole catalytic site offered by the catalyst became increasingly encapsulated by hydroxyl groups within the glucose. Consequently, this singular catalytic site became incapable of facilitating the subsequent dehydration of generated HMF, leading to the formation of LA and FA. The yield of HMF exhibited a gradual decline with the augmentation of catalyst dosage. This phenomenon could be attributed to the dynamics of glucose hydrolysis, wherein an increased quantity of catalysts supplied additional acidic sites. These sites played a pivotal role in sustaining the ongoing dehydration of HMF, leading to the production of LA. Additionally, as depicted in Figure 5C, there was nearly a 12% reduction in the LA yield when the substrate concentration escalated from 30 g/L to 50 g/L. This could be attributed to the heightened concentrations of glucose inducing more dynamic intermolecular movements and the generation of additional intermediates (1,2-enediols, etc.) These intermediates possessed a furan ring structure and were highly prone to generating undesirable by-products, such as humins. This occurred through intermolecular etherification reactions and electrophilic substitution of furfuryl alcohols [39].

Consequently, the optimal LA yield of 42.5% was achieved under conditions of 180 °C temperature, 5 h reaction time, 30 g/L substrate concentration, and 1 g catalyst dosage. The accompanying FA yield peaked at 52.1%. Calculations based on the equation revealed that the production of LA was concomitant with the production of an equivalent number of moles of FA. Throughout this reaction, the molar ratio of LA to FA was 13:16, where the surplus FA likely originated from the gradual degradation of glucose. In this reaction system, equilibrium was essentially attained in the reaction process, resulting in few by-products. Hence, the dosage of 15% Cr/HZSM-5 catalyst proved advantageous for the cascade conversion of glucose to LA.

2.4. *Effect of Different Co-Solvents on the Conversion of Glucose*

Figure 7 shows the effect of different co-solvents on glucose conversion and LA yield. The addition of GVL and THF as co-solvents had a boosting effect on the LA yield. In a pure aqueous system, the LA yield reached 42.5%. However, the incorporation of two polar nonprotonic solvents led to an enhancement in the LA yield, ca., 51.4% and 46.5%, respectively. The primary explanation for this lies in the ability of polar nonprotonic solvents to augment interaction with the reaction substrate, which can substantially increase the yield of LA. The introduction of acetonitrile as a co-solvent resulted in a reduction in LA yield, which is possibly attributed to the inherent differences in individual solvents. Furthermore, the inclusion of methanol and ethanol, serving as typical polar proton solvents, adversely impacted the yield of LA. A substantial distinction existed in the characteristics of these two solvent types. Polar nonprotonic solvents possessed the capability to accept electrons and exhibited a reduced propensity to form hydrogen bonds with the solute. Consequently, this accelerated the mass transfer rate and diminished the reaction energy barriers, thereby facilitating the reaction process. By contrast, polar protonated solvents were prone to accepting electrons, leading to the formation of hydrogen bonds with the solute and resulting in stronger interactions [40]. These interactions may potentially influence the catalytic efficiency of the overall reaction system. A comprehensive analysis conclusively identified GVL as the most effective solvent.

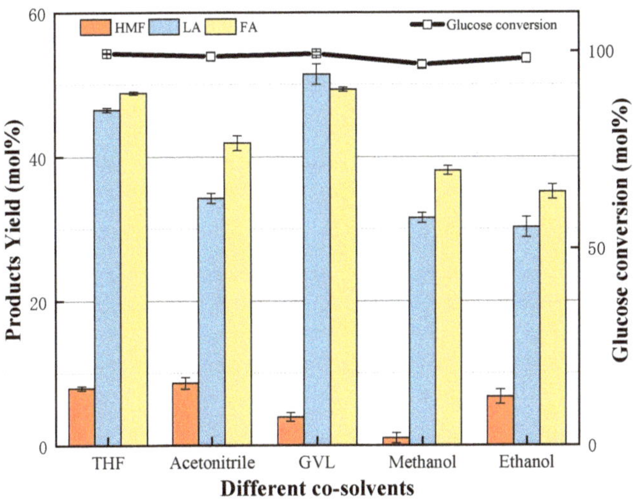

Figure 7. Effect of different solvent systems on glucose conversion and downstream product distribution.

2.5. Effect of Different GVL Contents in Solvent Systems on the Conversion of Glucose

Figure 8 shows the pattern of the effect of different GVL contents in solvent systems on the conversion of glucose to LA. As the GVL content increased, the glucose conversion consistently remained at approximately 99% (±1%). The yields of LA and FA exhibited an initial increase followed by a decline, whereas the yield of HMF steadily increased. At a GVL/H_2O ratio of 20/80, the LA yield reached 51.4%, approximately 10% higher than that in the pure water system. This finding suggests that the addition of GVL effectively mitigates side reactions caused by the condensation of the target product with the reaction substrate. However, further increasing the GVL proportion failed to enhance the LA yield, likely due to GVL's coating effect on HMF, which impedes the rehydration reaction. Moreover, due to glucose's higher solubility in water, the use of water as a co-solvent enhances substrate solubility and reduces the viscosity of the reaction medium (GVL/H_2O), thereby accelerating the mass transfer rate [41]. Based on a comprehensive analysis, the optimal GVL/H_2O ratio was identified as 20/80.

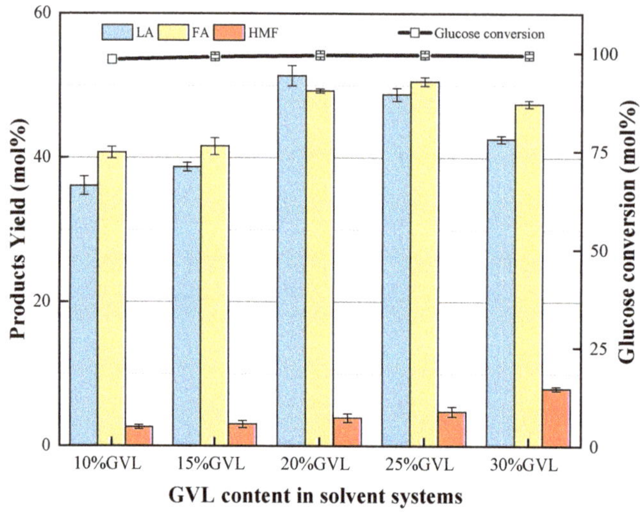

Figure 8. Effect of GVL content in solvent systems on glucose conversion and downstream product distribution.

2.6. Kinetic Study of the Response of GVL to Glucose to LA

Table 2 summarizes the kinetic parameters for glucose dehydration and HMF rehydration in various solvent systems over the 15% Cr/HZSM-5 catalyst. The reaction energy barrier (Ea) was subsequently determined from the Arrhenius equation. As indicated in Table 1, the reaction energy barrier for glucose dehydration in pure aqueous solvent was 59.19 kJ/mol, whereas that for HMF rehydration was 52.62 kJ/mol.

Table 2. Kinetic parameters of glucose dehydration and HMF rehydration in H_2O and GVL/H_2O (1:4) solvent systems.

Solvent Systems	Tem. (°C)	Glucose Dehydration K_G	R^2	Ea (kJ/mol)	HMF Rehydration K_H	R^2	Ea (kJ/mol)
H_2O	160	0.09578	0.96322	59.19	0.07813	0.95625	52.62
	180	0.20125	0.94285		0.15054	0.96437	
	200	0.38468	0.97468		0.26869	0.99518	
20%GVL/H_2O	160	0.11245	0.9788	27.02	0.08351	0.99087	26.27
	180	0.13284	0.94533		0.13683	0.91705	
	200	0.21312	0.97582		0.15417	0.98887	

The specific role of GVL as a co-solvent in the conversion of glucose to LA has not been adequately elucidated. Researchers have classified the impacts of incorporating co-solvents into two categories: physical type and chemical type [42]. The former aimed to safeguard biomass molecules from undesirable side effects, while the latter enabled the co-solvents as catalysts to modify the reaction energy barriers or directly engage in the reaction. To deeply understand the influence of GVL co-solvent on the reactions of glucose dehydration and HMF rehydration, kinetic analyses of the solvent system with a GVL/H_2O ratio of 20/80 were conducted. It is evident from Table 1 that the energy barriers for glucose dehydration and HMF rehydration in GVL/H_2O solvent exhibited a substantial reduction, in comparison to those in pure water (27.02 kJ/mol < 59.19 kJ/mol and 26.27 kJ/mol <

52.62 kJ/mol). This implied that the incorporation of GVL had a positive impact on both glucose dehydration and HMF rehydration, thereby augmenting the yield of the desired product LA. Under the same temperatures, the glucose dehydration rate constant (K_G) in the GVL/H_2O system exhibited a significant increase compared to that in the pure water system. This further emphasized the promotional impact of GVL on glucose dehydration. In a word, the inclusion of GVL reduced the energy barrier for this secondary reaction, signifying that the action of GVL as a co-solvent in the production of LA from glucose primarily fell into the chemical category.

2.7. MD Simulation Analysis of Glucose to LA Under Solvent Action

Molecular dynamics simulations provided additional insights into the interaction between the target product LA and the solvents (H_2O, GVL/H_2O). The examination of the distribution, position, and arrangement of solvent molecules around the product involved calculating spatial density distribution maps. These maps, represented in red for GVL and blue for water, were generated around glucose (Figure 9). Additionally, the radial distribution functions (RDF, Figure 10) of solvent molecules' centers of mass were calculated for further analysis.

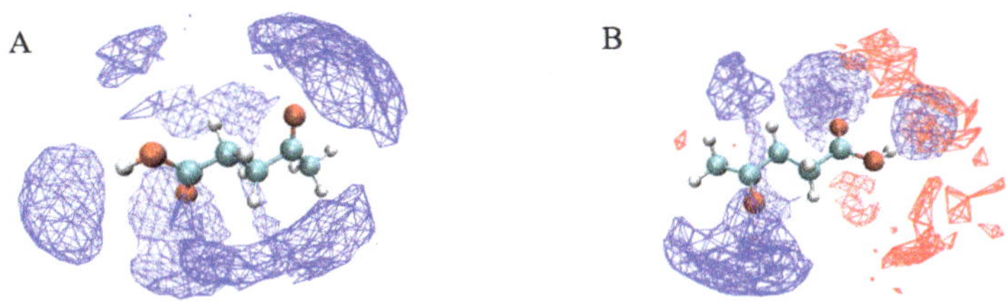

Figure 9. (**A**) Spatial density distribution of water (blue) around LA. (**B**) Spatial density distribution of GVL (red) and water (blue) around LA.

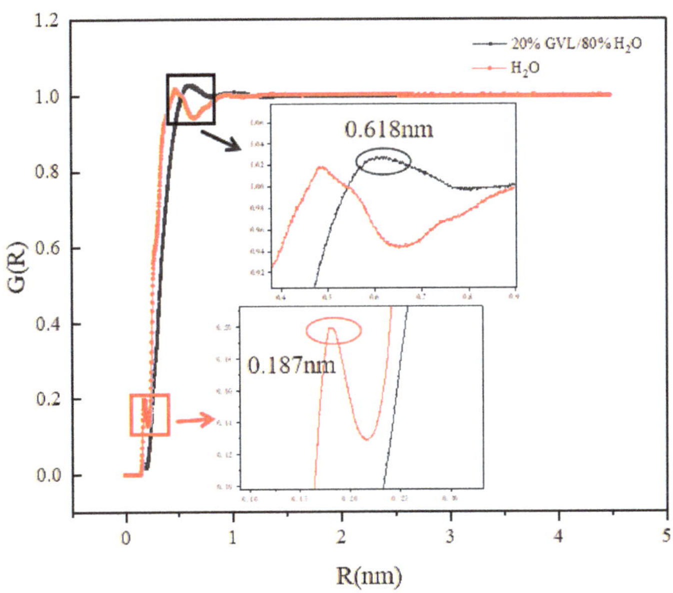

Figure 10. RDF of solvent molecules relative to LA in pure or mixed solvents.

Figure 9 illustrates the spatial density distribution of LA in both water and the mixed solvent system. As depicted in the figure, within the GVL/H_2O system, water molecules surrounding LA are relocated to the second solvation shell due to the competitive interactions between GVL and water. Within the first solvation shell, GVL primarily surrounds the carbonyl, hydroxyl, and carboxyl groups of LA, and this coordination plays a pivotal role in minimizing side reactions, such as esterification and condensation, which can occur with LA in the presence of excess water. Consequently, in the GVL/H_2O biphasic solvent system, the introduction of GVL provides a protective effect on the functional groups of LA while simultaneously reducing the mobility of glucose molecules within the reaction system. These synergistic effects facilitate the efficient catalytic conversion of glucose to LA. To elucidate the solvation shell layer more specifically, we simulated the solvation peaks across different solvent systems, as shown in Figure 10.

In a pure aqueous solvent, the solvation peak for water was 0.187 nm, whereas the solvation peak in the biphasic solvent reached 0.618 nm with the addition of 20% GVL to the system. The radial distribution function (RDF) curves between LA and the biphasic solvent exhibit a pronounced first solvation peak, indicating robust interactions between the biphasic solvent molecules and LA. This suggests that within the biphasic solvent system, GVL molecules actively compete with water to enter LA's first solvation shell.

2.8. Catalyst Reusability

To evaluate the stability of the 15%Cr/HZSM-5 catalyst, recovery tests were conducted. As depicted in Figure 11A, the 15%Cr/HZSM-5 catalyst exhibited excellent reusability. After four reaction cycles under identical conditions, the catalytic activity remained largely unchanged. In the fifth cycle, the target product LA achieved a yield of 45.7%, representing a 4.7% decline compared to the fresh 15%Cr/HZSM-5 catalyst. Figure 11B shows that the FT-IR characterization of the catalyst after five cycles revealed characteristic peaks at 456 cm^{-1}, 546 cm^{-1}, and 794 cm^{-1}. However, the weakened intensity of these peaks after five cycles suggests partial structural degradation of the catalyst, thereby reducing its stability.

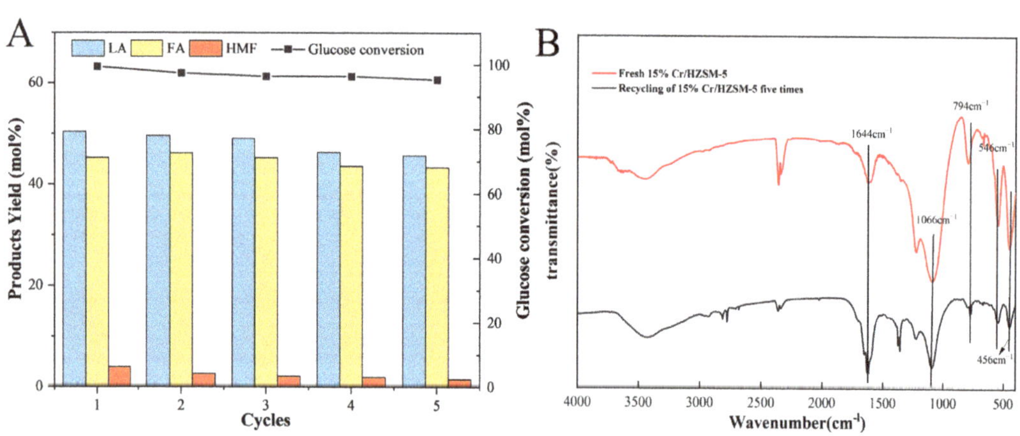

Figure 11. (**A**) 15% Cr/HZSM-5 catalyst reusability in the conversion of glucose to LA. Reaction conditions: 1.5 g glucose, 40 mL water and 10 mL GVL, 1 g catalyst, 180 °C reaction for 5 h. (**B**) FTIR spectra of fresh, and five times-recycled 15%Cr/HZSM-5 catalysts.

3. Experimental

3.1. Materials

Zinc Chloride ($ZnCl_2$, AR), Ferric Chloride ($FeCl_3$, AR), Chromium Chloride ($CrCl_3$, AR), Glucose ($C_6H_{12}O_6$, AR), Fructose ($C_6H_{12}O_6$, GR), 5-hydroxymethylfurfural ($C_6H_6O_3$, GR), Levulinic acid ($C_5H_7O_3$, GR), and HZSM-5 were obtained from Biotest Technology, Tianjin, China.

3.2. Catalyst Preparation

In this paper, a wet impregnation method was used to synthesize solid acid catalysts. For the preparation of the 15% Cr/HZSM-5 catalyst, 1.5 g of $CrCl_3 \cdot 6H_2O$ was accurately weighed and dissolved in 100 mL of deionized water to ensure complete dissolution. Subsequently, 10 g of HZSM-5 was precisely weighed and uniformly mixed with the solution. The resulting mixture was stirred at room temperature for 3 hours, followed by drying in an oven at 105 °C overnight. Finally, the dried mixture was calcined in air to yield a solid acid catalyst, which was heated to 550 °C at a rate of 2 °C/min and calcined for 6 h. The obtained solid acid catalyst was ground and passed through a 200-mesh sieve.

3.3. Catalytic Reactions

The conversion of glucose to LA was carried out in a 250 mL intermittent reactor equipped with a magnetic stirrer. At the beginning of the experiment, 1 g of glucose, 1 g of Cr/HZSM-5 catalyst, and 50 mL of water were added to the reactor. The optimal concentrations of glucose solution and catalyst dosage were subsequently determined under varying reaction conditions. The reaction was carried out at 140–220 °C for 1–7 h. After the reaction, the reactor was placed in ice water to stop the reaction. Each set of responses was repeated three times.

After the reaction, the catalyst was recovered by centrifugation and thoroughly washed with water and ethanol. The catalyst was subsequently dried overnight in an oven and reused in the next cycle.

3.4. Analytical Characterization

The product distribution in the reaction solution was detected by HPLC (LC-20A), with a refractive detector using a column (BioRad Aminex HPX-87H, 300 × 7.8 mm) at 65 °C and 0.05 M H_2SO_4 as the mobile phase at a flow rate of 0.6 mL/min.

The conversion of glucose, the yield of fructose, HMF, LA, and FA were calculated as following:

$$\text{Glucose conversion} = \left(\frac{\text{mole of glucose reacted}}{\text{mole of initial glucose}}\right) \times 100\%$$

$$\text{Products yield} = \left(\frac{\text{mole of fructose, HMF, LA, and FA}}{\text{mole of initial glucose}}\right) \times 100\%$$

3.5. Characterization Test

The morphology and elemental composition of Cr/HZSM-5 catalysts were examined using a high-resolution field-emission scanning electron microscope (SEM, TESCAN MIRA LMS) coupled with an energy-dispersive spectrometer (EDS, TESCAN MIRA LMS). The equipment was acquired by TESCAN, Brno, Czech Republic. The crystalline structure was analyzed using an X-ray diffractometer (XRD, Rigaku Smartlab 9KW) with a scanning range of 5° to 90° and a scanning speed of 10°/min. The facility was acquired by Nippon Rigaku Corporation, Akishima, Tokyo, Japan. Functional groups in the samples were identified using a Fourier-transform infrared spectrometer (FT-IR, Perkin Elmer Spectrum One) over a wavenumber range of 400–4000 cm^{-1}. The facility was acquired by PerkinElmer AG, Waltham, Massachusetts, USA. The crystalline structure was characterized using a Rigaku Smartlab 9KW X-ray diffractometer (XRD) with a scanning range of 5° to 90° at a speed of 10°/min. The X-ray source employed a rotating copper anode operating at 40 kV and 40 mA. The facility was acquired by Nippon Rigaku Corporation, Akishima, Tokyo, Japan. Metal concentrations in the catalysts were detected using an inductively coupled plasma emission spectrometer (ICP-OES, Agilent 5110). The device was acquired by Agilent Technologies, Inc. of California, USA. The specific surface area and pore volume of the prepared catalysts were detected using a specific surface area and porosity analyzer (ASAP 2020 PLUS). The device was acquired by Micromeritics of Norcross, GA, USA. Metal valence states in the catalysts were detected using x-ray photoelectron spectroscopy (XPS, Thermo Scientific K-Alpha). The facility was acquired by Thermo Fisher Scientific, Waltham, MA, USA.

3.6. Molecular Dynamics Simulation

Molecular dynamics (MD) simulations were executed employing the GROMACS 2021 package. Files pertaining to formic acid (FA) and LA molecules were acquired by Sobtop and employed in subsequent MD simulations. Periodicboundarycondition (PBC) was used for all three dimensions of the system. Modeling of FA and LA molecules involved the Generalized Amber Force Field (GAFF) force field, while the water molecule was optimized by the TIP4P model. Upon the completion of system construction, an energy minimization process was executed by using the steepest descent method for a duration of 5000 steps and the conjugate gradient method for an additional 5000 steps. This meticulous approach ensured the conventional structure and rational geometric configuration for molecules. After the system was at the energy minimization point, the real dynamics simulation could be started. A 1 ns NVT ramp-up kinetics phase was performed initially, followed by a 1 ns NPT balancing phase, and culminating in a 50 ns NPT synthesis simulation. The Leapfrog method with a time step of 2 fs was employed as the integral solution algorithm. Long-range electrostatic interactions were calculated using the Cut-Off method, employing a truncation distance of 1.0 nm for non-bonded interactions. Pressure coupling was achieved through the isotropic Parrinello–Rahman method, maintaining a controlled pressure of 1 bar. Temperature coupling was analyzed by the Velocity-rescale heat bath method. All bonds were constructed by the Linear Constraint Solver (LINCS) algorithm. The simulation trajectory data with an interval of 2 ps were analyzed by GROMACS 2021.6 software package, and visualized by VMD 1.9.3 software [43].

4. Conclusions

In this study, we synthesized Cr/HZSM-5 catalysts with varying Cr loadings to evaluate their catalytic performance in the conversion of glucose to LA. It was found that at a reaction temperature of 180 °C, a reaction time of 5 h, a catalyst dosage of 1 g, and a glucose concentration of 30 g/L, the glucose conversion reached 99% and the LA yield was 42.5%. Building on this foundation, the impact of solvent enhancement on the catalytic performance of Cr/HZSM-5 catalysts was investigated, identifying GVL as the most effective solvent. The addition of 20% GVL into the solvent system led to an LA yield of 51.4%, representing an 8.9% increase compared to using water as the solvent. Reaction kinetics and molecular dynamics simulations were conducted to further elucidate the mechanism of solvent potentiation. Based on reaction kinetics calculations, it was demonstrated that the solvent system composed of GVL and water reduced the activation energy barrier for the conversion of glucose to LA, compared to the pure water system (78.89 kJ/mol < 86.21 kJ/mol). Molecular dynamics simulation results indicate that the incorporation of GVL forces water molecules in the solvent system into the secondary solvation shell, thereby diminishing the presence of water around the LA. Concurrently, GVL predominantly localizes around the carbonyl, hydroxyl, and carboxyl groups of LA, thereby minimizing the likelihood of side reactions and consequently enhancing the LA yield.

Supplementary Materials: The following supporting information can be downloaded at: https://www.mdpi.com/article/10.3390/catal15020162/s1, Text S1: Kinetic study; Figure S1: Kinetic modelling of reactions; Figure S2. SEM, TEM images and EDS spectra of HZSM-5 and 5-20% Cr/HZSM-5; Figure S3. EDS elemental mapping images of 5%-20% Cr/HZSM-5; Figure S4. NH_3-TPD profiles of HZSM-5 and Cr/HZSM-5 catalysts; Figure S5. Effect of 0%-20% Fe/HZSM-5 catalyst on glucose conversion and downstream product distribution; Figure S6. Effect of 0%-20% Zn/HZSM-5 catalyst on glucose conversion and downstream product distribution; Figure S7. State of the reaction system before (A) and after (B) the test; Figure S8. Determination of energy barriers for glucose dehydration and HMF rehydration in water from the Arrhenius equation. (A) Plots of Ln ($C_{glucose}$) vs. time at different reaction temperature; (B) Arrhenius plot of Ln (K) vs. 1/T for glucose dehydration. (C) Plots of Ln (C_{HMF}) vs. time at different reaction temperature; (D) Arrhenius plot of Ln(K) vs. 1/T for HMF rehydration; Figure S9. Determination of energy barriers for glucose dehydration and HMF rehydration in GVL/H_2O from the Arrhenius equation. (A) Plots of Ln ($C_{glucose}$) vs. time at different reaction temperatures; (B) Arrhenius plot of Ln (K) vs. 1/T for glucose dehydration. (C) Plots of Ln (C_{HMF}) vs. time at different reaction temperatures; (D) Arrhenius plot of Ln(K) vs. 1/T for HMF rehydration.

Author Contributions: Conceptualization, X.L.; Formal analysis, J.C. (Jiale Chen); Funding acquisition, R.Z., Z.L., J.X., Z.Y. and X.L.; Methodology, J.C. (Jing Chang); Resources, R.Z. and Y.Q.; Writing—original draft, H.W.; Writing—review and editing, H.W. and J.L. All authors have read and agreed to the published version of the manuscript.

Funding: This work was supported by the National Natural Science Foundation of China (Grant No. 51908400, 52066017, 51876180, 52206293; 22308253), Tianjin Natural Science Foundation key project (23JCZDJC00430), Central Financial Support Special Funds for Local Universities (Tibet University) ([2023] No. 1, [2024] No. 1), Science and Technology Plan of Qinghai Province (2022-GX-C13).

Data Availability Statement: The data that support the findings of this study are available from the corresponding author initials, upon reasonable request.

Acknowledgments: The authors thank the financial support from the funding sources mentioned above.

Conflicts of Interest: The authors declare no conflicts of interest.

References

1. Guo, Z.; Pedersen, C.M.; Wang, P.; Ma, M.; Zhao, Y.; Qiao, Y.; Wang, Y. D-Glucose isomerization with PAMAM dendrimers as environmentally friendly catalysts. *J. Agric. Food Chem.* **2021**, *69*, 5105–5112. [CrossRef] [PubMed]
2. Mongkolpichayarak, I.; Jiraroj, D.; Anutrasakda, W.; Ngamcharussrivichai, C.; Samec, J.S.; Tungasmita, D.N. Cr/MCM-22 catalyst for the synthesis of levulinic acid from green hydrothermolysis of renewable biomass resources. *J. Catal.* **2022**, *405*, 373–384. [CrossRef]
3. Zhang, T.; Tang, Y.; Guo, S.; Cao, X.; Pan, A.; Fang, G.; Zhou, J.; Liang, S. Fundamentals and perspectives in developing zinc-ion battery electrolytes: A comprehensive review. *Energy Environ. Sci.* **2020**, *13*, 4625–4665. [CrossRef]
4. Shen, F.; Sun, S.; Zhang, X.; Yang, J.; Qiu, M.; Qi, X. Mechanochemical-assisted production of 5-hydroxymethylfurfural from high concentration of cellulose. *Cellulose* **2020**, *27*, 3013–3023. [CrossRef]
5. Dutta, S.; De, S.; Alam, M.I.; Abu-Omar, M.M.; Saha, B. Direct conversion of cellulose and lignocellulosic biomass into chemicals and biofuel with metal chloride catalysts. *J. Catal.* **2012**, *288*, 8–15. [CrossRef]
6. Hu, W.; Chi, Z.; Wan, Y.; Wang, S.; Lin, J.; Wan, S.; Wang, Y. Synergetic effect of Lewis acid and base in modified Sn-β on the direct conversion of levoglucosan to lactic acid. *Catal. Sci. Technol.* **2020**, *10*, 2986–2993. [CrossRef]
7. Intaramas, K.; Jonglertjunya, W.; Laosiripojana, N.; Sakdaronnarong, C. Selective conversion of cassava mash to glucose using solid acid catalysts by sequential solid state mixed-milling reaction and thermo-hydrolysis. *Energy* **2018**, *149*, 837–847. [CrossRef]

8. Chauhan, A.; Bal, R.; Srivastava, R. Synthesis of γ-Valerolactone from Levulinic Acid with Co/NC, and from Furfural via Cascade Reaction Using Co/NC and H-Beta. *Energy Fuels* **2024**, *38*, 5998–6011. [CrossRef]
9. Gu, C.; Chen, L.; Liu, Y.; Zhang, X.; Liu, J.; Zhang, Q.; Wang, C.; Ma, L. One-pot conversion of biomass-derived levulinic acid to furanic biofuel 2-methyltetrahydrofuran over bimetallic NiCo/γ-Al_2O_3 catalysts. *Mol. Catal.* **2022**, *524*, 112317. [CrossRef]
10. Tyagi, U.; Anand, N.; Kumar, D. Synergistic effect of modified activated carbon and ionic liquid in the conversion of microcrystalline cellulose to 5-Hydroxymethyl Furfural. *Bioresour. Technol.* **2018**, *267*, 326–332. [CrossRef] [PubMed]
11. Hak, C.; Panchai, P.; Nutongkaew, T.; Grisdanurak, N.; Tulaphol, S. One-pot levulinic acid production from rice straw by acid hydrolysis in deep eutectic solvent. *Chem. Eng. Commun.* **2024**, *211*, 366–378. [CrossRef]
12. Kumar, K.; Parveen, F.; Patra, T.; Upadhyayula, S. Hydrothermal conversion of glucose to levulinic acid using multifunctional ionic liquids: Effects of metal ion co-catalysts on the product yield. *New J. Chem.* **2018**, *42*, 228–236. [CrossRef]
13. Xu, X.; Liang, B.; Zhu, Y.; Chen, J.; Gan, T.; Hu, H.; Zhang, Y.; Huang, Z.; Qin, Y. Direct and efficient conversion of cellulose to levulinic acid catalyzed by carbon foam-supported heteropolyacid with Brønsted–Lewis dual-acidic sites. *Bioresour. Technol.* **2023**, *387*, 129600. [CrossRef] [PubMed]
14. Ya'aini, N.; Amin, N.A.S.; Endud, S. Characterization and performance of hybrid catalysts for levulinic acid production from glucose. *Microporous Mesoporous Mater.* **2013**, *171*, 14–23. [CrossRef]
15. Chu, S.; Guo, X.; Dong, L.; Chen, X.; Li, Y.; Mu, X. The influence of pore structure and Si/Al ratio of HZSM-5 zeolites on the product distributions of α-cellulose hydrolysis. *Mol. Catal.* **2018**, *445*, 240–247. [CrossRef]
16. Bounoukta, C.E.; Megías-Sayago, C.; Ivanova, S.; Ammari, F.; Centeno, M.A.; Odriozola, J.A. Pursuing efficient systems for glucose transformation to levulinic acid: Homogeneous vs. heterogeneous catalysts and the effect of their co-action. *Fuel* **2022**, *318*, 123712. [CrossRef]
17. Pidko, E.A.; Degirmenci, V.; van Santen, R.A.; Hensen, E. Glucose activation by transient Cr^{2+} dimers. *Angew Chem* **2010**, *49*, 2530–2534. [CrossRef] [PubMed]
18. Yang, S.; He, D.; Zhang, L.; Zhang, Y.; Lu, J.; Luo, Y. Toxic chromium treatment induce amino-assisted electrostatic adsorption for the synthesis of highly dispersed chromium catalyst. *J. Hazard. Mater.* **2021**, *417*, 126155. [CrossRef]
19. Ayari, F.; Mhamdi, M.; Álvarez-Rodríguez, J.; Ruiz, A.G.; Delahay, G.; Ghorbel, A. Selective catalytic reduction of NO with NH_3 over Cr-ZSM-5 catalysts: General characterization and catalysts screening. *Appl. Catal. B Environ.* **2013**, *134*, 367–380. [CrossRef]
20. Figueiredo, H.; Silva, B.; Quintelas, C.; Raposo, M.; Parpot, P.; Fonseca, A.; Lewandowska, A.; Bañares, M.; Neves, I.; Tavares, T. Immobilization of chromium complexes in zeolite Y obtained from biosorbents: Synthesis, characterization and catalytic behaviour. *Appl. Catal. B Environ.* **2010**, *94*, 1–7. [CrossRef]
21. Kułażyński, M.; Van Ommen, J.; Trawczyński, J.; Walendziewski, J. Catalytic combustion of trichloroethylene over TiO_2-SiO_2 supported catalysts. *Appl. Catal. B Environ.* **2002**, *36*, 239–247. [CrossRef]
22. Raguindin, R.Q.; Desalegn, B.Z.; Gebresillase, M.N.; Seo, J.G. Yolk-shell nickel–cobalt phosphides as bifunctional catalysts in the solvent-free hydrogenation of Levulinic acid to gamma-Valerolactone. *Renew. Energy* **2022**, *191*, 763–774. [CrossRef]
23. Mellmer, M.A.; Sanpitakseree, C.; Demir, B.; Ma, K.; Elliott, W.A.; Bai, P.; Johnson, R.L.; Walker, T.W.; Shanks, B.H.; Rioux, R.M. Effects of chloride ions in acid-catalyzed biomass dehydration reactions in polar aprotic solvents. *Nat. Commun.* **2019**, *10*, 1132. [CrossRef]
24. Mthembu, L.D.; Gupta, R.; Dziike, F.; Lokhat, D.; Deenadayalu, N. Conversion of Biomass-Derived Levulinic Acid into γ-Valerolactone Using Methanesulfonic Acid: An Optimization Study Using Response Surface Methodology. *Fermentation* **2023**, *9*, 288. [CrossRef]
25. Xu, R.; Liu, K.; Du, H.; Liu, H.; Cao, X.; Zhao, X.; Qu, G.; Li, X.; Li, B.; Si, C. Falling leaves return to their roots: A review on the preparation of γ-valerolactone from lignocellulose and its application in the conversion of lignocellulose. *ChemSusChem* **2020**, *13*, 6461–6476. [CrossRef] [PubMed]
26. Song, B.; Yu, Y.; Wu, H. Tuning glucose decomposition in hot-compressed gamma-valerolactone/water mixtures: From isomerization to dehydration reactions. *Fuel* **2019**, *238*, 225–231. [CrossRef]
27. Wu, H.; Zhang, R.; Zhai, Y.; Song, X.; Xiong, J.; Li, X.; Qiao, Y.; Lu, X.; Yu, Z. Solvent Effects Enable Efficient Tandem Conversion of Cellulose and Its Monosaccharides Towards 5-Hydroxymethylfurfural. *ChemSusChem* **2023**, *16*, e202201809. [CrossRef] [PubMed]
28. Li, X.; Liu, Q.; Si, C.; Lu, L.; Luo, C.; Gu, X.; Liu, W.; Lu, X. Green and efficient production of furfural from corn cob over H-ZSM-5 using γ-valerolactone as solvent. *Ind. Crops Prod.* **2018**, *120*, 343–350. [CrossRef]
29. Chen, H.; Chen, L.; Lin, J.; Tan, K.; Li, J. Copper sites in copper-exchanged ZSM-5 for CO activation and methanol synthesis: XPS and FTIR studies. *Inorg. Chem.* **1997**, *36*, 1417–1423. [CrossRef]
30. Pu, H.; Han, C.; Wang, H.; Xu, S.; Zhang, L.; Zhang, Y.; Luo, Y. Characterization and investigation mechanism of hydrothermal stability for micropore-free ordered mesoporous molecular sieves. *Appl. Surf. Sci.* **2012**, *258*, 8895–8901. [CrossRef]
31. Cihanoğlu, A.; Gündüz, G.; Dükkancı, M. Degradation of acetic acid by heterogeneous Fenton-like oxidation over iron-containing ZSM-5 zeolites. *Appl. Catal. B Environ.* **2015**, *165*, 687–699. [CrossRef]

32. Luan, P.; Chen, J.; Liu, T.; Wang, J.; Yan, B.; Li, N.; Cui, X.; Chen, G.; Cheng, Z. Ex-situ catalytic fast pyrolysis of waste polycarbonate for aromatic hydrocarbons: Utilizing HZSM-5/modified HZSM-5. *J. Anal. Appl. Pyrolysis* **2024**, *182*, 106667. [CrossRef]
33. Mushrif, S.H.; Varghese, J.J.; Vlachos, D.G. Insights into the Cr (III) catalyzed isomerization mechanism of glucose to fructose in the presence of water using ab initio molecular dynamics. *Phys. Chem. Chem. Phys.* **2014**, *16*, 19564–19572. [CrossRef]
34. Choudhary, V.; Mushrif, S.H.; Ho, C.; Anderko, A.; Nikolakis, V.; Marinkovic, N.S.; Frenkel, A.I.; Sandler, S.I.; Vlachos, D.G. Insights into the interplay of Lewis and Brønsted acid catalysts in glucose and fructose conversion to 5-(hydroxymethyl) furfural and levulinic acid in aqueous media. *J. Am. Chem. Soc.* **2013**, *135*, 3997–4006. [CrossRef]
35. Taghavi, S.; Ghedini, E.; Menegazzo, F.; Giordana, A.; Cerrato, G.; Cruciani, G.; Di Michele, A.; Zendehdel, M.; Signoretto, M. Balanced acidity by microwave-assisted ion-exchange of ZSM-5 zeolite as a catalyst for transformation of glucose to levulinic acid. *Biomass Convers. Biorefinery* **2024**, *14*, 8251–8269. [CrossRef]
36. Hu, L.; Wu, Z.; Xu, J.; Sun, Y.; Lin, L.; Liu, S. Zeolite-promoted transformation of glucose into 5-hydroxymethylfurfural in ionic liquid. *Chem. Eng. J.* **2014**, *244*, 137–144. [CrossRef]
37. Teong, S.P.; Yi, G.; Cao, X.; Zhang, Y. Poly-benzylic ammonium chloride resins as solid catalysts for fructose dehydration. *ChemSusChem* **2014**, *7*, 2120–2124. [CrossRef]
38. Shi, N.; Liu, Q.; Ju, R.; He, X.; Zhang, Y.; Tang, S.; Ma, L. Condensation of α-carbonyl aldehydes leads to the formation of solid humins during the hydrothermal degradation of carbohydrates. *Acs Omega* **2019**, *4*, 7330–7343. [CrossRef] [PubMed]
39. Ghosh, M.K.; Howard, M.S.; Dussan, K.; Dooley, S. Mechanism and theory of d-glucopyranose homogeneous acid catalysis in the aqueous solution phase. *Phys. Chem. Chem. Phys.* **2019**, *21*, 17993–18011. [CrossRef] [PubMed]
40. Guo, H.; Duereh, A.; Su, Y.; Hensen, E.J.; Qi, X.; Smith Jr, R.L. Mechanistic role of protonated polar additives in ethanol for selective transformation of biomass-related compounds. *Appl. Catal. B Environ.* **2020**, *264*, 118509. [CrossRef]
41. Antonetti, C.; Fulignati, S.; Licursi, D.; Raspolli Galletti, A.M. Turning point toward the sustainable production of 5-hydroxymethyl-2-furaldehyde in water: Metal salts for its synthesis from fructose and inulin. *ACS Sustain. Chem. Eng.* **2019**, *7*, 6830–6838. [CrossRef]
42. Vasudevan, V.; Mushrif, S.H. Insights into the solvation of glucose in water, dimethyl sulfoxide (DMSO), tetrahydrofuran (THF) and N, N-dimethylformamide (DMF) and its possible implications on the conversion of glucose to platform chemicals. *Rsc Adv.* **2015**, *5*, 20756–20763. [CrossRef]
43. Abraham, M.J.; Murtola, T.; Schulz, R.; Páll, S.; Smith, J.C.; Hess, B.; Lindahl, E. GROMACS: High performance molecular simulations through multi-level parallelism from laptops to supercomputers. *SoftwareX* **2015**, *1*, 19–25. [CrossRef]

Disclaimer/Publisher's Note: The statements, opinions and data contained in all publications are solely those of the individual author(s) and contributor(s) and not of MDPI and/or the editor(s). MDPI and/or the editor(s) disclaim responsibility for any injury to people or property resulting from any ideas, methods, instructions or products referred to in the content.

catalysts

MDPI

Review

Zeolite-Supported TiO_2 for Enhanced Photocatalytic Performance in Environmental Applications: A Review

Sanja J. Armaković [1,3,*] and Stevan Armaković [2,3,*]

1 University of Novi Sad, Faculty of Sciences, Department of Chemistry, Biochemistry and Environmental Protection, 21000 Novi Sad, Serbia
2 University of Novi Sad, Faculty of Sciences, Department of Physics, 21000 Novi Sad, Serbia
3 Association for the International Development of Academic and Scientific Collaboration (AIDASCO), 21000 Novi Sad, Serbia
* Correspondence: sanja.armakovic@dh.uns.ac.rs (S.J.A.); stevan.armakovic@df.uns.ac.rs (S.A.)

Abstract: The combination of TiO_2 with zeolites has emerged as a transformative strategy to enhance photocatalytic performance for environmental applications. The combination of zeolites' regular pore structure, high surface area, and adsorption capacity with the photocatalytic properties of TiO_2 allows synergistic effects, significantly improving the removal of organic pollutants and hazardous substances from water. This review provides a comprehensive analysis of TiO_2–zeolite composites, focusing on their synthesis, structural characteristics, and photocatalytic mechanisms. Advances in the characterization of material and computational analysis are applied to explain the relationship between structure and catalytic activity. Environmental applications such as water purification and renewable energy production are critically evaluated, highlighting their potential for addressing pressing global challenges. The review also addresses key challenges, including material stability, scalability of synthesis methods, and cost-effectiveness, while presenting future perspectives for the development and application of TiO_2–zeolite composites in sustainable catalysis.

Keywords: TiO_2–zeolite composites; environmental catalysis; photocatalysis; catalytic mechanisms; water purification; organic pollutant degradation; synergistic effects; computational modeling

1. Introduction

Environmental pollution has escalated as one of the most threatening challenges of the 21st century, demanding immediate attention and action [1–4].

Figure 1 takes Europe as an example to depict the ecological status of surface water resources from 2004 to 2024, categorized into three distinct groups: natural, non-natural, and seriously disturbed. Figure 1 highlights the different environmental challenges and achievements across Europe. Clean (natural) water, represented in blue, is predominantly found in northern Europe, reflecting minimal human impact and preserved ecosystems. In contrast, central and southern Europe show higher proportions of non-natural disturbed (green) and seriously disturbed (yellow) water bodies, demonstrating the significant influence of activities such as agriculture, urbanization, and industrialization on water quality [5].

Parameters such as pH value, dissolved oxygen concentration, and chemical oxygen demand are important for evaluating water quality and detecting pollutants that require advanced treatment. Information from global water quality monitoring highlights the growing need for efficient treatment technologies to remove different pollutants from water

 https://doi.org/10.3390/catal15020174

systems [5,6]. At the country and basin levels, the water quality indicators are calculated as weighted averages of station-level indicator values, with weights determined by the availability of monitoring data for each parameter at individual stations.

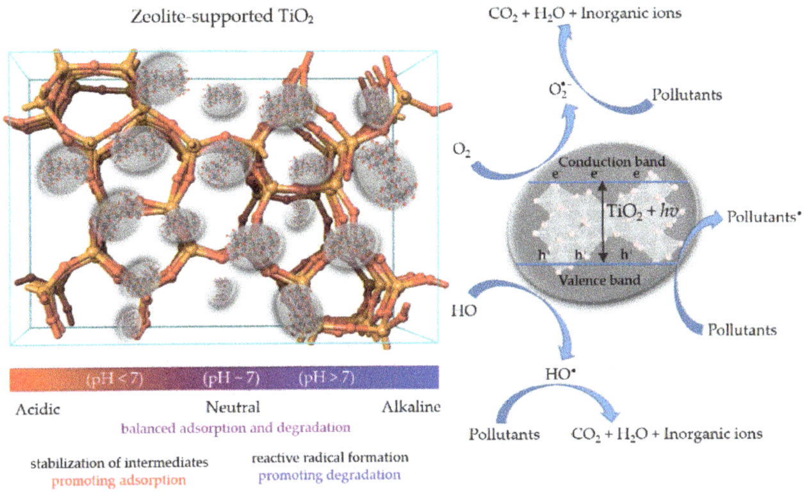

Figure 1. General schematic illustration of TiO_2–zeolite composite photocatalysis mechanisms.

This dataset reveals regional differences in ecological status and underscores the potential for targeted interventions and policies to deal with water pollution and restore environmental balance.

The need for efficient and sustainable water treatment technologies is more critical than ever. Photocatalysis is a promising solution, utilizing the energy from light to degrade pollutants into harmless products [7,8]. Titanium dioxide (TiO_2) has attracted significant attention because of its high photocatalytic activity, chemical stability, nontoxicity, and cost-effectiveness as one of the most used materials for photocatalysis. However, TiO_2 comes with certain limitations, such as a large bandgap that decreases its activity to UV light and a high recombination rate of photogenerated charge carriers, which reduce its efficiency under sunlight [3,4,9].

In recent years, various TiO_2-based composites, including TiO_2/graphene [10,11], TiO_2/carbon nanotubes [12,13], and TiO_2/metal oxides [14], have been extensively explored for environmental applications. These materials are known for improving photocatalytic performance by enhancing charge separation, increasing light absorption, and improving stability. However, zeolite-supported TiO_2 composites offer unique advantages such as the high surface area, suitable pore size, strong adsorption capacity, and ion-exchange properties of zeolites, which not only help in trapping pollutants but enhance the dispersion and stability of TiO_2 particles, leading to improved catalytic efficiency [15–17]. This combination makes TiO_2–zeolite composites particularly attractive for applications such as wastewater treatment and pollutant degradation [18–20].

This review first examines the synergistic mechanisms driving the enhanced photocatalytic performance of TiO_2–zeolite composites. It then deals with synthesis methods, advanced characterization techniques, and computational approaches, particularly density functional theory (DFT), to optimize their structure and performance in environmental applications. The findings of this review are of the highest importance, as they introduce the potential of TiO_2–zeolite composites to emerge as a sustainable solution for environmental catalysis. Additionally, recent advancements in material characterization and computational modeling are considered, aiming to explain the relationship between composite structure and catalytic activity. The environmental applications of these composites in water purification are critically evaluated. Furthermore, the challenges connected with material stability, scalability, and costs are emphasized as future perspectives for advancing TiO_2–zeolite composites.

2. Synergistic Mechanisms of TiO_2 and Zeolites

The combination of TiO_2 with zeolites has attracted attention in recent years due to the complementary properties of the components, which enhance photocatalytic performance [17,21]. Zeolites, with their highly microporous and mesoporous structures, have been recognized as effective support materials for TiO_2, providing a robust material to overcome limitations associated with independent TiO_2 photocatalysts [22,23]. This synergy appears from multiple physicochemical interactions between the two materials, including adsorption enhancement, charge carrier separation, and light absorption improvements [21,24].

2.1. Enhancing Photocatalytic Performance: Structural Benefits and Limitations Overcome by TiO_2–Zeolite Composites

The combination of TiO_2 with zeolites offers a promising and highly effective strategy for enhancing photocatalytic performance. This synergy plays a crucial role in improving charge transfer, pollutant adsorption, and photocatalytic reaction rates. Table 1 summarizes the key benefits of zeolite-supported TiO_2 composites. Recent research has shown the underlying mechanisms, structural interactions, and potential limitations of TiO_2–zeolite composites, highlighting their environmental applications [25–28].

2.1.1. Adsorption-Driven Catalysis and Pollutant Preconcentration

Zeolites are aluminosilicate materials that typically have high surface area, crystalline framework, and well-defined pore structure [29]. These properties allow the selective adsorption of organic and inorganic pollutants, effectively concentrating them near the photocatalytic sites of TiO_2. As a result, the reaction efficiency is significantly increased via the "preconcentration effect", which ensures that pollutants are more readily available for photocatalytic degradation (Figure 1).

2.1.2. Charge Carrier Separation and Electron–Hole Transport

Even with its strong oxidative capability, chemical stability, and nontoxicity, TiO_2 has fundamental limitations. One of the main limitations of TiO_2 is the rapid recombination of photogenerated electron–hole (e^-–h^+) pairs, which reduces its photocatalytic efficiency. The incorporation of zeolite into TiO_2 photocatalytic systems facilitates charge separation, as zeolites can act as electron sinks, delaying recombination and prolonging the lifetime of reactive charge carriers [30–35].

Zeolites with specific Si/Al ratios influence their electronic properties, affecting their ability to trap photogenerated e^- [17,36–38]. Cation-exchanged zeolites, such as M-ZSM-5 and H-Beta, have been reported to further enhance charge separation via their strong electrostatic interactions with TiO_2. These modifications contribute to the formation of an internal electric field, which assists in charge migration and enhances photocatalytic activity [7,27,39–42].

2.1.3. Reactive Oxygen Species Formation, Surface Characteristics, and Surface Acidity Effects

Zeolite-supported TiO_2 composites enhance the formation of reactive oxygen species (ROS), such as hydroxyl radicals ($^{\bullet}OH$), which are crucial for the degradation of organic pollutants (Table 1 and Figure 1). This enhancement is attributed to several synergistic effects, including improved charge carrier separation, enhanced adsorption of oxygen and pollutants, and the influence of zeolite's surface properties on ROS stabilization [19,28,36,43,44].

Studies by Piedra López et al. [17] demonstrated that metoprolol was degraded by over 95% under near-UV light, illustrating the effectiveness of this synergy.

The synergy between TiO_2 and zeolites has a strong effect on the surface characteristics and particle properties of the composite. While pure TiO_2 is primarily active under UV light because of its wide bandgap (~3.2 eV for anatase), the combination with zeolites can modify its optical properties. This can be achieved by preventing agglomeration and increasing active surface area, as well as doping TiO_2–zeolite composites with metals or nonmetals to induce.

Studies, including those by Sun et al. [45] and Torkian et al. [46], have demonstrated that these changes significantly improve photocatalytic efficiency by promoting ROS generation and extending the light absorption range of the TiO_2 phase into the visible spectrum.

Acidic and basic sites on the zeolite further direct degradation reactions [47–49]. Experimental results showed higher tetracycline adsorption on zeolites in an alkaline medium due to additional binding sites ($-NH^-$), while in acidic conditions, tetracycline interacted via hydroxyl groups ($-OH$). Photocatalytic degradation studies using radical scavengers confirmed h^+ as the dominant reactive species, with $^{\bullet}OH$ also contributing. The acidity and basicity of zeolites influence degradation pathways, with basic zeolites enhancing $^{\bullet}OH$ formation and acidic zeolites stabilizing intermediates [48].

Table 1. Roles of zeolite in improving the photocatalytic degradation mechanism of TiO_2 [19,21,28,38,44,50–54].

Role	What Happens	Why It Is Important	Example
Adsorption	Zeolite's porous structure traps and concentrates pollutants near TiO_2 active sites.	Increases pollutant concentration close to TiO_2, enhancing reaction efficiency.	Molecules adsorbed into zeolite pores are degraded efficiently by TiO_2.
Dispersion	Zeolite prevents TiO_2 particles from agglomerating by evenly distributing them on its surface.	Better TiO_2 active surface area, allowing more pollutants to interact with catalytic sites.	Without zeolite, TiO_2 particles might cluster, reducing the available surface area for catalysis.
Enhanced mass transfer	Zeolite's porous structure facilitates diffusion of pollutants to TiO_2 and removal of products.	Ensures efficient transport of reactants and intermediates, avoiding reaction block.	Pollutants diffuse through zeolite pores to TiO_2 for degradation by ROS.
pH regulation	Zeolite buffers change in pH during the reaction, maintaining optimal conditions for TiO_2 activity.	Stabilizes TiO_2 photocatalytic performance, especially under varying reaction conditions.	Acidic products are neutralized by zeolite, maintaining an ideal pH for pollutant degradation.
ROS	TiO_2 generates ROS under light, which degrades pollutants adsorbed on zeolite.	Zeolite ensures pollutants are close to TiO_2, improving their interaction with ROS and enhancing degradation.	Molecules are degraded into CO_2 and water by ROS generated on TiO_2.

Recent studies, including that by Sun et al. [45], have highlighted the critical role of cation-exchanged zeolites in further optimization of material's properties. For instance, the substitution of native cations in zeolites with $Ca2^+$ or K^+ enhances surface acidity and facilitates stronger interactions with TiO_2, resulting in improved charge separation and more efficient proton generation. This modification also contributes to a more significant reduction of e−–h+ recombination, which is vital for increased photocatalytic efficiency under visible light irradiation. Guang Hu et al. [21] emphasized the role of key zeolite parameters, such as the Si/Al ratio and surface acidity, in further enhancing catalytic efficiency.

2.2. Impact of TiO_2 Distribution, Morphology, and Crystallinity on Photocatalytic Performance

The distribution of TiO_2 on the zeolite surface plays a crucial role in photocatalytic efficiency. A homogeneous distribution of TiO_2 provides a higher availability of active sites, leading to better pollutant adsorption and charge separation. Different synthesis techniques, such as sol–gel, hydrothermal, and impregnation, directly influence the dispersion of TiO_2. Nonuniform TiO_2 distribution with agglomeration can lead to inactive zones and increased recombination.

The morphology of TiO_2 has a direct impact on its light absorption properties and charge carrier dynamics. Nanoparticles of TiO_2 offer a high specific surface area, but excessive packing may lead to faster electron–hole recombination. Nanorods and nanosheets promote directional electron transport, reducing recombination losses and enhancing photocatalytic activity. Studies have shown that TiO_2 nanorods supported on zeolite exhibited up to 40% higher photocatalytic efficiency in organic pollutant degradation compared with spherical nanoparticles. Mesoporous TiO_2 structures provide additional pathways for pollutant diffusion, increasing the degradation rate by facilitating mass transfer within the catalyst.

The crystal structure of TiO_2 significantly influences charge carrier dynamics and photocatalytic efficiency (Table 2).

Table 2. Comparison of different TiO_2 morphologies and crystalline phases on zeolites.

Parameter	Morphology	Crystal Phase	Effect on Photocatalysis
High specific surface area	Nanoparticles	Anatase	High adsorption, but prone to aggregation
Directional electron transport	Nanorods	Anatase–rutile	Improved charge carrier separation
Greater thermal stability	Nanotubes	Rutile	Lower activity but better stability
Enhanced pollutant diffusion	Mesoporous structures	Anatase	Faster degradation rate

The combination of natural zeolites as supports offers additional practical benefits, including cost-effectiveness, availability, and environmental compatibility. When combined with scalable synthesis methods such as sol–gel and hydrothermal techniques, these composites become highly applicable for comprehensive applications, emphasizing their potential as a sustainable solution for tackling emerging contaminants and other environmental challenges [27,40,55,56]. Additionally, the choice of synthesis method significantly influences the structural properties of zeolite-supported TiO_2. Some reasons for this could include affecting the distribution, morphology, and crystallinity of the TiO_2. This is crucial for optimizing its efficiency in water purification and pollutant degradation. Because of this, this review provides a more detailed analysis of this topic in the following section.

3. Strategies for Enhancing TiO_2–Zeolite Performance: Synthesis and Modification Techniques

The synthesis and modification techniques used for zeolite-supported TiO_2 play an important role in determining its structural characteristics, electronic behavior, and catalytic performance. Different synthesis methods and modifications can introduce variations in the particle dispersion, pore accessibility, crystallinity, and light absorption properties of TiO_2 [19,21,25,28,29,38,44,50–54,57]. By optimizing these techniques and balancing their advantages and limitations, it is possible to enhance the synergistic interactions between TiO_2 and the zeolite framework, thereby improving photocatalytic efficiency and expanding the range of environmental applications [27,29,56,58].

3.1. Synthesis Methods and Their Influence on Material Properties

3.1.1. Sol–Gel Method

The sol–gel method is a widely used technique for synthesizing TiO_2–zeolite composites because of its ability to produce uniform and high-quality materials [45,59,60]. This method involves the hydrolysis and condensation of titanium precursors, such as titanium isopropoxide or titanium tetrachloride, in the presence of zeolites. The resulting combination of TiO_2 with the zeolite framework enhances photocatalytic performance by employing the structural and adsorptive properties of zeolites [27,61].

In this process, titanium precursors are dissolved in a solvent, hydrolyzed with water or a catalyst, and condensed to form TiO_2. Zeolites are introduced during this stage, allowing the deposition of TiO_2 nanoparticles on and within their porous structure. Drying and subsequent calcination at high temperatures complete the synthesis, ensuring the formation of TiO_2, typically in the anatase phase, which is optimal for photocatalysis [62].

The sol–gel method offers key advantages, including uniform dispersion of TiO_2, precise control over particle size and morphology, and enhanced interaction between TiO_2 and zeolites. These characteristics improve charge transfer, prevent agglomeration, and maximize photocatalytic efficiency. However, challenges such as pore blockage from extreme TiO_2 loading and energy-intensive calcination require careful optimization [62,63].

TiO_2–zeolite composites synthesized through this approach have shown excellent efficacy in degrading pollutants such as dyes, pharmaceuticals, and pesticides. Future advancements in the use of greener solvents, energy-efficient calcination, and hybrid techniques are expected to further improve the scalability and environmental sustainability of the sol–gel process for large-scale applications [61–63].

3.1.2. Hydrothermal Synthesis

Hydrothermal synthesis is a widely adopted method for producing TiO_2–zeolite composites, offering advantages in material crystallinity and structural stability. This technique involves the crystallization of TiO_2 in the presence of zeolites under controlled high-temperature and high-pressure conditions, typically in an autoclave [24,49,64,65].

The hydrothermal process forms well-crystallized TiO_2 structures with strong adhesion to the zeolite framework, enhancing the thermal stability of the composites and making them suitable for applications in varying environmental conditions. Additionally, the close contact between the TiO_2 and the zeolites facilitates efficient charge transfer and separation, thereby improving photocatalytic performance [24,64].

Key parameters, such as reaction temperature, pressure, precursor concentration, and synthesis time, significantly influence the properties of the final composite. Proper optimization of these parameters confirms uniform TiO_2 distribution within the zeolite pores and prevents structural damage to the zeolite borders [49,65].

Hydrothermal synthesis demonstrates high photocatalytic efficiency, particularly in degrading organic pollutants in water. Their ability to create persistent and active photocatalysts makes it a competitive choice for environmental remediation. However, the process is energy-intensive and too long, requiring further investigation to improve scalability and cost-effectiveness. Future research should focus on the reduction of energy consumption and exploration of alternative synthesis conditions to make hydrothermal synthesis more sustainable for large-scale applications [49,54,64].

3.1.3. Impregnation Method

The impregnation method is a simple and cost-effective technique for synthesizing TiO_2–zeolite composites, making it highly suitable for industrial applications. In this approach, a solution of a titanium precursor, such as titanium isopropoxide or titanium tetrachloride, is introduced into the zeolite. The mixture is then dried to remove solvents and calcined at high temperatures to put TiO_2 within the zeolite's pores or on its surface [66–69].

This method is popular for its simplicity and scalability. It requires relatively inexpensive equipment and materials, making it accessible for practical production. Furthermore, the technique allows for flexibility in the amount of TiO_2 formed, which can be adjusted based on specific application requirements [70].

However, reaching uniform TiO_2 dispersion across the zeolite framework can be challenging. Unequal distribution may lead to aggregation or pore blockage, which would reduce the adsorption capacity and limit pollutant access to composite active sites. Therefore, optimizing synthesis parameters, such as precursor concentration, drying conditions, and calcination temperature, is crucial for ensuring structural and functional composite properties [68].

Regardless of these challenges, TiO_2–zeolite composites synthesized through the impregnation method have demonstrated significant photocatalytic activity in the decomposition of pollutants. Their cost-effectiveness makes this method an attractive option for environmental application technologies. Future research should focus on clarifying techniques by improving TiO_2 dispersion and minimizing structural issues, giving more efficient and robust composite materials for practical applications [70,71].

3.1.4. Atomic Layer Deposition

Atomic layer deposition (ALD) is an advanced technique for the synthesis of TiO_2–zeolite composites. It involves deposition of an ultrathin and uniform TiO_2 layer on zeolite surfaces. This method involves a sequential, self-limiting chemical vapor deposition process, ensuring precise control over the thickness and coverage of the deposited TiO_2 layer. The nanoscale precision achieved through ALD enhances the interaction between TiO_2 and the zeolite, resulting in much better composite performance [72].

One of the primary advantages of ALD is its ability to uniformly coat complex geometries and porous structures, making it ideal for zeolites with complex pore networks. This uniformity ensures effective TiO_2 distribution, avoiding agglomeration or uneven loading, which are familiar problems with other synthesis methods. To optimize photocatalytic activity, the controlled deposition allows for precise tuning of the crystallinity and phase composition of the TiO_2 [73].

ALD is particularly effective in enhancing light absorption and charge separation in TiO_2–zeolite composites. By placing a thin layer of TiO_2, the method minimizes the recombination of photogenerated e^-–h^+ pairs, improving the generation of ROS necessary for pollutant degradation. The precise control over layer thickness further optimizes light absorption, making the composites suitable for applications under both UV and visible light [74].

ALD has limitations, including high operational costs and the need for specialized equipment, that may restrict its industrial applications. However, its ability to produce high-performance photocatalytic materials makes it a valuable tool for research and development in practical environmental solutions [72–74].

3.1.5. Green and Sustainable Synthesis

Ecofriendly synthesis methods are becoming increasingly important in the development of TiO_2–zeolite composites. These approaches emphasize reduced environmental impact with sustainable development goals. They aim to minimize energy consumption and avoid the use of hazardous chemicals, making them safer and more sustainable alternatives to traditional methods [67,75–78].

One promising technique is solvent-free processing, which eliminates the need for organic solvents commonly used in conventional synthesis, reducing waste generation and toxicity risks [67,75]. Another promising strategy involves the use of biogenic precursors, such as plant extracts or bioderived materials. These methods not only lower the environmental footprint of the production process but introduce unique structural and chemical qualities to the resulting composites [75,77].

By focusing on sustainability, these green synthesis techniques contribute to the significant adoption of TiO_2–zeolite composites in environmental applications, particularly in water treatment and pollution remediation [67,76]. Future research should scale up these methods while keeping their ecofriendly nature to ensure their feasibility for industrial applications [75,77].

During the synthesis of TiO_2–zeolite composites, TiO_2 particles can cluster because of their high surface energy and attraction between particles. To prevent this, factors such as temperature, mixing methods, and the use of surfactants or dispersing agents are important for achieving even particle distribution, which improves photocatalytic performance.

3.2. Doping and Cocatalyst Combination

To overcome the limitations of TiO_2, its large bandgap and restricted activity under UV light, different doping strategies have been employed to extend its light absorption range into the visible and near-infrared spectrum. Near-infrared light-responsive TiO_2 materials have been extensively studied to improve solar energy utilization, with doping and structural modifications being key to extending light absorption [79–82]. Zeolite-supported TiO_2 composites contribute to this by enhancing charge separation and stabilizing the photocatalytic system [73,83,84].

Metal doping. Incorporating metal dopants, such as copper (Cu), silver (Ag), iron (Fe), nickel (Ni), or cobalt (Co), into TiO_2–zeolite composites introduces localized electronic states within the bandgap of TiO_2. These states enable the absorption of visible light, expanding the range of the photocatalyst out of the UV spectrum. For example, Cu and Fe dopants can act as charge carrier traps, promoting the separation of photogenerated e^-–h^+ pairs and extending their lifetimes. This mechanism enhances the production of ROS, which is critical for pollutant degradation [85–87].

Nonmetal doping. Nonmetal elements, such as nitrogen (N) or sulfur (S), are also effective dopants for TiO_2. These elements replace oxygen atoms in the TiO_2 lattice or form interstitial bonds, obtain the bandgap, and shift light absorption into the visible spectrum. Nonmetal doping is particularly advantageous for improving photocatalytic activity under solar irradiation, offering an environmentally friendly and cost-effective approach [88,89].

Cocatalyst combination. The addition of cocatalysts, such as graphene or noble metals such as palladium (Pd) and platinum (Pt), further enhances the performance of TiO_2–zeolite composites. Cocatalysts facilitate electron transfer by providing alternative pathways for charge carriers, reducing recombination. Graphene acts as an electron reservoir, improving charge separation and conductivity. Noble metals increase catalytic efficiency by creating Schottky barriers that promote electron flow and increase ROS generation [90,91].

Doping and cocatalyst combination lead to significant improvements, but challenges such as maintaining long-term stability, controlling dopant distribution, and optimizing cocatalyst loading remain. Future research activities should focus on understanding the interactions among dopants, cocatalysts, and the TiO_2–zeolite framework to maximize their synergistic effects. Additionally, the development of ecofriendly and applicable modification methods will be crucial for their practical implementation [58,89,91,92].

3.3. Surface Modification of Zeolites

Pretreatment and post-treatment of zeolites are effective strategies to optimize their structural and chemical properties, which improve their interaction with TiO_2 in photocatalytic composites. These modifications enhance the compatibility of zeolites with TiO_2 and enable the design of composites for specific applications [58–60,71,91,93,94].

Acid treatment involves exposing zeolites to acidic solutions, such as sulfuric or hydrochloric acid, to remove impurities and adjust their surface properties. This process can increase surface acidity, enhancing the adsorption of organic pollutants and promoting catalytic activity. Acid treatment also creates additional active sites on the zeolite surface, facilitating better combination with TiO_2 nanoparticles [60,94].

Dealumination is a postsynthesis modification technique that selectively removes aluminum atoms from the zeolite. This process increases the zeolite's Si/Al ratio, changes its hydrophobicity, and improves its thermal and chemical stability. Dealumination enhances the material adsorption characteristics, making it more effective for interacting with nonpolar organic pollutants in photocatalytic processes [27].

Ion exchange is another method to modify zeolites by replacing their native cations (e.g., Na^+ or K^+) with desired metal ions, such as Cu^{2+}, Fe^{3+}, or Zn^{2+}. This modification introduces catalytic active sites and improves the ability to interact with TiO_2 and photogenerated charge carriers. For example, ion-exchanged zeolites can exhibit enhanced redox properties, changing photocatalytic efficiency [59,95,96].

Modified zeolites have been successfully employed in TiO_2–zeolite composites to degrade organic pollutants, remove heavy metals, and facilitate advanced oxidation processes [59,93,95]. Future research should focus on combining multiple modification techniques to achieve synergistic effects and exploring ecofriendly modification methods to reduce environmental impact [60,94]. The development of scalable approaches for zeolite modification will be essential for its industrial application [93,94].

Table 3 provides a concise summary of each synthesis and modification method's strengths and weaknesses, offering a practical guide for researchers to select the most suitable method based on specific application requirements and resource availability.

Table 3. Advantages/limitations of synthesis and modification techniques for TiO_2–zeolite composites.

Synthesis Method	Advantages	Limitations	References
Sol–gel	Uniform dispersion of TiO_2 nanoparticles Control of particle size and morphology	Risk of pore blockage in zeolites Requirement of careful optimization of precursor ratios	[59–61]
Hydrothermal	Production of well-crystallized TiO_2 Strong adhesion to zeolite surfaces High thermal stability	High-temperature and high-pressure conditions required Longer synthesis duration	[49,64,65]
Impregnation	Simplicity and cost-effectiveness Scalability for industrial applications	Uneven TiO_2 distribution possible Potential for pore blockage in zeolites	[67–69]
Atomic layer deposition	Precise control over TiO_2 layer thickness and uniformity Enhanced interaction with the zeolite	Expensiveness and time-intensiveness Requires specialized equipment	[73,74]
Green and sustainable	Ecofriendly approach Reduces environmental impact and agrees with sustainability goals	Limited scalability in some methods Alternative precursors or equipment may be required	[75–77]
Doping and cocatalyst combination	Improvement of visible light absorption Better charge separation and ROS generation	Complex synthesis process Possible stability issues with dopants	[86,88,89]
Surface modification of zeolites	Enhanced adsorption properties Tailoring of zeolite structure for specific applications	Additional processing steps required Risk of structural degradation with aggressive treatments	[59,60,94]

4. Characterization Techniques

Understanding the structural, chemical, and photocatalytic properties of zeolite-supported TiO_2 composites is crucial for optimizing their performance in environmental applications. This section highlights key characterization techniques used to evaluate the materials, including scanning electron microscopy (SEM), transmission electron microscopy (TEM), X-ray diffraction (XRD), Brunauer–Emmett–Teller (BET) analysis, and various spectroscopy techniques. These techniques help identify important features such as particle distribution, potential clustering, surface area, and crystal structure. By understanding these properties, researchers can minimize TiO_2 particle aggregation, ensure uniform distribution on the zeolite surface, and enhance overall photocatalytic efficiency [16,18,19,25,36,42,45,46,56,63,66,72,84,97,98].

SEM is extensively used to evaluate the surface morphology of TiO_2–zeolite composites. It provides high-resolution images that reveal the dispersion and uniformity of TiO_2 particles on the zeolite matrix, insights into how synthesis methods influence agglomeration, and evidence of reduced porosity due to TiO_2 deposition [43,87,99].

SEM studies often highlight smoother surfaces after TiO_2 loading, indicating well-dispersed nanoparticles, critical for photocatalytic efficiency [16,46,97].

TEM offers detailed visualization of internal structures and interfaces at the nanoscale, measurement of the dimensions of TiO_2, information on the depth of TiO_2 deposition within zeolite pores, and identification of homogeneously distributed particles [25,34,45].

TEM studies frequently confirm the integration of TiO_2 into the zeolite, ensuring structural stability [71,84].

XRD is crucial for identifying crystalline phases and assessing the structural integrity of composites. Significant observations include differentiation among the anatase, rutile, and brookite phases of TiO_2 and confirmation that the zeolite remains intact postsynthesis [15,16,71,100].

For example, the presence of characteristic peaks for anatase at $2\theta = 25°$ confirms the photocatalytically active phase of TiO_2 [45].

BET analysis measures surface area, pore volume, and pore size distribution. TiO_2 loading often reduces the surface area because of pore blockage and evidence of mesoporous or microporous transformations [45,46].

These findings are critical for correlating structural properties with photocatalytic performance [36].

Spectroscopic techniques. UV-Vis, Raman, Fourier transform infrared (FTIR), and X-ray photoelectron spectroscopy (XPS) collectively provide critical insights into the optical, structural, and chemical properties of TiO_2–zeolite composites, enabling the assessment of their photocatalytic potential and interaction mechanisms with pollutants [15,17,43,45,66,87,98].

UV-Vis spectroscopy is used to analyze the optical properties and light absorption properties of TiO_2–zeolite composites. Observations include evaluating optical band gaps to predict photocatalytic activity and analyzing light absorption in modified composites [43,45,66].

Further, diffuse reflection spectroscopy (DRS) enables precise measurement of absorption behavior across the UV and visible ranges [84,101].

Raman spectroscopy provides information on vibrational modes, enabling the analysis of molecular and structural changes. Specific findings include the detection of anatase, rutile, or brookite phases through characteristic Raman peaks and changes in vibrational modes, indicating successful integration of TiO_2 into the zeolite [87,102].

FTIR spectroscopy is a technique for the characterization of the functional groups and chemical bonds within zeolite-supported TiO_2 composites. FTIR provides insights into the interaction of TiO_2 with the zeolite and adsorbed molecules by analyzing vibrational modes. FTIR spectra reveal characteristic bands for Ti-O-Ti and Ti-O-Si linkages, indicating the successful integration of TiO_2 onto the zeolite surface. Shifts or intensity changes in bands associated with Si-O-Si and Al-O-Si vibrations suggest structural alterations in the zeolite framework due to TiO_2 deposition. FTIR helps identify the types of organic pollutants adsorbed on the composite surface during photocatalytic reactions, offering insights into degradation mechanisms [17,46,97].

For example, the appearance of peaks at 950–1200 cm^{-1} corresponding to Si-O stretching vibrations confirms the zeolite's structural integrity, while new peaks around 550–800 cm^{-1} indicate TiO_2 incorporation. FTIR is particularly valuable for understanding the chemical interactions that drive photocatalytic activity, making it a complementary tool alongside SEM, TEM, and other techniques.

XPS studies surface chemistry and electronic states, revealing elemental analysis of Ti, O, Si, Al, etc.; insights into interactions between TiO_2 and pollutants; and shifts in binding energies during photocatalytic processes. XPS studies are fundamental in confirming the chemical interactions and stability of TiO_2–zeolite composites [36,72,87].

The characterization techniques discussed above are summarized in Table 4. The combination of these techniques provides a comprehensive understanding of the physicochemical properties and catalytic potential of TiO_2–zeolite composites. SEM and TEM offer morphological insights, XRD and Raman identify structural integrity and phases, BET assesses porosity and surface area, FTIR analyses functional groups and chemical interactions, and UV-Vis and XPS elucidate optical and chemical properties. The mentioned methods serve as a guide, directing the optimization of synthesis protocols and enhancing material performance for environmental applications.

Table 4. Overview of characterization techniques for zeolite-supported TiO_2.

Technique	Purpose	Key Insights	References
SEM	Surface morphology analysis	Dispersion, particle size and aggregation, and pore blocking by TiO_2	[43,87]
TEM	Internal structure visualization	Particle size distribution, layer thickness, and structural uniformity	[25,45]
XRD	Crystalline phase identification	Phase identification and structural stability	[16,71]
BET	Surface area and porosity measurement	Quantification of surface area reduction and pore size changes	[45,46]
UV-Vis	Optical property evaluation	Band gap determination and light absorption shifts	[43,45]
Raman	Molecular and structural vibration analysis	Identification of TiO_2 phases and structural modifications	[87]
FTIR	Characterization of the functional groups and chemical bonds	Identification of functional groups	[45,97]
XPS	Surface chemical composition and bonding state analysis	Chemical composition, bonding states, and catalytic mechanisms	[72,87]

The costs of characterization techniques differ based on equipment, maintenance, and operational expenses. Techniques such as TEM and XPS are the most expensive, with substantial costs that can overtop standard laboratory budgets because of their advanced capabilities and high-resolution output. SEM and Raman spectroscopy also fall into the high-cost category, requiring significant investment for quality imaging and spectral analysis. FTIR spectroscopy is a moderately priced technique that offers valuable insights into functional groups and chemical interactions, making it a practical choice for understanding structural and chemical modifications in composites. UV-Vis spectroscopy is a practical and affordable solution for routine optical analysis. Along with XRD, BET, and FTIR analysis, it is a moderate-cost technique that offers reliable structural and surface data at relatively accessible pricing levels.

Costs play a crucial role in the selection of characterization methods. More affordable methods such as UV-Vis spectroscopy are favored for quick assessments. At the same time, high-cost techniques such as TEM and XPS are reserved for specialized studies that demand precise nanoscale or chemical-state insights. Researchers carefully balance these costs against requirements to select the most appropriate methods for their needs.

5. Applications in Environmental Catalysis

Zeolite-supported TiO_2 composites are emerging as highly promising materials in environmental catalysis because of their dual functionality, the superior photocatalytic activity of TiO_2, and the strong adsorption and ion-exchange properties of zeolite. These composites offer practical solutions for addressing critical environmental issues, including the removal of organic pollutants, heavy metals, and gaseous pollutants from water or air [21,28,45,103,104].

5.1. *Removal of Organic Pollutants*

One of the most significant applications of zeolite-supported TiO_2 is the degradation of organic pollutants to nontoxic inorganic compounds. The composites effectively degrade dyes, pharmaceuticals, and pesticides into less dangerous products under irradiation. The literature data showed that these materials can achieve over 90% efficiency in methylene blue and methyl orange degradation under UV exposure, demonstrating their effectiveness in the purification of dye-contaminated wastewater. Similarly, their ability to degrade pharmaceutical pollutants makes them valuable in treating industrial and hospital effluents [44,105].

It can be stated that, although UV light exposure leads to high degradation efficiencies of dyes and pharmaceuticals because of the high energy available, sunlight-driven photocatalysis is more sustainable. With proper modifications, such as doping or cocatalysts, the efficiency of zeolite-supported TiO_2 under sunlight can approach that of UV systems.

5.2. *Heavy Metal Remediation*

In addition to organic pollutant removal, zeolite-supported TiO_2 composites are highly effective in heavy metal remediation. The strong ion-exchange capacity of zeolites allows for the adsorption of toxic metal from aqueous media. The photocatalytic activity of the zeolite-supported TiO_2 reduces these metals to less toxic or inert states. These properties are advantageous for improving the quality of drinking water in areas affected by industrial contamination [105].

5.3. *Air Purification*

Air purification represents another vital application of these composites [106]. Zeolite-supported TiO_2 excels in removing volatile organic compounds (VOCs) and nitrogen oxides (NOx) from polluted air. The porous structure of zeolites catches gaseous pollutants, while TiO_2 decomposes them into less dangerous compounds under UV light. Data from the literature have shown that these composites can achieve up to 80% removal of formaldehyde and acetaldehyde and significant reductions in NOx levels. This capability makes them highly relevant for improving urban air quality [71,107].

Sunlight can also drive VOC and NOx removal, but achieving efficiency comparable to that of UV systems requires structural optimization or the addition of cocatalysts.

5.4. *Comparison with Other Photocatalytic Materials*

When compared with other photocatalytic materials, zeolite-supported TiO_2 materials offer several advantages. Unlike pure TiO_2, these composites enhance stability and improve pollutant catch due to the zeolite matrix. Compared with graphene–TiO_2 hybrids, zeolite-supported TiO_2 is more cost-effective because of the abundance of affordable natural zeolites. Its performance under sunlight reduces costs, making it an attractive option for large-scale environmental applications [27,46,96].

5.5. Sustainability and Green Applications

The sustainability of TiO_2–zeolite composites stems from the availability of natural zeolites, and when coupled with green synthesis techniques, these composites align with green chemistry principles. Sunlight-driven photocatalysis further minimizes energy consumption, making TiO_2–zeolite composites particularly suited for environmentally sensitive applications [21]. These factors make the composites suitable for use in resource-limited and environmentally sensitive regions, underscoring their potential to address pressing global pollution challenges.

Table 5 summarizes the most commonly used zeolite-supported TiO_2 composites, highlighting their key properties, applications, and benefits.

Table 5. Commonly used zeolites in combination with TiO_2 and their applications.

Zeolite	Properties	Applications	Benefits with TiO_2	References
ZSM-5	High silica content, thermal stability, adjustable acidity	Dyes: methylene blue, methyl orange VOCs: toluene, formaldehyde	Enhances adsorption and TiO_2 dispersion	[34,71,99,101,108,109]
Y Zeolite (Faujasite)	Large pore size, high cation-exchange capacity	Heavy metals: Pb^{2+}, Cd^{2+} Air purification: NOx, VOCs	Provides large surface area for TiO_2 deposition	[26,68,110]
Beta Zeolite	High hydrophilicity, large surface area	Pharmaceuticals: antibiotics, effluents Dyes: rhodamine B	Improves photocatalytic efficiency	[35,43,91,111]
Mordenite	High thermal stability, channel-like structure	Heavy metal removal: Cr^{6+}, Ni^{2+} Organic pollutants: pesticides, hydrocarbons	Synergizes with TiO_2 under UV light	[112–114]
Clinoptilolite	Natural zeolite, high cation-exchange capacity	Heavy metals Pesticides: atrazine	Cost-effective and widely available	[102,115–118]
Zeolite A	Small pore size, strong ion-exchange capacity	Water softening: hardness removal Selective adsorption: arsenic	Ensures close interaction with TiO_2 active sites	[103,119,120]

5.6. Influence of Energy Sources on Photocatalytic Efficiency

The performance of zeolite-supported TiO_2 composites is highly dependent on the energy source used to activate the photocatalyst. While UV light is traditionally preferred for its high-energy photons and strong ROS generation, sunlight and visible light offer more sustainable alternatives for large-scale applications. The following Table 6 summarizes the key advantages, challenges, and optimization strategies for different energy sources to improve photocatalytic efficiency.

Table 6. Influence of energy sources on photocatalytic efficiency of zeolite-supported TiO_2.

Energy Source	Advantages	Challenges	Optimization Strategies	References
UV light	High-energy photons directly excite TiO_2, leading to high ROS generation and efficient pollutant degradation.	Limited to small-scale applications because of high energy consumption.	No significant optimization is needed, but ensuring good TiO_2 dispersion and light penetration maximizes efficiency.	[14,31,63]
Sunlight (natural/simulated)	Sustainable, cost-effective, and suitable for large-scale applications. Abundant natural energy source.	Sunlight contains a limited fraction of UV light (~5%), resulting in lower photocatalysis.	Doping with metals or nonmetals to extend light absorption into the visible spectrum.	[3,8,19,54]
Visible light (simulated light)	Expands the application of photocatalysis to regions with limited sunlight availability.	TiO_2 has a large bandgap (~3.2 eV) and is inactive under primarily visible light.	Cocatalysts (graphene, noble metals) or heterojunctions enhance charge separation and light absorption.	[10,14,21,86]

6. Computational Modeling in TiO_2–Zeolite Research

Computational modeling by atomistic calculations plays a crucial role in understanding composites by providing detailed insights into their structural and electronic properties [24,53]. Techniques such as density functional theory (DFT) and molecular dynamics (MD) simulations have been employed in various aspects of materials science to explain the interaction between structures, optimize structural parameters, and explore the charge transfer dynamics crucial for photocatalysis. In the context of zeolites, computational studies have shown the significant influence of the Si/Al proportion on the surface acidity of zeolites and its interaction with photogenerated charge. This insight is crucial in improving pollutant degradation efficiency. Researchers have also used atomistic calculations to simulate the adsorption of different pollutants on new materials and allow the design of composites with specific properties such as high adsorption capacity, rapid degradation, and long-term stability.

Despite the extensive research on zeolites and TiO_2 individually, for their exceptional adsorption and photocatalytic properties, computational studies focusing on their interaction remain scarce. While zeolites are renowned for their high adsorption capacities and tunable acidity, and TiO_2 for its unparalleled photocatalytic activity, computational investigations into the synergistic mechanisms between these materials are notably underexplored. The lack of such studies is a consequence of the complexity of modeling hybrid systems, which require accurate representation of both the periodic nature of zeolites and the electronic properties of TiO_2. Additionally, challenges arise from the need to account for diverse interaction mechanisms, such as charge transfer, surface binding, and coupled adsorption–photocatalysis processes. This gap in computational research presents a significant limitation, as understanding these interactions could unlock new routes for designing advanced materials with enhanced environmental remediation capabilities. In this chapter, we aim to address this challenge by reviewing the available database, computational tools, and methodologies.

6.1. The Role of Atomistic Calculations in Understanding How Photocatalyst Structure Influences Removal Efficiency

The structure of a photocatalyst plays a crucial role in determining its efficiency in pollutant degradation. As already mentioned, some key structural parameters, such as surface area, crystal phase, particle size, and morphology, directly influence photocatalytic activity. Modifying the photocatalyst's structure through doping can significantly enhance its pollutant removal efficiency. Doping involves introducing foreign atoms into the photocatalyst's lattice, which can alter its electronic structure and improve light absorption.

For example, doping TiO_2 with nonmetals such as nitrogen can narrow its band gap, enabling it to utilize visible light more effectively [121,122]. Similar results have also been obtained with boron doping [123,124]. This modification enhances the generation of reactive oxygen species that degrade pollutants. Furthermore, creating composite structures by combining different semiconductors can facilitate efficient charge separation, reducing recombination losses.

Regarding photocatalyst structure, there has also been significant interest in doping TiO_2 with transition metal ions due to their partially filled d-orbitals [125,126]. Specifically, studies have established that doping TiO_2 with Al, Cu, Mo, and W leads to a narrowed band gap, improving its photocatalytic efficiency [127]. Additionally, the introduction of Al^{3+} into TiO_2 generates oxygen vacancies, which further enhance photocatalytic activity [127,128].

Determining how modifications to a photocatalyst's structure influence its photocatalytic properties is an expensive and time-consuming process. Fortunately, atomistic calculations provide a powerful tool for predicting how structural changes affect key photocatalytic properties, with the band gap being a primary focus. This is because a material's electronic band structure directly governs its optical and electronic properties, which, in turn, dictate its photocatalytic efficiency. In this context, DFT offers a reliable computational framework for predicting crucial electronic properties, such as band gap energy, DOS, and band edge positions. These factors are essential in determining a material's ability to absorb light and generate charge carriers (electrons and holes), which drive photocatalytic reactions.

When discussing the application of DFT for obtaining band structures, it is important to acknowledge its limitations in the context of photocatalysis. One of the most critical properties, the bandgap, is often severely underestimated when standard density functionals, such as the widely used Perdew–Burke–Ernzerhof (PBE) functional [129,130], are applied. To address this challenge, researchers have pursued two main approaches: (i) the development of novel density functionals that provide a more accurate description of the electronic subsystem and (ii) the implementation of corrections to existing density functionals that have already demonstrated reliability across various applications.

Both approaches are equally important, though the first is generally associated with higher computational costs. Among recently developed functionals, one of the most noteworthy is the Heyd–Scuseria–Ernzerhof (HSE06) hybrid functional [131], which has shown remarkable accuracy in predicting band gaps across a wide range of materials, as demonstrated in several studies [132,133].

To further address the underestimation of band gaps, the DFT+U method [134,135] has been introduced. This approach incorporates a Hubbard-like term into the Hamiltonian, which improves the description of on-site electron–electron interactions, offering a computationally affordable correction that significantly enhances the accuracy of bandgap predictions.

Another widely used bandgap correction technique is the DFT-1/2 method [136,137], which mitigates self-interaction errors in standard DFT by introducing an atomic self-energy potential to counteract electron–hole self-interaction energy. This correction effectively refines bandgap predictions while maintaining computational efficiency.

Therefore, researchers working with DFT-based band structure calculations have multiple strategies to improve accuracy. They may use the computationally inexpensive PBE functional—often considered the workhorse for periodic structures—and enhance its performance using DFT+U or DFT-1/2 corrections. Alternatively, when higher precision is required and computational resources allow, they can apply hybrid functionals such as HSE06, which offer improved predictive power at a higher computational cost.

6.2. Database of Zeolite Structures

The accurate modeling of zeolite structures is fundamental in computational studies of TiO_2–zeolite composites. While the periodic structures of all TiO_2 polymorphs are well-established and readily available in various databases, zeolites present a much more significant challenge because of their vast diversity in forms and derivatives. TiO_2 structures can often be generated with relative ease using crystal builders in various software tools. In contrast, the modeling of zeolites is far more complex, primarily because of their specific frameworks, their significant variability, and the large sizes of their unit cells. This complexity necessitates advanced computational strategies and resources to create reliable models that reflect their structural and chemical properties accurately.

In these regards, it is important to mention the Database of Zeolite Structures is available at https://www.iza-structure.org/databases/ (accessed on 25 December 2024). The Database of Zeolite Structures, curated by the Structure Commission of the International

Zeolite Association (IZA-SC), serves as a comprehensive resource for researchers and professionals in the field of zeolite science. Established in 1996 by Christian Baerlocher, the database was initially an online version of two critical publications: the "Atlas of Zeolite Structure Types" and the "Collection of Simulated XRD Powder Patterns for Zeolites" Over the years, it has evolved significantly, incorporating dynamic data generation, enhanced search capabilities, and interactive features to facilitate user engagement.

This extensive repository provides detailed structural information on all zeolite framework types approved by the IZA-SC. Users can access crystallographic data, simulated powder diffraction patterns, and descriptions for each framework type. The database also includes user-controlled animated displays, measured powder patterns from verified syntheses, and NMR spectra for various zeolites. Additionally, it offers detailed instructions for building models, facilitating the accurate representation of zeolite structures in computational studies. In Figure 2, we present the unit cells of two well-known zeolite frameworks that can be found in the Database of Zeolite Structures. Structures were made with the Maestro molecular visualization program [127] using .cif files from the Database of Zeolite Structures.

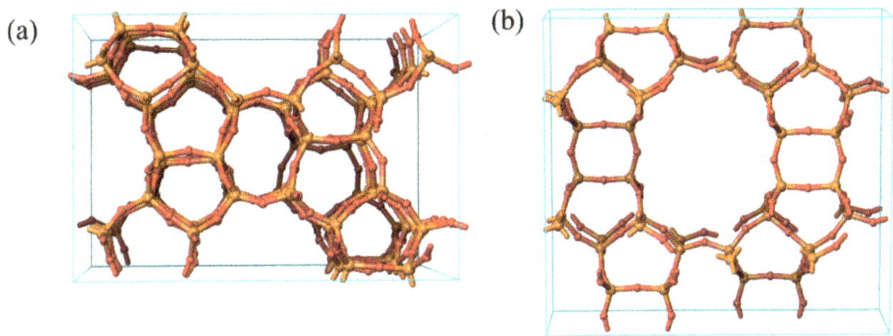

Figure 2. Unit cell structures of two well-known versions of ZSM-5 zeolites: (**a**) calcined ZSM-5 and (**b**) mordenite.

Zeolite structures in the database are systematically categorized by their unique three-letter framework type codes (FTCs), assigned by the IZA-SC. These codes are derived from the names of the type materials and are essential for distinguishing between different framework topologies. The database's search functionality allows users to filter structures based on various criteria, including unit cell parameters, chemical composition, and pore characteristics. This organized distribution enables researchers to efficiently locate and download molecular structures of zeolites and subject those structures to atomistic calculations.

In addition to the specific structures of zeolites provided in .cif format, it is important to note that all pristine zeolite frameworks are also readily available in this format. This feature of the database is extremely valuable, as it allows users to fine-tune these structures to suit their specific computational and experimental needs.

6.3. *Tools for Atomistic Calculations—Types and Capabilities*

As is already known, atomistic calculations provide invaluable insights into the properties of molecular and periodic structures and their interactions with other systems. With advancements in science and technology, a wide range of computational codes and software packages have been developed. These tools have now been optimized for efficiency, enabling even complex simulations to be run on standard personal computers without technical issues. Computational tools for atomistic calculations can be broadly categorized into three main groups: open-source, free-for-academics, and commercial software.

Open-source software is freely available to everyone and offers complete access to the source code. This allows users to modify, improve, and adapt the software for specific needs, making it highly flexible and suitable for a wide range of research applications. Notable examples of open-source codes for atomistic calculations based on quantum mechanical approaches include Quantum ESPRESSO [138–141], PSI4 [142–144], Multiwfn [145,146], xTB [147–150], and others.

Given the periodic nature of both zeolites and TiO_2, Quantum ESPRESSO is particularly important, as it supports periodic DFT calculations, enabling users to obtain critical information such as band structures, projected density of states (PDOS), charge distributions, and charge transfer properties.

On the other hand, PSI4 offers a variety of computational methods for studying the electronic properties of molecules. At the same time, Multiwfn provides exceptional capabilities for analyzing output files from modern atomistic codes and calculating thousands of molecular properties. The xTB code, developed by the group of Professor Stefan Grimme, is renowned for its efficient semiempirical calculations based on the extended tight-binding method. It also offers the capability to perform MD simulations using the methods it covers. It includes the GFN-FF force field [151], a highly accurate and fast force field that bridges the gap between semiempirical calculations and standard classical force field MD simulations.

For force-field-based MD simulations, LAMMPS [152,153] and GROMACS [154–161] are among the most widely used open-source tools. Since LAMMPS is more commonly applied in materials science, with a variety of force fields available, GROMACS is primarily designed for simulating the dynamics of biological molecules, such as proteins and lipids.

Free-for-academic tools are typically provided at no cost to researchers affiliated with academic institutions. While they may not be open-source, they are designed to support educational and research efforts without the financial burden of licensing fees. One of the most prominent examples is ORCA [162–170], which offers an extensive range of possibilities for atomistic calculations and MD simulations. The latest version, ORCA 6, introduced significant improvements, including unparalleled speed due to code optimization and an expanded array of computational methods. While ORCA is predominantly focused on molecular systems, its diverse capabilities also make it suitable for studying cluster or molecular configurations of zeolites and TiO_2 nanoparticles.

In this context, we also refer to our project, atomistica.online (https://atomistica.online (accessed on 25 December 2024)) [171,172]. Atomistica.online is a free-for-academics, online molecular modeling platform that provides a range of web-based and desktop tools for atomistic calculations on molecules. Among these tools is an online xTB tool, which acts as an interface connecting users to the xTB code hosted on a remote server. This allows users to perform semiempirical calculations directly from their web browsers without the need for installation or maintenance. The platform's fast semiempirical calculations are particularly valuable for (pre)optimizing large molecular and cluster models of combined zeolite–TiO_2 systems, significantly accelerating computational workflows for modeling such complex systems.

Commercial software packages are fully developed, highly polished products offered by companies for a fee. These tools typically provide extensive customer support, regular updates, and advanced features tailored for industrial and research applications. Examples include Materials and Biologics Science Suites by Schrödinger Inc. (https://www.schrodinger.com (accessed on 25 December 2024)), Amsterdam Density Functional (ADF) by Software for Chemistry and Materials B.V. (https://www.scm.com (accessed on 25 December 2024)), QuantumATK by Synopsys (formerly Atomistix Toolkit, https://www.synopsys.com/manufacturing/quantumatk.html (accessed on 25 December

2024)), and others. These packages are comprehensive, offering tools for DFT calculations, MD simulations, advanced visualization, graphical user interfaces (GUI), and postprocessing capabilities.

For example, both Schrödinger's Materials Science Suite and ADF integrate Quantum ESPRESSO with advanced GUIs for setting up and postprocessing calculations. ADF also includes its own periodic DFT program, BAND, while Schrödinger's suite incorporates Desmond for MD simulations with the latest force fields. QuantumATK provides its codes for both molecular and periodic DFT calculations, as well as advanced system builders and tools for force field generation.

Each category of computational tools has its unique strengths and limitations. Open-source tools foster innovation and collaboration, offering flexibility and accessibility. Free-for-academic software lowers barriers for students and researchers, enabling high-quality calculations without financial constraints. Commercial tools, while requiring significant investment, deliver unmatched reliability, support, and advanced features, making them indispensable for modern applications. The choice of tool depends on the specific needs, expertise, and budget of the user, ensuring that researchers can access appropriate resources for their studies.

6.4. Computational Studies of Zeolites and Zeolite–TiO_2 Structures

As is already known, first-principles calculations, particularly those based on the DFT approach, are highly valuable because of their favorable balance between computational cost and accuracy. However, performing calculations at the DFT level for large structures, such as zeolites, can be very challenging. Nonetheless, with a careful selection of density functionals, basis sets, pseudopotentials, and other computational parameters, the literature has demonstrated that it is indeed possible to effectively investigate large zeolite unit cells and their interactions with TiO_2 nanostructures. Additionally, MD simulations based on force fields have proven to be a valuable computational tool for studying the dynamical properties of various structures, primarily because of their ability to handle huge systems with satisfactory accuracy. However, the availability of force fields capable of obtaining accurate results for zeolites is limited, and the development of new, improved force fields remains an active area of research. As noted earlier, a literature search revealed a limited number of computational studies addressing combined zeolite and TiO_2 structures. Nevertheless, the studies discussed in this chapter may serve as valuable inspiration for further investigations.

Historically, one of the first first-principles studies of zeolites and some titania-based compounds is found in the paper by Zicovich and Dovesi [173]. In their study of the so-called titanozeolites, the authors employed the periodic quantum mechanical program CRYSTAL [174,175] to study the equilibrium geometry, electronic structure, and relative stability of these materials. Their study encompassed nine crystalline models with compositions of $(TiO_2)x(SiO_2)1$-x, incorporating various framework structures such as chabazite, sodalite, and α-quartz, as well as rutile as a TiO_2 reference. By employing a split-valence double-zeta basis set, the geometries of the models were fully optimized, and the influence of extended basis sets on electronic properties and relative stabilities was investigated. A key finding highlighted the lower stability of fourfold-coordinated titanium in titanozeolites compared with sixfold-coordinated titanium in rutile.

The study by Zhang et al. [176] highlights the structural and electronic synergies achieved by modifying zeolites with TiO_2, offering insights for designing advanced catalysts for hydrocarbon cracking and other applications. The authors combined DFT calculations and experimental approaches to investigate the structural, electronic, and acidic properties of TiO_2-modified USY zeolite (TiO_2/USY). The DMol^3 module of the Materials

Studio software package was used for the DFT calculations, employing the BLYP [177–181] functional under the GGA approximation. Adsorption and deprotonation energies were key metrics for evaluating interaction strength and acidity changes. The results revealed two primary configurations for TiO_2 adsorption on USY zeolite, the ring-form and bridge-form configurations, with the bridge-form being more stable because of stronger Ti-O and Al-O bonding. The modifications introduced Ti-OH acid sites, enhancing medium-strength acidity while slightly weakening Brønsted acidity at Al-OH-Si sites.

Atomistic calculations were particularly useful in the study by Li et al. [182], where computational investigation of TiO_2-modified zeolite 13X provided important additional insights into the synergy between these materials and their enhanced adsorption capabilities. Using a combination of DFT, Monte Carlo (MC), and MD simulations, the authors elucidated the adsorption mechanisms of stearyl alcohol (C18H37OH) on zeolite 13X modified with TiO_2 via atomic layer deposition (ALD). Both periodic and molecular/cluster DFT calculations were applied using tools such as Dmol3 and Gaussian 09. In particular, it was found that physisorption occurred within the zeolite micropores, driven by van der Waals interactions, and that chemisorption sites were provided by ALD-deposited TiO_2 nanoislands, which formed stable Ti-O bonds with stearyl alcohol.

In the context of computational studies, a review by Salashoori et al. [183] is also beneficial, although it does not directly consider combined zeolite and TiO_2 structures. However, it references zeolites and TiO_2 in the context of adsorption and their roles as potential materials for pollutant removal, which is important, as it can provide ideas for researchers on how to combine these materials. From the computational aspect, among other methods, this paper also highlights the application of MD simulations with GROMACS and LAMMPS codes. Since zeolites are challenging to model because of their size, this is interesting. Some of the notable case studies mentioned involve the study of the adsorption properties of zeolites towards pharmaceutical pollutants. Taking into account the size of the zeolite structures, we reference the study by Fischer [184], which utilized DFT calculations to investigate the adsorption properties of eleven all-silica zeolites with varying pore sizes for carbamazepine. This research highlighted the potential for selective removal of carbamazepine using tailored zeolite frameworks.

7. Future Perspectives

The future development and application of TiO_2–zeolite composites offer promising directions for addressing environmental challenges. Several research areas hold potential for enhancing the efficiency and sustainability of these materials. Table 7 outlines key focus areas for future research, their objectives, and corresponding references that support or inspire these perspectives. This summarized table serves as a roadmap for advancing TiO_2–zeolite composites to meet emerging environmental and technological demands.

Table 7. Future perspectives on zeolite-supported TiO_2.

Focus Area	Description	Goal	References
Visible light activation	Develop doped TiO_2–zeolite composites to extend activity into the visible spectrum.	Enhance efficiency under natural sunlight.	[45,46,86,88,89]
Advanced synthesis	Employ ecofriendly methods such as atomic layer deposition and green synthesis.	Improve scalability and sustainability.	[61,63,73,74]
Emerging pollutants	Target pharmaceuticals, microplastics, and persistent organic pollutants.	Address complex pollution scenarios.	[39,44,105]

Table 7. *Cont.*

Focus Area	Description	Goal	References
Dual-functionality systems	Combine adsorption and photocatalysis for selective pollutant removal and degradation.	Maximize pollutant remediation.	[43,87,97]
Computational modeling	Use tools such as DFT and molecular dynamics to optimize material interactions and design.	Guide the development of tailored composites.	[171,172,176]
Real-world applications	Test stability under varying environmental conditions and conduct lifecycle assessments.	Validate practicality and durability.	[7,37,55,71]
Smart system integration	Embed composites into purification devices, filters, and membranes.	Enable sustainable and innovative environmental solutions.	[66,71,110]
Circular economy	Explore reusability and regeneration techniques for materials.	Promote sustainability and waste reduction.	[67,93,94]
Multidisciplinary research	Foster collaboration among chemistry, engineering, and environmental sciences.	Accelerate innovation and practical applications.	[7,21,29]

8. Conclusions

Zeolite-supported TiO_2 has demonstrated potential for environmental applications, offering a sustainable solution for removing pollutants from the environment. The synergistic properties of zeolites, such as high adsorption capacity and robust structural stability, in combination with the photocatalytic efficiency of TiO_2, enable the effective degradation of a wide range of pollutants, including organic contaminants, heavy metals, and gaseous pollutants such as VOCs and NOx. These materials have shown excellent efficiency under UV light, with promising advancements toward extending the activity to the visible spectrum.

Improvements in material stability and optimization of synthesis methods remain challenges. Progress in material characterization techniques and computational modeling has provided more insights into the structure of materials. Adapting material properties plays a crucial role in influencing catalytic activity. Green synthesis techniques, along with the exploration of multifunctional systems, present promising opportunities for future research.

Zeolite-supported TiO_2 composites represent a sustainable platform for environmental application. Interdisciplinary collaboration and innovations in this field have the potential to solve global pollution challenges and contribute to the development of ecofriendly technologies for air and water purification.

Author Contributions: Conceptualization, S.J.A. and S.A.; methodology, S.J.A. and S.A.; software, S.J.A. and S.A.; formal analysis, S.J.A. and S.A.; data curation, S.J.A. and S.A.; writing—original draft preparation, S.J.A. and S.A.; visualization, S.J.A. and S.A.; supervision, S.J.A.; project administration, S.J.A. and S.A.; funding acquisition, S.J.A. and S.A. All authors have read and agreed to the published version of the manuscript.

Funding: This work is supported by the Ministry of Education, Science and Technological Development of the Republic of Serbia (Grant No. 451-03-68/2022-14/200125) and the Association for the International Development of Academic and Scientific Collaboration (AIDASCO).

Data Availability Statement: Data are contained within the article.

Conflicts of Interest: The authors declare no conflict of interest.

References

1. Vasseghian, Y.; Hosseinzadeh, S.; Khataee, A.; Dragoi, E.-N. The Concentration of Persistent Organic Pollutants in Water Resources: A Global Systematic Review, Meta-Analysis and Probabilistic Risk Assessment. *Sci. Total Environ.* **2021**, *796*, 149000. [CrossRef] [PubMed]
2. Jones, K.C. Persistent Organic Pollutants (POPs) and Related Chemicals in the Global Environment: Some Personal Reflections. *Environ. Sci. Technol.* **2021**, *55*, 9400–9412. [CrossRef]
3. Chen, D.; Cheng, Y.; Zhou, N.; Chen, P.; Wang, Y.; Li, K.; Huo, S.; Cheng, P.; Peng, P.; Zhang, R.; et al. Photocatalytic Degradation of Organic Pollutants Using TiO_2-Based Photocatalysts: A Review. *J. Clean. Prod.* **2020**, *268*, 121725. [CrossRef]
4. Tee, S.Y.; Kong, J.; Koh, J.J.; Teng, C.P.; Wang, X.; Wang, X.; Teo, S.L.; Thitsartarn, W.; Han, M.-Y.; Seh, Z.W. Structurally and Surficially Activated TiO_2 Nanomaterials for Photochemical Reactions. *Nanoscale* **2024**, *16*, 18165–18212. [CrossRef]
5. GEMStat. GEMStat Database 2024. Available online: https://bafg.maps.arcgis.com/apps/webappviewer/index.html?id=2366605a186e4e7db42d9c35a2b64ba4 (accessed on 5 December 2024).
6. UNEP. A Framework for Freshwater Ecosystem Management. In *Technical Guide for Classification and Target-Setting*; UNEP: Nairobi, Kenya, 2017; Volume 2. Available online: https://www.unenvironment.org/resources/publication/framework-freshwater-ecosystem-management (accessed on 20 December 2024).
7. Zhang, J.; Wu, H.; Shi, L.; Wu, Z.; Zhang, S.; Wang, S.; Sun, H. Photocatalysis Coupling with Membrane Technology for Sustainable and Continuous Purification of Wastewater. *Sep. Purif. Technol.* **2024**, *329*, 125225. [CrossRef]
8. Muscetta, M.; Ganguly, P.; Clarizia, L. Solar-Powered Photocatalysis in Water Purification: Applications and Commercialization Challenges. *J. Environ. Chem. Eng.* **2024**, *12*, 113073. [CrossRef]
9. Lian, P.; Qin, A.; Liao, L.; Zhang, K. Progress on the Nanoscale Spherical TiO_2 Photocatalysts: Mechanisms, Synthesis and Degradation Applications. *Nano Sel.* **2021**, *2*, 447–467. [CrossRef]
10. Padmanabhan, N.T.; Thomas, N.; Louis, J.; Mathew, D.T.; Ganguly, P.; John, H.; Pillai, S.C. Graphene Coupled TiO_2 Photocatalysts for Environmental Applications: A Review. *Chemosphere* **2021**, *271*, 129506. [CrossRef]
11. Shaik, B.B.; Katari, N.K.; Raghupathi, J.K.; Jonnalagadda, S.B.; Rana, S. Titanium Dioxide/Graphene-Based Nanocomposites as Photocatalyst for Environmental Applications: A Review. *ChemistrySelect* **2024**, *9*, e202403521. [CrossRef]
12. Kaveh, R.; Alijani, H.; Falletta, E.; Bianchi, C.L.; Mokhtarifar, M.; Boffito, D.C. Advancements in Superhydrophilic Titanium Dioxide/Graphene Oxide Composites Coatings for Self-Cleaning Applications on Glass Substrates: A Comprehensive Review. *Prog. Org. Coat.* **2024**, *190*, 108347. [CrossRef]
13. K., A.; Devarajan, Y. Nanomaterials-Based Wastewater Treatment: Addressing Challenges and Advancing Sustainable Solutions. *BioNanoScience* **2024**, *15*, 149. [CrossRef]
14. Song, C.; Xiao, L.; Chen, Y.; Yang, F.; Meng, H.; Zhang, W.; Zhang, Y.; Wu, Y. TiO_2-Based Catalysts with Various Structures for Photocatalytic Application: A Review. *Catalysts* **2024**, *14*, 366. [CrossRef]
15. Diban, N.; Pacuła, A.; Kumakiri, I.; Barquín, C.; Rivero, M.J.; Urtiaga, A.; Ortiz, I. TiO_2–Zeolite Metal Composites for Photocatalytic Degradation of Organic Pollutants in Water. *Catalysts* **2021**, *11*, 1367. [CrossRef]
16. Roongraung, K.; Chuangchote, S.; Laosiripojana, N. Enhancement of Photocatalytic Oxidation of Glucose to Value-Added Chemicals on TiO_2 Photocatalysts by A Zeolite (Type Y) Support and Metal Loading. *Catalysts* **2020**, *10*, 423. [CrossRef]
17. Piedra López, J.G.; González Pichardo, O.H.; Pinedo Escobar, J.A.; De Haro Del Río, D.A.; Inchaurregui Méndez, H.; González Rodríguez, L.M. Photocatalytic Degradation of Metoprolol in Aqueous Medium Using a TiO_2/Natural Zeolite Composite. *Fuel* **2021**, *284*, 119030. [CrossRef]
18. Petcu, G.; Papa, F.; Anghel, E.M.; Atkinson, I.; Preda, S.; Somacescu, S.; Culita, D.C.; Baran, A.; Ciobanu, E.M.; Jecu, L.M.; et al. Effects of Aluminosilicate Gel Treatment and TiO_2 Loading on Photocatalytic Properties of Au–TiO_2/Zeolite Y. *Gels* **2023**, *9*, 503. [CrossRef] [PubMed]
19. Jalloul, G.; Hijazi, N.; Boyadjian, C.; Awala, H.; Albadarin, A.B.; Ahmad, M.N. Titania-Zeolite Composite for Tetracycline Photocatalytic Degradation under Visible Light: A Comparison between Doping and Ion Exchange. *Heliyon* **2024**, *10*, e31854. [CrossRef] [PubMed]
20. Arifin, S.N.H.; Radin Mohamed, R.M.S.; Al-Gheethi, A.A.; Wei, L.C.; Yashni, G.; Fitriani, N.; Naushad, M.; Albadarin, A.B. Modified TiO_2 Nanotubes-Zeolite Composite Photocatalyst: Characteristics, Microstructure and Applicability for Degrading Triclocarban. *Chemosphere* **2022**, *287*, 132278. [CrossRef]
21. Hu, G.; Yang, J.; Duan, X.; Farnood, R.; Yang, C.; Yang, J.; Liu, W.; Liu, Q. Recent Developments and Challenges in Zeolite-Based Composite Photocatalysts for Environmental Applications. *Chem. Eng. J.* **2021**, *417*, 129209. [CrossRef]
22. Lazau, C.; Ratiu, C.; Orha, C.; Pode, R.; Manea, F. Photocatalytic Activity of Undoped and Ag-Doped TiO_2-Supported Zeolite for Humic Acid Degradation and Mineralization. *Mater. Res. Bull.* **2011**, *46*, 1916–1921. [CrossRef]
23. Khatamian, M.; Hashemian, S.; Yavari, A.; Saket, M. Preparation of Metal Ion (Fe^{3+} and Ni^{2+}) Doped TiO_2 Nanoparticles Supported on ZSM-5 Zeolite and Investigation of Its Photocatalytic Activity. *Mater. Sci. Eng. B* **2012**, *177*, 1623–1627. [CrossRef]

24. Tsaplin, D.E.; Ostroumova, V.A.; Kulikov, L.A.; Zolotukhina, A.V.; Sadovnikov, A.A.; Kryuchkov, M.D.; Egazaryants, S.V.; Maksimov, A.L.; Wang, K.; Luo, Z.; et al. Synthesis and Investigation of Zeolite TiO_2/Al-ZSM-12 Structure and Properties. *Catalysts* **2023**, *13*, 216. [CrossRef]
25. Mubarak, M.F.; Mohamed, A.M.G.; Keshawy, M.; elMoghny, T.A.; Shehata, N. Adsorption of Heavy Metals and Hardness Ions from Groundwater onto Modified Zeolite: Batch and Column Studies. *Alex. Eng. J.* **2022**, *61*, 4189–4207. [CrossRef]
26. Bello, M.O.; Shelake, S.P.; Abdus-Salam, N.; Adekola, F.A.; Vennapoosa, C.S.; Sesha Sainath, A.V.; Pal, U. Na-Y Zeolite Supported TiO_2/Pd Nanoparticles for Enhanced Photoredox Catalytic Properties and Green Hydrogen Generation. *Catal. Commun.* **2024**, *186*, 106817. [CrossRef]
27. Liu, N.; Qi, R.; Sun, X.; Kawazoe, N.; Chen, G.; Yang, Y. Synthesis and Characterization of 3D-Zeolite–Modified TiO_2-Based Photocatalyst with Synergistic Effect for Elimination of Organic Pollutant in Wastewater Treatment. *Front. Environ. Sci.* **2022**, *10*, 1009045. [CrossRef]
28. Ruíz-Baltazar, Á.D.J. Advancements in Nanoparticle-Modified Zeolites for Sustainable Water Treatment: An Interdisciplinary Review. *Sci. Total Environ.* **2024**, *946*, 174373. [CrossRef]
29. Milojević-Rakić, M.; Bajuk-Bogdanović, D. Recent Advances in Zeolites and Porous Materials Applications in Catalysis and Adsorption Processes. *Catalysts* **2023**, *13*, 863. [CrossRef]
30. Navidpour, A.H.; Abbasi, S.; Li, D.; Mojiri, A.; Zhou, J.L. Investigation of Advanced Oxidation Process in the Presence of TiO_2 Semiconductor as Photocatalyst: Property, Principle, Kinetic Analysis, and Photocatalytic Activity. *Catalysts* **2023**, *13*, 232. [CrossRef]
31. Ochiai, T.; Fujishima, A. Photoelectrochemical Properties of TiO_2 Photocatalyst and Its Applications for Environmental Purification. *J. Photochem. Photobiol. C Photochem. Rev.* **2012**, *13*, 247–262. [CrossRef]
32. Kang, X.; Liu, S.; Dai, Z.; He, Y.; Song, X.; Tan, Z. Titanium Dioxide: From Engineering to Applications. *Catalysts* **2019**, *9*, 191. [CrossRef]
33. Eddy, D.R.; Permana, M.D.; Sakti, L.K.; Sheha, G.A.N.; Solihudin; Hidayat, S.; Takei, T.; Kumada, N.; Rahayu, I. Heterophase Polymorph of TiO_2 (Anatase, Rutile, Brookite, TiO_2 (B)) for Efficient Photocatalyst: Fabrication and Activity. *Nanomaterials* **2023**, *13*, 704. [CrossRef]
34. Zainudin, N.F.; Abdullah, A.Z.; Mohamed, A.R. Characteristics of Supported Nano-TiO_2/ZSM-5/Silica Gel (SNTZS): Photocatalytic Degradation of Phenol. *J. Hazard. Mater.* **2010**, *174*, 299–306. [CrossRef] [PubMed]
35. Shen, W.; Xiang, D.; Yang, J.; Tang, Y.; Xin, C.; Guo, Q.; Yu, X. Fabrication of Beta Zeolite Supported Ti^{3+}-TiO_2/CdS Composite for Ultrahigh-Performance Photodegradation of Tetracycline under Visible-Light Illumination. *Colloids Surf. A Physicochem. Eng. Asp.* **2022**, *653*, 129965. [CrossRef]
36. Taheri Najafabadi, A.; Taghipour, F. Physicochemical Impact of Zeolites as the Support for Photocatalytic Hydrogen Production Using Solar-Activated TiO_2-Based Nanoparticles. *Energy Convers. Manag.* **2014**, *82*, 106–113. [CrossRef]
37. Sun, Q.; Hu, X.; Zheng, S.; Sun, Z.; Liu, S.; Li, H. Influence of Calcination Temperature on the Structural, Adsorption and Photocatalytic Properties of TiO_2 Nanoparticles Supported on Natural Zeolite. *Powder Technol.* **2015**, *274*, 88–97. [CrossRef]
38. Corma, A.; Garcia, H. Zeolite-Based Photocatalysts. *Chem. Commun.* **2004**, 1443–1459. [CrossRef] [PubMed]
39. Liu, X.; Liu, Y.; Lu, S.; Guo, W.; Xi, B. Performance and Mechanism into TiO_2/Zeolite Composites for Sulfadiazine Adsorption and Photodegradation. *Chem. Eng. J.* **2018**, *350*, 131–147. [CrossRef]
40. Chen, K.; Qu, F.; Huang, Y.; Cai, J.; Wu, F.; Li, W. Advancing Photocatalytic Concrete Technologies in Design, Performance and Application for a Sustainable Future. *Adv. Nanocomposites* **2024**, *1*, 180–200. [CrossRef]
41. Kamegawa, T.; Kido, R.; Yamahana, D.; Yamashita, H. Design of TiO_2-Zeolite Composites with Enhanced Photocatalytic Performances under Irradiation of UV and Visible Light. *Microporous Mesoporous Mater.* **2013**, *165*, 142–147. [CrossRef]
42. González Rodríguez, L.M.; Pinedo Escobar, J.A.; Piedra López, J.G.; De Haro Del Rio, D.A.; Suarez Vázquez, S.I.; Carrillo Martínez, C.J.; Solis, C.G.; Chulim, A.C. Synthesis, Characterization and Photocatalytic Activity Evaluation of WO_3, TiO_2 and WO_3/TiO_2 Supported on Zeolite Faujasite. *Int. J. Chem. React. Eng.* **2020**, *18*, 20200095. [CrossRef]
43. Jalloul, G.; Al-Mousawi, A.; Chocr, F.; Merhi, A.; Awala, H.; Boyadjian, C. Fe-Sensitized Zeolite Supported TiO_2 for the Degradation of Tetracycline Using Blue LED Irradiation. *Front. Environ. Sci.* **2022**, *10*, 873257. [CrossRef]
44. Behravesh, S.; Mirghaffari, N.; Alemrajabi, A.A.; Davar, F.; Soleimani, M. Photocatalytic Degradation of Acetaminophen and Codeine Medicines Using a Novel Zeolite-Supported TiO_2 and ZnO under UV and Sunlight Irradiation. *Environ. Sci. Pollut. Res.* **2020**, *27*, 26929–26942. [CrossRef]
45. Sun, T.; Wei, J.; Zhou, C.; Wang, Y.; Shu, Z.; Zhou, J.; Chen, J. Facile Preparation and Enhanced Photocatalytic Hydrogen Evolution of Cation-Exchanged Zeolite LTA Supported TiO_2 Photocatalysts. *Int. J. Hydrog. Energy* **2023**, *48*, 13851–13863. [CrossRef]
46. Torkian, N.; Bahrami, A.; Hosseini-Abari, A.; Momeni, M.M.; Abdolkarimi-Mahabadi, M.; Bayat, A.; Hajipour, P.; Amini Rourani, H.; Abbasi, M.S.; Torkian, S.; et al. Synthesis and Characterization of Ag-Ion-Exchanged Zeolite/TiO_2 Nanocomposites for Antibacterial Applications and Photocatalytic Degradation of Antibiotics. *Environ. Res.* **2022**, *207*, 112157. [CrossRef] [PubMed]

47. Collado, L.; Reñones, P.; Fermoso, J.; Fresno, F.; Garrido, L.; Pérez-Dieste, V.; Escudero, C.; Hernández-Alonso, M.D.; Coronado, J.M.; Serrano, D.P.; et al. The Role of the Surface Acidic/Basic Centers and Redox Sites on TiO_2 in the Photocatalytic CO_2 Reduction. *Appl. Catal. B Environ.* **2022**, *303*, 120931. [CrossRef]
48. John, K.I.; Adeleye, A.T.; Adeniyi, A.G.; Sani, L.A.; Abesa, S.; Orege, I.J.; Adenle, A.A.; Elawad, M.; Omorogie, M.O. Screening of Zeolites Series: H-β/H-MOR/H-ZSM-5 as Potential Templates for Photocatalyst Heterostructure Composites through Photocatalytic Degradation of Tetracycline. *J. Mol. Struct.* **2023**, *1277*, 134852. [CrossRef]
49. Fernández-Catalá, J.; Sánchez-Rubio, M.; Navlani-García, M.; Berenguer-Murcia, Á.; Cazorla-Amorós, D. Synthesis of TiO_2/Nanozeolite Composites for Highly Efficient Photocatalytic Oxidation of Propene in the Gas Phase. *ACS Omega* **2020**, *5*, 31323–31331. [CrossRef]
50. Xie, Y.; Wang, J.; Ren, F.; Shuai, H.; Du, G. Nonmetallic Mineral as the Carrier of TiO_2 Photocatalyst: A Review. *Front. Catal.* **2022**, *2*, 806316. [CrossRef]
51. Mergenbayeva, S.; Abitayev, Z.; Batyrbayeva, M.; Vakros, J.; Mantzavinos, D.; Atabaev, T.S.; Poulopoulos, S.G. TiO_2/Zeolite Composites for SMX Degradation under UV Irradiation. *Catalysts* **2024**, *14*, 147. [CrossRef]
52. Ghasemi, Z.; Younesi, H.; Zinatizadeh, A.A. Kinetics and Thermodynamics of Photocatalytic Degradation of Organic Pollutants in Petroleum Refinery Wastewater over Nano-TiO_2 Supported on Fe-ZSM-5. *J. Taiwan Inst. Chem. Eng.* **2016**, *65*, 357–366. [CrossRef]
53. Zhang, W.; Nomura, Y.; Fukahori, S.; Kiso, T.; Myoujin, K.; Fujiwara, T. Modeling, Validation, and Optimization of a Rotating Advanced Oxidation Contactor Equipped with TiO_2-Zeolite Composite Sheets for Sulfamethazine Removal. *Process Saf. Environ. Prot.* **2024**, *187*, 1248–1258. [CrossRef]
54. Swaminaathan, P.; Saravanan, A.; Yaashikaa, P.R.; Vickram, A.S. Recent Advances in Photocatalytic Degradation of Persistent Organic Pollutants: Mechanisms, Challenges, and Modification Strategies. *Sustain. Chem. Environ.* **2024**, *8*, 100171. [CrossRef]
55. Cano-Casanova, L.; Amorós-Pérez, A.; Lillo-Ródenas, M.Á.; Román-Martínez, M.D.C. Effect of the Preparation Method (Sol-Gel or Hydrothermal) and Conditions on the TiO_2 Properties and Activity for Propene Oxidation. *Materials* **2018**, *11*, 2227. [CrossRef] [PubMed]
56. Shin, S.; Kim, M.-J. Hydrothermal Synthesis of Zeolites from Residual Waste Generated via Indirect Carbonation of Coal Fly Ash. *Sustain. Environ. Res.* **2024**, *34*, 1. [CrossRef]
57. Jayasree, P.; Remya, N. Photocatalytic Degradation of Paracetamol Using Aluminosilicate Supported TiO_2. *Water Sci. Technol.* **2020**, *82*, 2114–2124. [CrossRef] [PubMed]
58. Elghniji, K.; Elaloui, E.; Moussaoui, Y. Coating of Anatase Titania on Clinoptilolite by Metal Organic Chemical Vapor Deposition Method: Enhanced Mesoporosity and Photocatalytic Activity. *Chem. Pap.* **2018**, *72*, 1159–1168. [CrossRef]
59. Gomez, S.; Marchena, C.L.; Pizzio, L.; Pierella, L. Preparation and Characterization of TiO_2/HZSM-11 Zeolite for Photodegradation of Dichlorvos in Aqueous Solution. *J. Hazard. Mater.* **2013**, *258–259*, 19–26. [CrossRef]
60. Chong, M.N.; Tneu, Z.Y.; Poh, P.E.; Jin, B.; Aryal, R. Synthesis, Characterisation and Application of TiO_2–Zeolite Nanocomposites for the Advanced Treatment of Industrial Dye Wastewater. *J. Taiwan Inst. Chem. Eng.* **2015**, *50*, 288–296. [CrossRef]
61. Wang, M.; Zhang, W.; Tao, Y.; Yang, L. Calcination Effects on Sol–Gel Preparation of Porous TiO_2/Zeolite Composites for Acid Red 1 Degradation. *Desalination Water Treat.* **2022**, *254*, 251–259. [CrossRef]
62. Gatou, M.-A.; Syrrakou, A.; Lagopati, N.; Pavlatou, E.A. Photocatalytic TiO_2-Based Nanostructures as a Promising Material for Diverse Environmental Applications: A Review. *Reactions* **2024**, *5*, 135–194. [CrossRef]
63. Ramesh, N.; Lai, C.W.; Johan, M.R.B.; Mousavi, S.M.; Badruddin, I.A.; Kumar, A.; Sharma, G.; Gapsari, F. Progress in Photocatalytic Degradation of Industrial Organic Dye by Utilising the Silver Doped Titanium Dioxide Nanocomposite. *Heliyon* **2024**, *10*, e40998. [CrossRef]
64. Zhou, P.; Shen, Y.; Zhao, S.; Chen, Y.; Gao, S.; Liu, W.; Wei, D. Hydrothermal Synthesis of Novel Ternary Hierarchical MoS_2/TiO_2/Clinoptilolite Nanocomposites with Remarkably Enhanced Visible Light Response towards Xanthates. *Appl. Surf. Sci.* **2021**, *542*, 148578. [CrossRef]
65. Ullah, R.; Sun, J.; Gul, A.; Bai, S. One-Step Hydrothermal Synthesis of TiO_2-Supported Clinoptilolite: An Integrated Photocatalytic Adsorbent for Removal of Crystal Violet Dye from Aqueous Media. *J. Environ. Chem. Eng.* **2020**, *8*, 103852. [CrossRef]
66. Suárez, S.; Jansson, I.; Ohtani, B.; Sánchez, B. From Titania Nanoparticles to Decahedral Anatase Particles: Photocatalytic Activity of TiO_2/Zeolite Hybrids for VOCs Oxidation. *Catal. Today* **2019**, *326*, 2–7. [CrossRef]
67. Fatkhasari, Y.; Rouf, N.A.; Ermadayanti, W.A.; Kurniawan, R.Y.; Bagastyo, A.Y. Synthesis of TiO_2/Zeolite-A Composite for the Removal of Methylene Blue on Direct Sunlight. *J. Tek. ITS* **2019**, *8*, F115–F120. [CrossRef]
68. Younes, N. NaY Zeolite and TiO_2 Impregnated NaY Zeolite for the Adsorption and Photocatalytic Degradation of Methylene Blue under Sunlight. *Iran. J. Chem. Chem. Eng.* **2022**, *41*, 6.
69. Iqbal, R.M.; Susanti, I.; Rachman, R.A.; Pradana, T.A.; Toepak, E.P. Synthesis, Characterization, and Photocatalytic Activity of N-Doped TiO_2/Zeolite-NaY for Methylene Blue Removal. *J. Pure App. Chem. Res.* **2021**, *10*, 132–139. [CrossRef]
70. Saqib, N.U.; Adnan, R.; Shah, I. Zeolite Supported TiO_2 with Enhanced Degradation Efficiency for Organic Dye under Household Compact Fluorescent Light. *Mater. Res. Express* **2019**, *6*, 095506. [CrossRef]

71. Jansson, I.; Suárez, S.; Garcia-Garcia, F.J.; Sánchez, B. Zeolite–TiO 2 Hybrid Composites for Pollutant Degradation in Gas Phase. *Appl. Catal. B Environ.* **2015**, *178*, 100–107. [CrossRef]
72. Adebayo, B.O.; Yu, H.; Rownaghi, A.A.; Liang, X.; Rezaei, F. Atomic Layer Deposited Pt/TiO_2-SiO_2 and Pt/ZrO_2-SiO_2 for Sequential Adsorption and Oxidation of VOCs. *Chem. Eng. J.* **2022**, *444*, 136603. [CrossRef]
73. Xu, D.; Yin, J.; Gao, Y.; Zhu, D.; Wang, S. Atomic-Scale Designing of Zeolite Based Catalysts by Atomic Layer Deposition. *ChemPhysChem* **2021**, *22*, 1287–1301. [CrossRef]
74. Ke, W.; Qin, X.; Palomino, R.M.; Simonovis, J.P.; Senanayake, S.D.; Rodriguez, J.A.; Zaera, F. Redox Properties of TiO_2 Thin Films Grown on Mesoporous Silica by Atomic Layer Deposition. *J. Phys. Chem. Lett.* **2023**, *14*, 4696–4703. [CrossRef]
75. Bogdan, L.; Palčić, A.; Duplančić, M.; Leskovac, M.; Tomašić, V. Eco-Friendly Synthesis of TiO_2/ZIF-8 Composites: Characterization and Application for the Removal of Imidacloprid from Wastewater. *Processes* **2023**, *11*, 963. [CrossRef]
76. Zendehdel, M.; Cruciani, G.; Barghi, B. Micro-Meso Structure NaP Zeolite @TiO_2 Nanocomposite: Eco-Friendly Photocatalyst for Simultaneous Removal COD and Degradation of Methylene Blue under Solar Irradiation. *Photochem. Photobiol. Sci.* **2022**, *21*, 1011–1029. [CrossRef] [PubMed]
77. Ratnakaram, V.N.; Ummadi, R.R.; Devineni, S.R.; Ch, S.; Pavuluri, C.M. Eco-Friendly TiO_2-Zno Composite Nano-Catalysts: Synthesis, Characterization, Catalytic Performance in Synthesis of A-Amino/Hydroxyphosphonates and Their Adme, Docking and Antimicrobial Studies. 2024. Available online: https://papers.ssrn.com/sol3/papers.cfm?abstract_id=5047867 (accessed on 20 December 2024).
78. Ferreira, R.; Morales-Torres, S.; Pastrana-Martinez, L.M.; Maldonado-Hódar, F.J.; Lourenço, J.P.; Silva, J.M.; João, I.M.; Ribeiro, M.F.; Fernandes, A. On the Characterization and Performance of Highly Dispersed TiO_2 Nanoparticles onto BEA Zeolite in the Continuous Gas-Phase Photooxidation of Ethylene. *RCS Appl. Interfaces 2025* 2, 150. [CrossRef]
79. Han, C.; Kundu, B.K.; Liang, Y.; Sun, Y. Near-Infrared Light-Driven Photocatalysis with an Emphasis on Two-Photon Excitation: Concepts, Materials, and Applications. *Adv. Mater.* **2024**, *36*, 2307759. [CrossRef]
80. Jiang, L.; Zhou, S.; Yang, J.; Wang, H.; Yu, H.; Chen, H.; Zhao, Y.; Yuan, X.; Chu, W.; Li, H. Near-Infrared Light Responsive TiO_2 for Efficient Solar Energy Utilization. *Adv. Funct. Mater.* **2022**, *32*, 2108977. [CrossRef]
81. Wang, M.; Hou, Z.; Al Kheraif, A.A.; Xing, B.; Lin, J. Mini Review of TiO_2-Based Multifunctional Nanocomposites for Near-Infrared Light–Responsive Phototherapy. *Adv. Healthc. Mater.* **2018**, *7*, 1800351. [CrossRef]
82. Zhao, J.; Xu, J.; Jian, X.; Xu, J.; Gao, Z.; Song, Y.-Y. NIR Light-Driven Photocatalysis on Amphiphilic TiO_2 Nanotubes for Controllable Drug Release. *ACS Appl. Mater. Interfaces* **2020**, *12*, 23606–23616. [CrossRef]
83. Enzweiler, H.; Yassue-Cordeiro, P.H.; Schwaab, M.; Barbosa-Coutinho, E.; Olsen Scaliante, M.H.N.; Fernandes, N.R.C. Evaluation of Pd-TiO_2/ZSM-5 Catalysts Composition Effects on Hydrogen Production by Photocatalytic Water Splitting. *Int. J. Hydrog. Energy* **2018**, *43*, 6515–6525. [CrossRef]
84. El-Bery, H.M.; Abdelhamid, H.N. Photocatalytic Hydrogen Generation via Water Splitting Using ZIF-67 Derived Co_3O_4@C/TiO_2. *J. Environ. Chem. Eng.* **2021**, *9*, 105702. [CrossRef]
85. Kanakaraju, D.; Kockler, J.; Motti, C.A.; Glass, B.D.; Oelgemöller, M. Titanium Dioxide/Zeolite Integrated Photocatalytic Adsorbents for the Degradation of Amoxicillin. *Appl. Catal. B: Environ.* **2015**, *166–167*, 45–55. [CrossRef]
86. Kanakaraju, D.; Anak Kutiang, F.D.; Lim, Y.C.; Goh, P.S. Recent Progress of Ag/TiO_2 Photocatalyst for Wastewater Treatment: Doping, Co-Doping, and Green Materials Functionalization. *Appl. Mater. Today* **2022**, *27*, 101500. [CrossRef]
87. Petcu, G.; Papa, F.; Atkinson, I.; Baran, A.; Apostol, N.G.; Petrescu, S.; Richaudeau, L.; Blin, J.-L.; Parvulescu, V. Co- and Ni-Doped TiO_2 Nanoparticles Supported on Zeolite Y with Photocatalytic Properties. *Nanomaterials* **2023**, *13*, 2200. [CrossRef]
88. Akhter, P.; Arshad, A.; Saleem, A.; Hussain, M. Recent Development in Non-Metal-Doped Titanium Dioxide Photocatalysts for Different Dyes Degradation and the Study of Their Strategic Factors: A Review. *Catalysts* **2022**, *12*, 1331. [CrossRef]
89. Cheng, Z.-L.; Han, S. Preparation and Photoelectrocatalytic Performance of N-Doped TiO_2/NaY Zeolite Membrane Composite Electrode Material. *Water Sci. Technol.* **2016**, *73*, 486–492. [CrossRef] [PubMed]
90. Feng, N.; Xu, J.; Deng, F. Solid-State NMR of Active Sites in TiO_2 Photocatalysis: A Critical Review. *Chem. Synth.* **2024**, *4*, 43. [CrossRef]
91. Yan, D.; Li, Q.; Zhang, H.; Zhou, X.; Chen, H. A Highly Dispersed Mesoporous zeolite@TiO_2—Supported Pt for Enhanced Sulfur-Resistance Catalytic CO Oxidation. *Catal. Commun.* **2020**, *142*, 106042. [CrossRef]
92. Morales-Torres, S.; Pastrana-Martínez, L.M.; Figueiredo, J.L.; Faria, J.L.; Silva, A.M.T. Design of Graphene-Based TiO_2 Photocatalysts—A Review. *Environ. Sci. Pollut. Res.* **2012**, *19*, 3676–3687. [CrossRef]
93. Neppolian, B.; Mine, S.; Horiuchi, Y.; Bianchi, C.L.; Matsuoka, M.; Dionysiou, D.D.; Anpo, M. Efficient Photocatalytic Degradation of Organics Present in Gas and Liquid Phases Using Pt-TiO_2/Zeolite (H-ZSM). *Chemosphere* **2016**, *153*, 237–243. [CrossRef] [PubMed]
94. Wojciechowska, K. The Influence of Desilication/Dealumination Processes on the Physicochemical Properties of Clinoptilolite. *Clay Min.* **2019**, *54*, 111–119. [CrossRef]

95. Huang, M.; Xu, C.; Wu, Z.; Huang, Y.; Lin, J.; Wu, J. Photocatalytic Discolorization of Methyl Orange Solution by Pt Modified TiO_2 Loaded on Natural Zeolite. *Dye. Pigment.* **2008**, *77*, 327–334. [CrossRef]
96. Zhang, G.; Song, A.; Duan, Y.; Zheng, S. Enhanced Photocatalytic Activity of TiO_2/Zeolite Composite for Abatement of Pollutants. *Microporous Mesoporous Mater.* **2018**, *255*, 61–68. [CrossRef]
97. Imessaoudene, A.; Mechraoui, O.; Aberkane, B.; Benabbas, A.; Manseri, A.; Moussaoui, Y.; Bollinger, J.-C.; Amrane, A.; Zoukel, A.; Mouni, L. Synthesis of a TiO_2/Zeolite Composite: Evaluation of Adsorption-Photodegradation Synergy for the Removal of Malachite Green. *Nano-Struct. Nano-Objects* **2024**, *38*, 101191. [CrossRef]
98. Sarabyar, S.; Farahbakhsh, A.; Tahmasebi, H.A.; Mahmoodzadeh Vaziri, B.; Khosroyar, S. Enhancing Photocatalytic Degradation of Beta-Blocker Drugs Using TiO_2 NPs/Zeolite and ZnO NPs/Zeolite as Photocatalysts: Optimization and Kinetic Investigations. *Sci. Rep.* **2024**, *14*, 27390. [CrossRef] [PubMed]
99. Chang, C.-T.; Wang, J.-J.; Ouyang, T.; Zhang, Q.; Jing, Y.-H. Photocatalytic Degradation of Acetaminophen in Aqueous Solutions by TiO_2/ZSM-5 Zeolite with Low Energy Irradiation. *Mater. Sci. Eng. B* **2015**, *196*, 53–60. [CrossRef]
100. Pan, Y.; Zhang, Y.; Huang, Y.; Jia, Y.; Chen, L. Enhanced Photocatalytic Oxidation Degradability for Real Cyanide Wastewater by Designing Photocatalyst GO/TiO_2/ZSM-5: Performance and Mechanism Research. *Chem. Eng. J.* **2022**, *428*, 131257. [CrossRef]
101. Badvi, K.; Javanbakht, V. Enhanced Photocatalytic Degradation of Dye Contaminants with TiO_2 Immobilized on ZSM-5 Zeolite Modified with Nickel Nanoparticles. *J. Clean. Prod.* **2021**, *280*, 124518. [CrossRef]
102. Hosseinzadeh, G.; Ghasemian, N.; Zinatloo-Ajabshir, S. TiO_2/Graphene Nanocomposite Supported on Clinoptilolite Nanoplate and Its Enhanced Visible Light Photocatalytic Activity. *Inorg. Chem. Commun.* **2022**, *136*, 109144. [CrossRef]
103. Luzzi, E.; Ambrogi, V.; Filippone, G.; Liguori, B.; Salzano De Luna, M.; Aprea, P. Increasing Awareness of Materials and the Environment: Hands-On Outreach Activity Presenting Water Purification Materials and Concepts. *J. Chem. Educ.* **2021**, *98*, 1296–1301. [CrossRef]
104. Mohadesi, M.; Sanavi Fard, M.; Shokri, A. The Application of Modified Nano-TiO_2 Photocatalyst for Wastewater Treatment: A Review. *Int. J. Environ. Anal. Chem.* **2024**, *104*, 2571–2592. [CrossRef]
105. Liaquat, I.; Munir, R.; Abbasi, N.A.; Sadia, B.; Muneer, A.; Younas, F.; Sardar, M.F.; Zahid, M.; Noreen, S. Exploring Zeolite-Based Composites in Adsorption and Photocatalysis for Toxic Wastewater Treatment: Preparation, Mechanisms, and Future Perspectives. *Environ. Pollut.* **2024**, *349*, 123922. [CrossRef] [PubMed]
106. Cheng, H.; Ren, X.; Yao, Y.; Tang, X.; Yi, H.; Gao, F.; Zhou, Y.; Yu, Q. Application of Unconventional External-Field Treatments in Air Pollutants Removal over Zeolite-Based Adsorbents/Catalysts. *Catalysts* **2023**, *13*, 1461. [CrossRef]
107. Mendoza, J.A.; Lee, D.H.; Kang, J.-H. Photocatalytic Removal of NOx Using TiO_2-Coated Zeolite. *Environ. Eng. Res.* **2016**, *21*, 291–296. [CrossRef]
108. Kim, S.; Kim, S.; Park, H.-J.; Park, S.; Kim, J.Y.; Jeong, Y.W.; Yang, H.H.; Choi, Y.; Yeom, M.; Song, D.; et al. Practical Scale Evaluation of a Photocatalytic Air Purifier Equipped with a Titania-Zeolite Composite Bead Filter for VOC Removal and Viral Inactivation. *Environ. Res.* **2022**, *204*, 112036. [CrossRef]
109. Zhang, W.; Bi, F.; Yu, Y.; He, H. Phosphoric Acid Treating of ZSM-5 Zeolite for the Enhanced Photocatalytic Activity of TiO_2/HZSM-5. *J. Mol. Catal. A Chem.* **2013**, *372*, 6–12. [CrossRef]
110. Rajan, M.S.; John, A.; Yoon, M.; Thomas, J. Zeolite Y-Supported Carbon-Doped TiO_2 Nanocomposites: Efficient Solar Photocatalysts for the Purification of Medicinal Wastewater. *Environ. Sci. Pollut. Res.* **2023**, *30*, 60638–60653. [CrossRef]
111. G., G.; D'Souza, J.Q.; Sundaram, N.G. UV Induced Photocatalytic Degradation of Caffeine Using TiO_2–H-Beta Zeolite Composite. *Minerals* **2023**, *13*, 465. [CrossRef]
112. Ahmed, A.M.; Jarullah, A.T.; Hussein, H.M.; Ahmed, A.N. Mordenite-Type Zeolite from Iraq Sand: Synthesis and Characterization. *J. Pet. Res. Stud.* **2023**, *13*, 126–142. [CrossRef]
113. Shelyapina, M.G.; Yocupicio-Gaxiola, R.I.; Val'kovskii, G.A.; Kotolevich, Y.; Petranovskii, V. Pillaring 2D Mordenite with TiO_2 Nanoparticles. *Russ. J. Phys. Chem. A* **2024**, *98*, 2298–2301. [CrossRef]
114. Shelyapina, M.G.; Mazur, A.; Yocupicio-Gaxiola, R.I.; Caudillo-Flores, U.; Urtaza, A.; Rodionov, I.A.; Zvereva, I.A.; Petranovskii, V. Local Structure of TiO_2/2D Mordenite Mesoporous Nanocomposites Probed by NMR. *Appl. Magn. Reson.* **2022**, *53*, 1609–1620. [CrossRef]
115. Ullah, R.; Sun, J.; Gul, A.; Munir, T.; Wu, X. Evaluations of Physico-Chemical Properties of TiO_2/Clinoptilolite Synthesized via Three Methods on Photocatalytic Degradation of Crystal Violet. *Chin. J. Chem. Eng.* **2021**, *33*, 181–189. [CrossRef]
116. Shen, Y.; Zhou, P.; Zhao, S.; Li, A.; Chen, Y.; Bai, J.; Han, C.; Wei, D.; Ao, Y. Synthesis of High-Efficient TiO_2/Clinoptilolite Photocatalyst for Complete Degradation of Xanthate. *Miner. Eng.* **2020**, *159*, 106640. [CrossRef]
117. Zhou, P.; Shen, Y.; Zhao, S.; Li, G.; Cui, B.; Wei, D.; Shen, Y. Synthesis of Clinoptilolite-Supported BiOCl/TiO_2 Heterojunction Nanocomposites with Highly-Enhanced Photocatalytic Activity for the Complete Degradation of Xanthates under Visible Light. *Chem. Eng. J.* **2021**, *407*, 126697. [CrossRef]
118. Amiri, S.; Anbia, M. Insights into the Effect of Parameters and Pathway of Visible-Light Photodegradation of Glyphosate and Diazinon by C-TiO_2/Clinoptilolite Nanocomposite. *J. Photochem. Photobiol. A Chem.* **2024**, *446*, 115146. [CrossRef]

119. Fang, X.-H.; Fang, F.; Lu, C.-H.; Zheng, L. Removal of Cs^+, Sr^{2+}, and Co^{2+} Ions from the Mixture of Organics and Suspended Solids Aqueous Solutions by Zeolites. *Nucl. Eng. Technol.* **2017**, *49*, 556–561. [CrossRef]
120. Kusrini, E.; Mualim, N.M.; Rahman, A.; Usman, A. Application of Activated Na-Zeolite a as a Water Softening Agent to Remove Ca^{2+} and Mg^{2+} Ions from Water. *AIP Conf. Proc.* **2020**, *2255*, 060012.
121. Lim, D.-H.; Ali Maitlo, H.; Younis, S.A.; Kim, K.-H. The Practical Utility of Nitrogen Doped TiO_2 as a Photocatalyst for the Oxidative Removal of Gaseous Formaldehyde. *Mater. Today Nano* **2024**, *27*, 100499. [CrossRef]
122. Foo, C.; Li, Y.; Lebedev, K.; Chen, T.; Day, S.; Tang, C.; Tsang, S.C.E. Characterisation of Oxygen Defects and Nitrogen Impurities in TiO_2 Photocatalysts Using Variable-Temperature X-Ray Powder Diffraction. *Nat. Commun.* **2021**, *12*, 661. [CrossRef]
123. Cano-Casanova, L.; Ansón-Casaos, A.; Hernández-Ferrer, J.; Benito, A.M.; Maser, W.K.; Garro, N.; Lillo-Ródenas, M.A.; Román-Martínez, M.C. Surface-Enriched Boron-Doped TiO_2 Nanoparticles as Photocatalysts for Propene Oxidation. *ACS Appl. Nano Mater.* **2022**, *5*, 12527–12539. [CrossRef]
124. Niu, P.; Wu, G.; Chen, P.; Zheng, H.; Cao, Q.; Jiang, H. Optimization of Boron Doped TiO_2 as an Efficient Visible Light-Driven Photocatalyst for Organic Dye Degradation With High Reusability. *Front. Chem.* **2020**, *8*, 172. [CrossRef]
125. Liu, B.; Chen, H.M.; Liu, C.; Andrews, S.C.; Hahn, C.; Yang, P. Large-Scale Synthesis of Transition-Metal-Doped TiO_2 Nanowires with Controllable Overpotential. *J. Am. Chem. Soc.* **2013**, *135*, 9995–9998. [CrossRef]
126. Weng, H.; Dong, J. First-principles investigation of transition-metal-doped group-IV semiconductors: R_xY_{1-x} (R = Cr, Mn, Fe; Y = Si, Ge). *Phys. Rev. B* **2005**, *71*, 035201. [CrossRef]
127. Khlyustova, A.; Sirotkin, N.; Kusova, T.; Kraev, A.; Titov, V.; Agafonov, A. Doped TiO_2: The Effect of Doping Elements on Photocatalytic Activity. *Mater. Adv.* **2020**, *1*, 1193–1201. [CrossRef]
128. Choi, Y.J.; Seeley, Z.; Bandyopadhyay, A.; Bose, S.; Akbar, S.A. Aluminum-Doped TiO_2 Nano-Powders for Gas Sensors. *Sens. Actuators B Chem.* **2007**, *124*, 111–117. [CrossRef]
129. Perdew, J.P.; Burke, K.; Ernzerhof, M. Generalized Gradient Approximation Made Simple. *Phys. Rev. Lett.* **1996**, *77*, 3865–3868. [CrossRef] [PubMed]
130. Perdew, J.P.; Burke, K.; Ernzerhof, M. Generalized Gradient Approximation Made Simple. *Phys. Rev. Lett.* **1997**, *78*, 1396. [CrossRef]
131. Heyd, J.; Scuseria, G.E.; Ernzerhof, M. Hybrid Functionals Based on a Screened Coulomb Potential. *J. Chem. Phys.* **2003**, *118*, 8207–8215. [CrossRef]
132. Borlido, P.; Schmidt, J.; Huran, A.W.; Tran, F.; Marques, M.A.L.; Botti, S. Exchange-Correlation Functionals for Band Gaps of Solids: Benchmark, Reparametrization and Machine Learning. *Npj Comput. Mater.* **2020**, *6*, 96. [CrossRef]
133. Borlido, P.; Aull, T.; Huran, A.W.; Tran, F.; Marques, M.A.L.; Botti, S. Large-Scale Benchmark of Exchange–Correlation Functionals for the Determination of Electronic Band Gaps of Solids. *J. Chem. Theory Comput.* **2019**, *15*, 5069–5079. [CrossRef]
134. Dudarev, S.L.; Botton, G.A.; Savrasov, S.Y.; Humphreys, C.J.; Sutton, A.P. Electron-Energy-Loss Spectra and the Structural Stability of Nickel Oxide: An LSDA+U Study. *Phys. Rev. B* **1998**, *57*, 1505–1509. [CrossRef]
135. Cococcioni, M.; de Gironcoli, S. Linear Response Approach to the Calculation of the Effective Interaction Parameters in the $\mathrm{LDA}+\mathrm{U}$ Method. *Phys. Rev. B* **2005**, *71*, 035105. [CrossRef]
136. Ferreira, L.G.; Marques, M.; Teles, L.K. Approximation to Density Functional Theory for the Calculation of Band Gaps of Semiconductors. *Phys. Rev. B* **2008**, *78*, 125116. [CrossRef]
137. Ferreira, L.G.; Marques, M.; Teles, L.K. Slater Half-Occupation Technique Revisited: The LDA-1/2 and GGA-1/2 Approaches for Atomic Ionization Energies and Band Gaps in Semiconductors. *AIP Adv.* **2011**, *1*, 032119. [CrossRef]
138. Scandolo, S.; Giannozzi, P.; Cavazzoni, C.; de Gironcoli, S.; Pasquarello, A.; Baroni, S. First-Principles Codes for Computational Crystallography in the Quantum-ESPRESSO Package. *Z. Für Krist. Cryst. Mater.* **2005**, *220*, 574–579. [CrossRef]
139. Giannozzi, P.; Baseggio, O.; Bonfà, P.; Brunato, D.; Car, R.; Carnimeo, I.; Cavazzoni, C.; de Gironcoli, S.; Delugas, P.; Ferrari Ruffino, F.; et al. Quantum ESPRESSO toward the Exascale. *J. Chem. Phys.* **2020**, *152*, 154105. [CrossRef] [PubMed]
140. Giannozzi, P.; Andreussi, O.; Brumme, T.; Bunau, O.; Buongiorno Nardelli, M.; Calandra, M.; Car, R.; Cavazzoni, C.; Ceresoli, D.; Cococcioni, M.; et al. Advanced Capabilities for Materials Modelling with Quantum ESPRESSO. *J. Phys. Condens. Matter* **2017**, *29*, 465901. [CrossRef] [PubMed]
141. Giannozzi, P.; Baroni, S.; Bonini, N.; Calandra, M.; Car, R.; Cavazzoni, C.; Ceresoli, D.; Chiarotti, G.L.; Cococcioni, M.; Dabo, I.; et al. QUANTUM ESPRESSO: A Modular and Open-Source Software Project for Quantum Simulations of Materials. *J. Phys. Condens. Matter* **2009**, *21*, 395502. [CrossRef]
142. Turney, J.M.; Simmonett, A.C.; Parrish, R.M.; Hohenstein, E.G.; Evangelista, F.A.; Fermann, J.T.; Mintz, B.J.; Burns, L.A.; Wilke, J.J.; Abrams, M.L.; et al. Psi4: An Open-Source Ab Initio Electronic Structure Program. *WIREs Comput. Mol. Sci.* **2012**, *2*, 556–565. [CrossRef]
143. Parrish, R.M.; Burns, L.A.; Smith, D.G.A.; Simmonett, A.C.; DePrince, A.E.I.; Hohenstein, E.G.; Bozkaya, U.; Sokolov, A.Y.; Di Remigio, R.; Richard, R.M.; et al. Psi4 1.1: An Open-Source Electronic Structure Program Emphasizing Automation, Advanced Libraries, and Interoperability. *J. Chem. Theory Comput.* **2017**, *13*, 3185–3197. [CrossRef]

144. Smith, D.G.A.; Burns, L.A.; Simmonett, A.C.; Parrish, R.M.; Schieber, M.C.; Galvelis, R.; Kraus, P.; Kruse, H.; Di Remigio, R.; Alenaizan, A.; et al. PSI4 1.4: Open-Source Software for High-Throughput Quantum Chemistry. *J. Chem. Phys.* **2020**, *152*, 184108. [CrossRef]
145. Lu, T. A Comprehensive Electron Wavefunction Analysis Toolbox for Chemists, Multiwfn. *J. Chem. Phys.* **2024**, *161*, 082503. [CrossRef]
146. Lu, T.; Chen, F. Multiwfn: A Multifunctional Wavefunction Analyzer. *J. Comput. Chem.* **2012**, *33*, 580–592. [CrossRef]
147. Bannwarth, C.; Caldeweyher, E.; Ehlert, S.; Hansen, A.; Pracht, P.; Seibert, J.; Spicher, S.; Grimme, S. Extended Tight-Binding Quantum Chemistry Methods. *WIREs Comput. Mol. Sci.* **2021**, *11*, e1493. [CrossRef]
148. Ehlert, S.; Stahn, M.; Spicher, S.; Grimme, S. Robust and Efficient Implicit Solvation Model for Fast Semiempirical Methods. *J. Chem. Theory Comput.* **2021**, *17*, 4250–4261. [CrossRef]
149. Grimme, S.; Bannwarth, C.; Shushkov, P. A Robust and Accurate Tight-Binding Quantum Chemical Method for Structures, Vibrational Frequencies, and Noncovalent Interactions of Large Molecular Systems Parametrized for All Spd-Block Elements (Z = 1–86). *J. Chem. Theory Comput.* **2017**, *13*, 1989–2009. [CrossRef] [PubMed]
150. Bannwarth, C.; Ehlert, S.; Grimme, S. GFN2-xTB—An Accurate and Broadly Parametrized Self-Consistent Tight-Binding Quantum Chemical Method with Multipole Electrostatics and Density-Dependent Dispersion Contributions. *J. Chem. Theory Comput.* **2019**, *15*, 1652–1671. [CrossRef]
151. Spicher, S.; Grimme, S. Robust Atomistic Modeling of Materials, Organometallic, and Biochemical Systems. *Angew. Chem. Int. Ed.* **2020**, *59*, 15665–15673. [CrossRef] [PubMed]
152. Plimpton, S. Fast Parallel Algorithms for Short-Range Molecular Dynamics. *J. Comput. Phys.* **1995**, *117*, 1–19. [CrossRef]
153. Thompson, A.P.; Aktulga, H.M.; Berger, R.; Bolintineanu, D.S.; Brown, W.M.; Crozier, P.S.; in 't Veld, P.J.; Kohlmeyer, A.; Moore, S.G.; Nguyen, T.D.; et al. LAMMPS—A Flexible Simulation Tool for Particle-Based Materials Modeling at the Atomic, Meso, and Continuum Scales. *Comput. Phys. Commun.* **2022**, *271*, 108171. [CrossRef]
154. Páll, S.; Abraham, M.J.; Kutzner, C.; Hess, B.; Lindahl, E. Tackling Exascale Software Challenges in Molecular Dynamics Simulations with GROMACS. In *Solving Software Challenges for Exascale*; Markidis, S., Laure, E., Eds.; Springer International Publishing: Cham, Switzerland, 2015; pp. 3–27.
155. Bekker, H.; Berendsen, H.; Dijkstra, E.; Achterop, S.; Vondrumen, R.; Vanderspoel, D.; Sijbers, A.; Keegstra, H.; Renardus, M. Gromacs—A Parallel Computer For Molecular-Dynamics Simulations. In Proceedings of the 4th International Conference on Computational Physics, Praha, Czechoslovakia, 24–28 August 1992; pp. 252–256.
156. Abraham, M.J.; Murtola, T.; Schulz, R.; Páll, S.; Smith, J.C.; Hess, B.; Lindahl, E. GROMACS: High Performance Molecular Simulations through Multi-Level Parallelism from Laptops to Supercomputers. *SoftwareX* **2015**, *1–2*, 19–25. [CrossRef]
157. Pronk, S.; Páll, S.; Schulz, R.; Larsson, P.; Bjelkmar, P.; Apostolov, R.; Shirts, M.R.; Smith, J.C.; Kasson, P.M.; van der Spoel, D.; et al. GROMACS 4.5: A High-Throughput and Highly Parallel Open Source Molecular Simulation Toolkit. *Bioinformatics* **2013**, *29*, 845–854. [CrossRef]
158. Hess, B.; Kutzner, C.; van der Spoel, D.; Lindahl, E. GROMACS 4: Algorithms for Highly Efficient, Load-Balanced, and Scalable Molecular Simulation. *J. Chem. Theory Comput.* **2008**, *4*, 435–447. [CrossRef]
159. Van Der Spoel, D.; Lindahl, E.; Hess, B.; Groenhof, G.; Mark, A.E.; Berendsen, H.J.C. GROMACS: Fast, Flexible, and Free. *J. Comput. Chem.* **2005**, *26*, 1701–1718. [CrossRef]
160. Lindahl, E.; Hess, B.; Van Der Spoel, D. GROMACS 3.0: A Package for Molecular Simulation and Trajectory Analysis. *J. Mol. Model.* **2001**, *7*, 306–317. [CrossRef]
161. Berendsen, H.J.C.; van der Spoel, D.; van Drunen, R. GROMACS: A Message-Passing Parallel Molecular Dynamics Implementation. *Comput. Phys. Commun.* **1995**, *91*, 43–56. [CrossRef]
162. Liakos, D.G.; Guo, Y.; Neese, F. Comprehensive Benchmark Results for the Domain Based Local Pair Natural Orbital Coupled Cluster Method (DLPNO-CCSD(T)) for Closed- and Open-Shell Systems. *J. Phys. Chem. A* **2020**, *124*, 90–100. [CrossRef] [PubMed]
163. Guo, Y.; Riplinger, C.; Liakos, D.G.; Becker, U.; Saitow, M.; Neese, F. Linear Scaling Perturbative Triples Correction Approximations for Open-Shell Domain-Based Local Pair Natural Orbital Coupled Cluster Singles and Doubles Theory [DLPNO-CCSD(T0/T)]. *J. Chem. Phys.* **2020**, *152*, 024116. [CrossRef]
164. Neese, F. The SHARK Integral Generation and Digestion System. *J. Comput. Chem.* **2022**, *44*, 381–396. [CrossRef] [PubMed]
165. Teale, A.M.; Helgaker, T.; Savin, A.; Adamo, C.; Aradi, B.; Arbuznikov, A.V.; Ayers, P.W.; Baerends, E.J.; Barone, V.; Calaminici, P.; et al. DFT Exchange: Sharing Perspectives on the Workhorse of Quantum Chemistry and Materials Science. *Phys. Chem. Chem. Phys.* **2022**, *24*, 28700–28781. [CrossRef]
166. Neese, F.; Wennmohs, F.; Hansen, A.; Becker, U. Efficient, Approximate and Parallel Hartree–Fock and Hybrid DFT Calculations. A 'Chain-of-Spheres' Algorithm for the Hartree–Fock Exchange. *Chem. Phys.* **2009**, *356*, 98–109. [CrossRef]
167. Neese, F. Software Update: The ORCA Program System—Version 5.0. *WIREs Comput. Mol. Sci.* **2022**, *12*, e1606. [CrossRef]
168. Neese, F. Software Update: The ORCA Program System, Version 4.0. *WIREs Comput. Mol. Sci.* **2018**, *8*, e1327. [CrossRef]
169. Neese, F. The ORCA Program System. *WIREs Comput. Mol. Sci.* **2012**, *2*, 73–78. [CrossRef]

170. Neese, F.; Wennmohs, F.; Becker, U.; Riplinger, C. The ORCA Quantum Chemistry Program Package. *J. Chem. Phys.* **2020**, *152*, 224108. [CrossRef]
171. Armaković, S.; Armaković, S.J. Atomistica.Online—Web Application for Generating Input Files for ORCA Molecular Modelling Package Made with the Anvil Platform. *Mol. Simul.* **2023**, *49*, 117–123. [CrossRef]
172. Armaković, S.; Armaković, S.J. Online and Desktop Graphical User Interfaces for Xtb Programme from Atomistica.Online Platform. *Mol. Simul.* **2024**, *50*, 560–570. [CrossRef]
173. Zicovich-Wilson, C.M.; Dovesi, R. Titanium-Containing Zeolites. A Periodic Ab Initio Hartree−Fock Characterization. *J. Phys. Chem. B* **1998**, *102*, 1411–1417. [CrossRef]
174. Dovesi, R.; Erba, A.; Orlando, R.; Zicovich-Wilson, C.M.; Civalleri, B.; Maschio, L.; Rérat, M.; Casassa, S.; Baima, J.; Salustro, S.; et al. Quantum-Mechanical Condensed Matter Simulations with CRYSTAL. *WIREs Comput. Mol. Sci.* **2018**, *8*, e1360. [CrossRef]
175. Erba, A.; Desmarais, J.K.; Casassa, S.; Civalleri, B.; Donà, L.; Bush, I.J.; Searle, B.; Maschio, L.; Edith-Daga, L.; Cossard, A.; et al. CRYSTAL23: A Program for Computational Solid State Physics and Chemistry. *J. Chem. Theory Comput.* **2023**, *19*, 6891–6932. [CrossRef]
176. Zhang, M.; Yang, K.; Wang, N.; Yu, Y. Combined DFT and Experimental Studies of Properties of TiO_2 Modified USY Zeolite. *J. Porous Mater.* **2015**, *22*, 595–602. [CrossRef]
177. Miehlich, B.; Savin, A.; Stoll, H.; Preuss, H. Results Obtained with the Correlation Energy Density Functionals of Becke and Lee, Yang and Parr. *Chem. Phys. Lett.* **1989**, *157*, 200–206. [CrossRef]
178. Lee, C.; Yang, W.; Parr, R.G. Development of the Colle-Salvetti Correlation-Energy Formula into a Functional of the Electron Density. *Phys. Rev. B* **1988**, *37*, 785–789. [CrossRef]
179. Vosko, S.H.; Wilk, L.; Nusair, M. Accurate Spin-Dependent Electron Liquid Correlation Energies for Local Spin Density Calculations: A Critical Analysis. *Can. J. Phys.* **1980**, *58*, 1200–1211. [CrossRef]
180. Becke, A.D. Density-Functional Exchange-Energy Approximation with Correct Asymptotic Behavior. *Phys. Rev. A* **1988**, *38*, 3098–3100. [CrossRef]
181. Slater, J.C.; Phillips, J.C. Quantum Theory of Molecules and Solids Vol. 4: The Self-Consistent Field for Molecules and Solids. *Phys. Today* **1974**, *27*, 49–50. [CrossRef]
182. Li, Y.; Wei, W.; Chen, J.; Zu, L.; Zhou, Y.; Wu, J.; Wu, X. Atomic Layer Deposition Precisely Modified Zeolite 13X: Physicochemical Synergistic Adsorption of Space Molecular Contaminants. *Appl. Surf. Sci.* **2022**, *590*, 153084. [CrossRef]
183. Salahshoori, I.; Mahdavi, S.; Moradi, Z.; Otadi, M.; Zare Kazemabadi, F.; Nobre, M.A.L.; Ali Khonakdar, H.; Baghban, A.; Wang, Q.; Mohammadi, A.H. Advancements in Molecular Simulation for Understanding Pharmaceutical Pollutant Adsorption: A State-of-the-Art Review. *J. Mol. Liq.* **2024**, *410*, 125513. [CrossRef]
184. Fischer, M. Adsorption of Carbamazepine in All-Silica Zeolites Studied with Density Functional Theory Calculations. *ChemPhysChem* **2023**, *24*, e202300022. [CrossRef]

Disclaimer/Publisher's Note: The statements, opinions and data contained in all publications are solely those of the individual author(s) and contributor(s) and not of MDPI and/or the editor(s). MDPI and/or the editor(s) disclaim responsibility for any injury to people or property resulting from any ideas, methods, instructions or products referred to in the content.

catalysts

MDPI

Article

Different Routes for the Hierarchization of *BEA Zeolite, Followed by Impregnation with Niobium and Application in Ethanol and 1-Propanol Dehydration

Deborah da Silva Valadares, Willian Henrique Ribeiro de Carvalho, Ana Lívia Fernandes Fonseca, Guilherme de França Machado, Matheus Ramos Silva, Pablo Teles Aragão Campos, José Alves Dias * and Sílvia Cláudia Loureiro Dias *

Laboratory of Catalysis, Chemistry Institute (IQ-UnB), University of Brasília, Campus Universitário Darcy Ribeiro–Asa Norte, Brasília 70910-900, DF, Brazil; deborahsvaladares@gmail.com (D.d.S.V.); williancarvalho3372@gmail.com (W.H.R.d.C.); analiviaffonseca@gmail.com (A.L.F.F.); guilhermedefrancamachado@gmail.com (G.d.F.M.); matheus.ramos.unb@gmail.com (M.R.S.); pablo.teles1998@gmail.com (P.T.A.C.)

* Correspondence: jdias@unb.br or josediasunb@gmail.com (J.A.D.); scdias@unb.br or silviadiasunb@gmail.com (S.C.L.D.); Tel.: +55-61-3107-3846 (J.A.D.)

Abstract: This study examined the hierarchical structuring of *BEA zeolite using two distinct approaches: double aluminum removal with solid ammonium hexafluorosilicate (2x-AHFS) and a solution of 0.2 M sodium hydroxide followed by 0.5 M hydrochloric acid (T-NaOH). Additionally, niobium pentoxide (Nb_2O_5) was impregnated at different loadings (5, 10, 15, and 20 wt.%) onto the hierarchized materials. Both treatments increased the SiO_2/Al_2O_3 ratio and produced crystals with domains of about the same size. The hierarchization methods generated secondary mesopores and reduced the micropores in the treated HB zeolite. The solid-state NMR analysis by ^{27}Al and ^{29}Si indicated that the 2x-AHFS treatment increased the hydrophobic character of the zeolite, while the treatment with NaOH/HCl resulted in a less hydrophobic material. A balanced quantity of Brønsted and Lewis sites was observed for all treated zeolites. Thus, these combined physicochemical characteristics of the new catalysts may explain their superior performance in the dehydration reactions. In the case of ethanol dehydration at 230 °C, the 20 wt.% Nb_2O_5 supported on the T-NaOH catalyst produced an 84% conversion and 86% selectivity for ethylene (EE), with 14% diethyl ether (DEE) as the only products. Conversely, in the 1-propanol dehydration reaction, the 20 wt.% Nb_2O_5 supported on 2x-AHFS achieved 99% conversion, producing 99% propene.

Keywords: BEA zeolite; hierarchization; niobium pentoxide; ethanol dehydration; 1-propanol dehydration

1. Introduction

Catalysis involved in the conversion of renewable resources to green chemicals has attracted significant attention over the past decade [1]. Among heterogeneous catalysts, zeolites account for more than 40% of solid catalysts used in the chemical industry due to their superior properties, such as modulated acidity, high surface area, thermal and chemical stability, and the ability to confine active metal species within their pores [2]. Monte Carlo simulation studies have shown that the structure of zeolites is inherently altered during adsorption processes, and the conversion (yield) largely depends on both the crystalline structure (e.g., 10- and 12-ring three-dimensional channel systems) and framework compositions (e.g., Si/Al ratios and the associated acidity and hydrophobicity/hydrophilicity) [3].

Zeolites are crystalline materials with pores and cavities that exhibit excellent thermal and chemical stability, Brønsted/Lewis acidity, and high catalytic activity [4]. Among the five zeolites with the greatest applicability (FER, MOR, MFI, FAU, and *BEA), *BEA zeolite stands out due to its wide micropores, intersecting channels, and 12-membered rings [5]. Despite its large pores, some reagents have difficulty accessing the active catalytic sites, and chemical reactions are limited by diffusion issues [6]. One way to overcome this problem is through the hierarchization of the zeolite, where the material acquires different types of pores. This procedure can be performed either before or after zeolite synthesis [7–9]. Several studies in the literature show that catalytic reactions exhibit higher conversion and selectivity for the products of interest after the hierarchization of the zeolite, including *BEA [10–12].

Hierarchical zeolites have interconnected pores of different sizes (bimodal micro-mesoporous structure), which reduce the steric limitations of bulky reagents and their access to catalytic sites, increase the intracrystalline diffusion rate, minimize catalyst deactivation caused by coke formation, maximize catalyst utilization, and modulate selectivity towards target products [13]. Among post-synthesis treatments of zeolites, solid-state dealumination has shown promise, as Al is removed from the structure with the reinsertion of Si into the vacancy left by Al. As a result, the Si/Al ratio increases, thereby enhancing the resistance of the zeolitic structure to extreme thermal and hydrothermal treatments [14]. On the other hand, basic leaching is a technique that removes Si from the framework (desilication) through aqueous alkaline treatment (NaOH). In addition to being promising, this technique is particularly suitable for zeolites with a high Si/Al ratio. Through the hydrolysis of Si-O-Si and Si-O-Al bonds, the created mesopores offer higher accessibility to the zeolite's outer surface, as they are primarily generated on the outer surface or in defective areas within the structure [15,16]. According to Groen et al. [17], the framework Si/Al ratio plays a crucial role, as lattice aluminum species control the silicon extraction from the zeolite framework due to the suppressed extraction of neighboring Si species.

One of the applications of hierarchical zeolites is the conversion of biomass-derived feedstock, such as bioethanol, into renewable ethylene via a dehydration reaction operated at lower temperatures (200–500 °C). Strong solid acid sites facilitate ethylene formation, while weak acid sites promote the formation of diethyl ether [17,18]. Additionally, the use of renewable sources like propanol is important for industrial processes, such as propene production via dehydration, since propene is a building block to produce a wide range of chemicals [19]. In a previous study by Valadares et al. [14], it was demonstrated that the catalytic properties of *BEA zeolite treated with ammonium hexafluorosilicate (1×-70 mol%) were effectively tuned by incorporating niobium species, which favored the production of diethyl ether (DEE) from ethanol dehydration and furfural from xylose dehydration. The addition of niobium enhanced the selectivity for these products, providing the possibility to control them through different hierarchical processes.

Considering the above, the objective of the present study is to produce hierarchical *BEA zeolite using two different methodologies: a double dealumination with ammonium hexafluorosilicate (2×-AHFS) in the solid state and treatment in solution with sodium hydroxide, followed by treatment with hydrochloric acid (NaOH/HCl). To evaluate the modifications of the materials, they were characterized structurally and texturally using several techniques: XRD, EDXRF, FT-IR, ^{27}Al and ^{29}Si MAS NMR, SEM, nitrogen sorption at low temperature, and gaseous adsorption of pyridine followed by FT-IR analysis. All modified catalysts were tested in ethanol and 1-propanol dehydration reactions to assess the effect of desilication/dealumination on the performance of the *BEA materials.

2. Results and Discussion

2.1. Powder XRD and XRF Characterizations

The different *BEA polymorphs present similar peaks in the powder X-ray diffractograms but with different positions, which are characteristic, particularly in the low-angle region, widely accepted as the *BEA fingerprint region (2θ = 5–10°). Both sharp and broad diffraction peaks suggest structural disorder (stacking faults) of the framework due to an intergrowth of distinct but closely related polymorphs (A and B). In all samples, the main diffraction peaks were identified at 2θ = 7.80° (broad) and 22.50° (sharp), which correspond to the reflections (101) and (116) of the *BEA zeolite, respectively (Figure 1) [20,21]. The provided *BEA zeolite showed characteristics of both polymorphs in nearly equal proportion (44% A-chiral and 56% B-achiral) [22]. The modified materials showed no significant deviation in angles compared to the starting protonic zeolite (HB), indicating the preservation of the crystalline structure and no phase change. Even a double amount of ammonium hexafluorsilicate (2×-AHFS) did not compromise the zeolite network. Relative crystallinity calculations were performed using the total areas of the characteristic diffraction peaks, with the parent zeolite designated as 100% crystalline (Table 1). An increase of 2% in crystallinity was observed for HB treated with 2×-AHFS, and a decrease of 12% was noted when treated with NaOH/HCl (T-NaOH). For the HB treated with NaOH, it was observed that when an extra metal source (e.g., niobium) is added above 10 wt.%, this metal will act as a healing atom, initially occupying the silanol nests to prevent further collapse of the zeolite [9,23]. The peak at 2θ = 27.1° (crystal plane 008), characteristic of the *BEA structure, decreased after NaOH treatment but not after 2x-AHFS treatment [24]. The diffractograms for the samples after the impregnation of Nb_2O_5 (Figure 1) for both treatments indicated a relative decrease in peak intensities for both peaks at 2θ = 22.5° (plane 116) and 7.8° (plane 101), since the supported Nb_2O_5 was amorphous when treated at 550 °C.

Table 1. Physicochemical properties of the catalysts.

Catalyst	$\frac{SiO_2}{Al_2O_3}$	%C [a]	2θ (°) [b]	β (rad) [b]	D (nm) [c]	L (nm) [d]	Strain [e]
HB	25	100	22.52	0.0105	13.1	4.6	0.0007
2x-AHFS	44	102	22.63	0.0115	11.9	5.2	0.0009
5Nb-AHFS	35	100	22.60	0.0119	11.6	6.5	0.0009
10Nb-AHFS	33	100	22.50	0.0119	11.6	6.7	0.0009
15Nb-AHFS	35	95	22.50	0.0108	12.7	6.6	0.0008
20Nb-AHFS	36	96	22.50	0.0112	12.3	6.6	0.0008
T-NaOH	39	88	22.57	0.0136	10.1	5.0	0.0012
5Nb-NaOH	42	87	22.77	0.0199	6.9	7.7	0.0023
10Nb-NaOH	41	89	22.69	0.0182	7.6	8.6	0.0020
15Nb-NaOH	41	100	22.67	0.0140	9.8	6.2	0.0013
20Nb-NaOH	38	106	22.72	0.0157	8.7	5.8	0.0015

[a] Crystallinity (%) calculated by Equation (1). [b] 2θ angle and the respective full width at half maximum (FWHM). [c] Average size of crystal domain, calculated by Scherrer's equation: $D = 0.9\ \lambda/\beta\ \cos\theta$. [d] Average particle diameter (L), calculated by Equation (2). [e] Microstrain (ε) calculated by $\varepsilon = \beta\ \sin(\theta)/4$.

The average crystallite domain calculated by XRD was 12 nm for the 2×-AHFS and 9 nm for T-NaOH, whereas the average particle diameter obtained using BET was 7 nm and 6 nm for 2×-AHFS and T-NaOH, respectively (Table 1). In general, the values did not significantly differ in magnitude for both methods and may reflect differences in calculations [25]. However, it is interesting to note that the dealumination using NaOH resulted in smaller particles, regardless of the calculation method. In addition, The T-NaOH caused higher microstrains, from 0.0022 to 0.0029, while the AHFS treatment produced microstrains from 0.0017 to 0.0019. The treatment with NaOH probably resulted in greater intergrowth and stacking faults, potentially impacting the physicochemical properties, particularly diffusivity, due to modifications in pore connectivity.

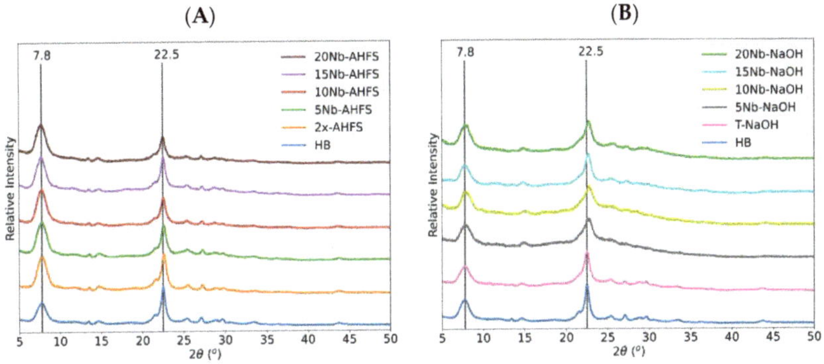

Figure 1. XRD patterns of HB, 2x-AHFS, and T-NaOH and the respective Nb_2O_5 supported on (**A**) 2x-AHFS and (**B**) T-NaOH treated zeolites.

All materials showed an increase in the SiO_2/Al_2O_3 ratio, as both treatments can remove framework Al (Table 1). A larger SiO_2/Al_2O_3 ratio corresponds to a smaller unit cell size and, thus, larger XRD angles, from 2θ = 22.52° for HB (SiO_2/Al_2O_3 = 25) to 22.57° for T-NaOH (SiO_2/Al_2O_3 = 39) and 22.63° for 2×-AHFS (SiO_2/Al_2O_3 = 44). Acid leaching can preferentially remove Al from the network and create mesopores within the zeolitic crystals. Defective sites are formed through the hydrolysis of Si-O-Al bonds, forming Si-O-defects and giving rise to extra-lattice Al species (EFAl) [24].

The Nb_2O_5 loading was obtained based on Nb analysis determined by energy dispersive X-ray fluorescence (EDXRF), which used the QualiQuant method of fundamental standards (Version 1.2, 2006, Shimadzu, Kyoto, Japan). The theoretical values were close to the real ones (deviations of 2–3%). Therefore, the nominal values were used here for simplicity.

2.2. FT-IR Spectroscopy

In relation to the FT-IR spectra, all catalysts showed characteristic bands at approximately 1220 cm^{-1}, which corresponds to the external symmetric vibration of the SiO_4 tetrahedron, and at approximately 1089 cm^{-1} (Figure 2), which corresponds to the asymmetric vibration between the tetrahedral atom (T, Si or Al) and the oxygen atoms (O-T-O bonds) [26]. After the modifications, these bands shifted to higher wavenumbers, which is an expected phenomenon in hierarchization due to the increase in the SiO_2/Al_2O_3 ratio [27]. The band around 946 cm^{-1} was also identified, corresponding to the vibration of the Si-O- bond [14,27]. This band was more prominent in materials treated with NaOH than in those treated with 2x-AHFS. Additionally, the band at 946 cm^{-1} is sensitive to the incorporation of niobium into the framework. When Nb is incorporated into the zeolite structure, the wavenumber shifts to higher values (in the range of 960–970 cm^{-1}), according to the literature [14,28]. This shift was not observed, reinforcing the presence of only niobium pentoxide species. Finally, the absorption bands observed around 628, 572, and 526 cm^{-1} correspond to the vibrations of the 4-, 5-, and 6-membered rings of the *BEA zeolite, respecitvely, indicating that the modifications did not significantly compromise the zeolite framework [14,28,29]. Additionally, a large band from 3550 to 3750 cm^{-1} was observed for the samples prepared under an inert atmosphere with Nujol (Figure S1). A band at 3610 cm^{-1} is assigned to bridging hydroxyl groups [Si-O(H)-Al], and another at approximately 3650 cm^{-1} is related to the presence of extra-framework Al-OH, as reported in the literature [9,30].

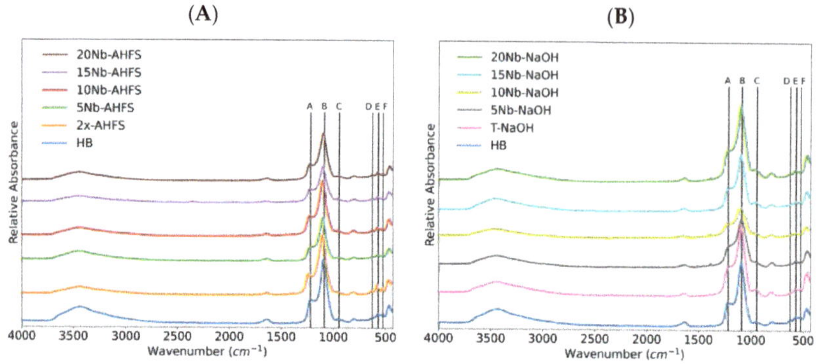

Figure 2. FT-IR spectra of HB, 2x-AHFS, and T-NaOH and the respective Nb_2O_5 supported on (**A**) 2x-AHFS and (**B**) T-NaOH treated zeolites. A, B, C, D, E, and F corresponds to 1220, 1089, 946, 628, 572, and 526 cm^{-1}, respectively, to the vibrations described in the text.

2.3. *Textural Properties Using N_2 Adsorption/Desorption Isotherms at −196 °C*

Figure S2 shows the N_2 physisorption isotherms (adsorption/desorption). All samples showed a combination of type I(a) and IV(a) isotherms with a hysteresis loop at higher relative pressures ($p/p_0 > 0.7$) that indicated the presence of a secondary mesoporous structure. The specific surface area and pore volume data are presented in Table 2. The specific surface area (S_{BET}), mesopore area (S_{Meso}), total pore volume (V_p), and mesopore volume (V_{Meso}) increased after both treatments in the following order: T-NaOH > 2x-AHFS > HB. This indicates an effective process of hierarchization of the zeolite BEA using both treatments. However, the addition of niobium forming Nb_2O_5 did not produce a crescent increase in all those values. For the Nb-NaOH catalysts, low loading of Nb_2O_5 (5 to 10 wt.%) initially decreased the BET area and caused a decrease in V_p and hence in the micropore and mesopore volumes. This can be attributed to the inclusion of small niobium clusters inside the zeolite pores, which precluded the complete filling of the original pores of T-NaOH zeolite. It is known that low niobium loading forms preferentially smaller clusters or isolated species on the surface of different supports [31,32]. Conversely, when the loading was increased to 15 to 20 wt.%, there was a subsequent increase in both the BET surface area and the total surface area of the catalysts. The development of larger clusters of Nb_2O_5 on the external surface creates an overlayer that effectively contributes to the total area of the material. This has been observed for Nb_2O_5 supported on silica-alumina catalysts [33]. The trend on the textural data observed for loading Nb on 2x-AHFS was generally in parallel with that observed on T-NaOH, which is consistent with the obtained isotherms. Some of those cited trends were plotted as a function of the treatment for easier comparison and are presented in Figure S3.

Thus, it can be inferred that the impregnation of HB zeolite after these hierarchization processes primarily generates extra-framework niobium oxide species on its surface, with the overlayer thickness depending on the niobium loading, as pointed out in the literature [34]. Moreover, the textural data showed that the diffusional constraints were alleviated by coupling the intrinsic microporosity with an auxiliary mesopore framework of inter- and intracrystalline structures. The model of silicon and aluminum dissolution is the main mechanism behind the chemical Si environments and mesopores formation [35–38].

Table 2. Textural parameters for the catalysts.

Catalyst	S_{BET} [a] (m^2/g)	S_{Ext} [b] (m^2/g)	S_{Micro} [c] (m^2/g)	S_{Meso} [d] (m^2/g)	V_{Micro} [e] (cm^3/g)	V_p [f] (cm^3/g)	V_{Meso} [g] (cm^3/g)
HB	649	190	459	188	0.19	0.85	0.66
2x-AHFS	577	193	384	211	0.16	0.87	0.71
5Nb-AHFS	462	170	292	193	0.12	0.67	0.55
10Nb-AHFS	448	156	292	179	0.12	0.63	0.51
15Nb-AHFS	484	163	321	199	0.13	0.67	0.54
20Nb-AHFS	512	183	330	211	0.13	0.77	0.64
T-NaOH	607	220	387	239	0.16	0.97	0.81
5Nb-NaOH	391	222	169	151	0.07	0.68	0.61
10Nb-NaOH	347	138	209	137	0.08	0.51	0.43
15Nb-NaOH	484	163	320	198	0.13	0.76	0.63
20Nb-NaOH	488	168	321	199	0.13	0.76	0.63

[a] Specific surface area obtained by BET method. The standard error (2σ) was ± 2 m^2 g^{-1}. [b] External surface obtained by t-plot method. [c] Microporous area obtained by t-plot method. [d] Mesoporous area obtained by BJH method. [e] Microporous volume obtained by t-plot method. [f] Total pore volume obtained by the amount of gas adsorbed at $p/p_0 = 0.98$. [g] Mesoporous volume (Vp − VMicro).

2.4. SEM Images and Analysis

The SEM micrographs of the treated zeolites and their niobium pentoxide (Nb_2O_5) impregnation are shown (Figure 3, and with EDX analyses in Figure S4) and compared to those of the HB zeolite. The protonic *BEA zeolite exhibited a distinct morphology of nanosized aggregates, as previously described in the literature [9]. After treatment with 2x-AHFS, the formation of a nearly spherical ball morphology with a uniform size was observed. The treatment using NaOH/HCl resulted in a material with morphological characteristics similar to those of HB. These images are consistent with the XRD patterns, which showed no significant differences among the materials. The impregnation of Nb_2O_5 onto both treated zeolites resulted in a rougher particle surface, but with characteristics analogous to those of the parent HB zeolite.

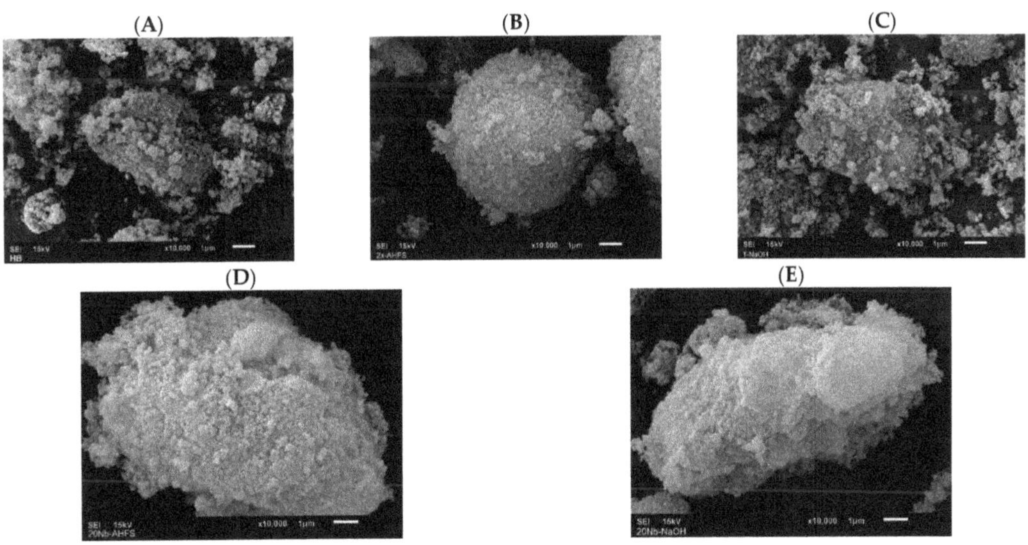

Figure 3. SEM images of (**A**) HB; (**B**) 2x-AHFS; (**C**) T-NaOH; (**D**) 20Nb-AHFS; and (**E**) 20Nb-NaOH.

2.5. ^{27}Al and ^{29}Si MAS NMR Spectroscopy

In the ^{27}Al MAS NMR spectra, all catalysts showed signals around −22 to 22 ppm for octahedral Al and 40 to 80 ppm for tetrahedral Al in the framework (Figure 4). The intense signal at 1.4 ppm present in the T-NaOH zeolite has been assigned to octahedral species that have fewer adsorbed water molecules, while the broadening of the signals in this region indicates a greater degree of hydration of the catalysts [39,40]. According to Groen et al. [17], the formation of mesopores in *BEA requires the dissolution of both Si and Al, with part of the removed Al being re-incorporated into the solid. This re-incorporation may be facilitated in the presence of niobium. For the treatment with 2x-AHFS, the ^{27}Al MAS NMR spectra showed a similar distribution of tetrahedral and octahedral Al as the HB zeolite. Niobium impregnation led to an increase in the relative area of tetrahedral Al for both treatments, in the range of 5 to 15% Nb_2O_5 (Table 3). For samples with 20% Nb_2O_5, the amount of tetrahedral Al reached a plateau around 80%. These behaviors suggest that niobium plays an important role in the reorganization of the zeolitic structure after hierarchical treatments.

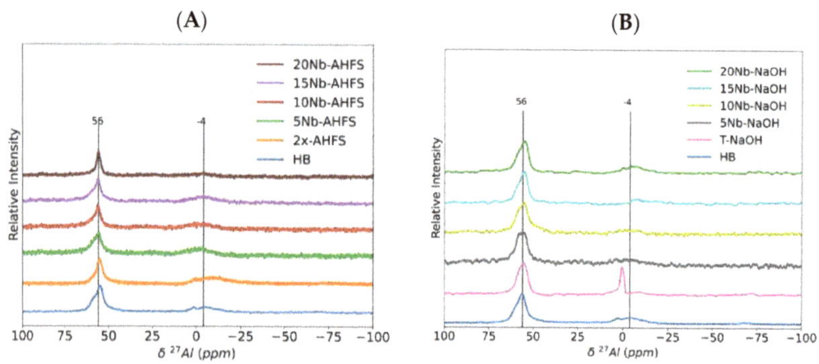

Figure 4. ^{27}Al MAS NMR spectra of HB, 2x-AHFS, and T-NaOH and the respective Nb_2O_5 supported on (**A**) 2x-AHFS and (**B**) T-NaOH treated zeolites.

Table 3. Relative distribution of tetrahedral (Al-T_d) and octahedral (Al-O_h) aluminum, according to the ^{27}Al MAS NMR spectra.

Catalyst	δ Al-T_d (ppm)	Area (%)	δ Al-O_h (ppm)	Area (%)
HB	56	62	−4.0	38
2x-AHFS	56	65	−4.0	35
5Nb-AHFS	57	70	−0.9	30
10Nb-AHFS	56	70	−6.8	30
15Nb-AHFS	57	69	−5.2	31
20Nb-AHFS	56	81	−4.3	19
T-NaOH	57	76	1.2	24
5Nb-NaOH	57	89	−1.4	11
10Nb-NaOH	57	91	−2.2	9
15Nb-NaOH	57	91	−7.2	9
20Nb-NaOH	57	79	−4.9	21

In the ^{29}Si MAS NMR spectra (Figure 5), the HB zeolite showed a signal at −102 ppm, which is related to the Q^3 environment (Si(1Al)). This signal almost disappears in the sample treated with 2x-AHFS. The signals at −111 and −115 ppm correspond to the Q^4 environments (Si(0Al)), with the latter (−115 ppm) not present in the HB sample but appearing markedly in the 2x-AHFS treated sample and in those impregnated with niobium. This suggests that Al removal from the zeolite promotes the reorganization of Si in the sample, leading to the formation of more Q^4-type environments [14,30,39]. The signal at −115 ppm can be attributed to the formation of silica (nest silanol) [14,30,35], which has a chemical shift in the same region and is more prominent in the treatment with 2x-AHFS. On the other hand, for the T-NaOH sample, the Q^3 environment remained practically unchanged relative to the HB sample. In addition, the Q^4 environment at −111 ppm remained constant, and the signal at −115 ppm did not change even with the addition of niobium. Therefore, in the treatment with 2x-AHFS, nest silanol is incorporated into the framework, which increases the Q^4 environment at −115 ppm, corroborating the increase in crystallinity observed by XRD and the higher SiO_2/Al_2O_3. Table 4 and Figure S5 provide a detailed presentation of the distribution environments derived from the deconvolution of the spectra.

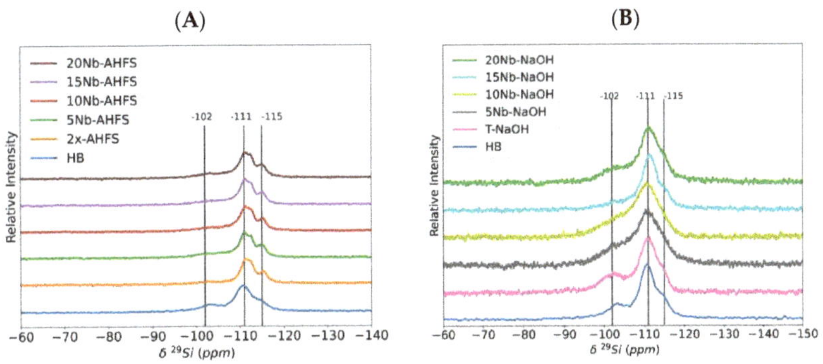

Figure 5. ^{29}Si MAS NMR spectra of HB, 2x-AHFS, and T-NaOH and the respective Nb_2O_5 supported on (**A**) 2x-AHFS and (**B**) T-NaOH treated zeolites.

Table 4. Relative distribution of Q^4 (Si(0Al)) and Q^3 (Si(1Al)) silicon environments based on deconvolution of ^{29}Si MAS NMR spectra.

Catalyst	Q^3 (ppm)	Area (%)	Q^4 (ppm)	Area (%)	Q^4 (ppm)	Area (%)
HB	−103	20	−111	67	−115	13
2x-AHFS	−102	8	−112	83	−115	9
5Nb-AHFS	−103	15	−112	74	−115	11
10Nb-AHFS	−103	20	−112	72	−115	8
15Nb-AHFS	−103	13	−112	78	−115	9
20Nb-AHFS	−103	14	−112	76	−114	10
T-NaOH	−102	26	−111	67	−114	7
5Nb-NaOH	−103	23	−112	66	−115	12
10Nb-NaOH	−104	29	−110	50	−113	21
15Nb-NaOH	−103	10	−112	84	−115	6
20Nb-NaOH	−103	18	−112	73	−115	10

2.6. Acidity of the Catalysts

Figure 6 presents the FT-IR spectra of pyridine adsorbed on the catalysts. The HB zeolite showed stronger intensity in the three bands between 1580 and 1420 cm^{-1}. These bands correspond to the interaction between pyridine and Brønsted sites (1540 cm^{-1}), the interaction with Lewis sites (1445 cm^{-1}), and a band at 1490 cm^{-1}, which is assigned to the interaction with both types of acid sites [41,42]. Qualitative analysis revealed that spectra from 2x-AHFS and T-NaOH treatments had the same bands, but with lower intensity compared to HB. After Nb_2O_5 impregnation, the 5% and 10% loadings showed a minor decrease in intensity, whereas the 15% and 20% loadings exhibited a relative increase. This may be related to the quantity of acid sites in these catalysts. The addition of niobium to the treated zeolites initially decreases the total amount of acid sites due to partial blocking access to the original acid sites of HB, which are in the micropores [14]. However, as the niobium loading increases, the formed Nb_2O_5 overlayer increases the total accessible sites (including the niobium contribution) on the new surface. The key point is that these acid sites facilitate the catalytic activity of these materials in dehydration reactions.

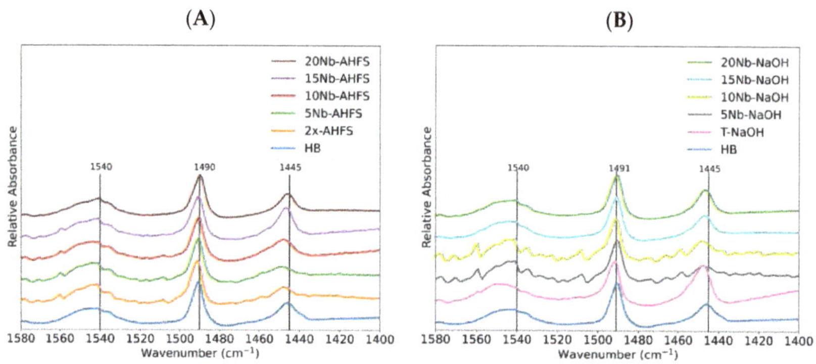

Figure 6. FT-IR spectra of pyridine adsorbed on HB, 2x-AHFS, and T-NaOH and the respective Nb_2O_5 supported on (**A**) 2x-AHFS and (**B**) T-NaOH treated zeolites.

Based on the FT-IR spectra, it was also possible to calculate the ratio between the areas corresponding to the Brønsted and Lewis sites (Table 5). As a result, the 2x-AHFS treatment showed a higher ratio value than T-NaOH. It is possible that the AHFS treatment removed only the extra-lattice Al and introduced Si atoms into the material. The T-NaOH treatment broke the Si-O bonds in the framework, followed by the removal of Al using an HCl solution, which led to a decrease in the Brønsted sites of the zeolite. Impregnation with Nb_2O_5 kept the ratio almost constant. This behavior provides evidence that Nb_2O_5 can influence the restructuring of Brønsted and Lewis sites. Niobium, with electronegativity similar to that of Al, may promote the interaction of pyridine with Brønsted sites and act as a coordinative center for pyridine.

Table 5. The relative Brønsted to Lewis ratio (B/L) obtained by FT-IR using Py adsorbed on the catalysts.

Catalyst	B/L
HB	1.1
2x-AHFS	1.2
5Nb-AHFS	1.1
10Nb-AHFS	1.1
15Nb-AHFS	1.1
20Nb-AHFS	1.1
T-NaOH	1.1
5Nb-NaOH	1.2
10Nb-NaOH	1.2
15Nb-NaOH	1.1
20Nb-NaOH	1.2

2.7. Ethanol Catalytic Dehydration

As pointed out in the literature, olefin production from (bio)alcohol dehydration is an important industrial process for the potential use of raw materials derived from biomass, substituting petroleum-based materials [43,44]. For the complete transformation (conversion) of ethanol, there are four types of reactions: (i) dehydration, (ii) dehydrogenation, (iii) dehydrogenation coupling, and (iv) hydrogenolysis [45–48]. The final selectivity can be controlled by a catalyst with different active sites (e.g., solid acid, bifunctional). When using a solid acid, the number, strength, and distribution of Brønsted and Lewis sites are critical, as is the accessibility to these sites.

As previously described, the dehydration reaction of ethanol was performed at 230 and 250 °C. The results are described from two perspectives: alcohol conversion and selectivity. Figure 7 shows the results at 230 °C. Compared to the parent zeolite (HB), which showed 91% conversion, both treated zeolites showed lower conversion (2x-AHFS = 51% and T-NaOH = 64%). Modifying the surface of treated zeolites with Nb_2O_5, it was observed a distinct behavior. The 2x-AHFS had an average conversion of around 47%, whereas the T-NaOH samples generally increased the conversion with the niobium loading, reaching a maximum of 84% with 20Nb-NaOH. Looking at the selectivity (Figure 7), we noted that both treated zeolites loaded with Nb_2O_5 showed only ethylene (EE) and diethyl ether (DEE) as the products. Although the HB zeolite showed the highest conversion and selectivity for EE and DEE, there is a minor formation (about 1%) of other products. Probably, because of the higher strength of Brønsted sites in HB, there is also an increase in parallel reactions compared to the modified zeolites. The most active catalysts were 20Nb-AHFS which showed 55% conversion, with selectivity of 56% (EE) and 41% (DEE), while 20Nb-NaOH had 84% conversion and selectivity of 86% (EE) and 14% (DEE).

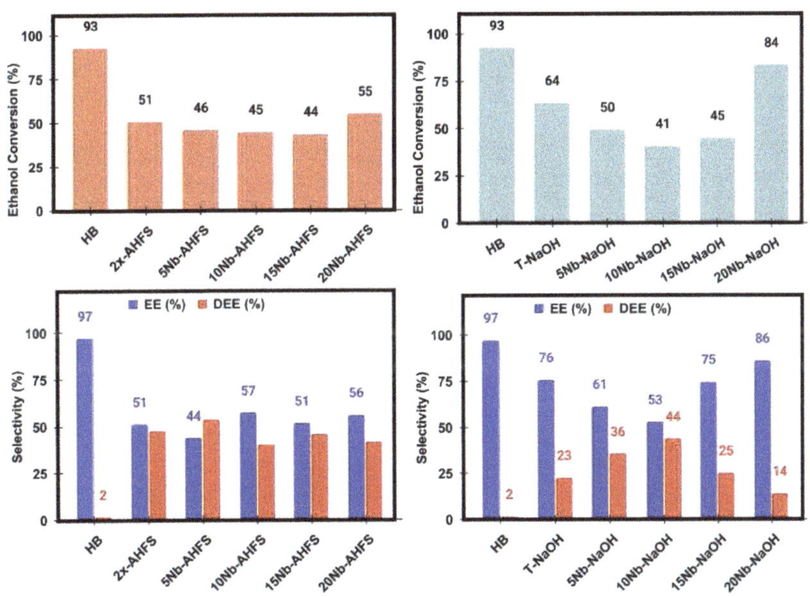

Figure 7. Results of conversion and selectivity for ethanol dehydration at 230 °C using the catalysts of HB, 2x-AHFS, and T-NaOH, and the respective x%Nb_2O_5 supported.

The same analysis approach was employed to run the reaction at 250 °C. The results for conversion and selectivity are in Figure 8. In general, all the catalytic conversions increased with prominent results for HB, which reached 96%, but again with 3% selectivity for other products. The treated HB zeolite displayed an increased conversion (52% for 2x-AHFS and 74% for T-NaOH). The insertion of niobium leads to the maximum conversion reaching 84%, with 82% selectivity for EE and 18% for DEE using a 20Nb-NaOH catalyst. Therefore, it can be concluded that although the highest conversion was achieved with HB, the exclusive products (EE and DEE) with high selectivity towards EE are more advantageous for the treated zeolites with Nb_2O_5, particularly the 20Nb-NaOH sample. As the reaction temperature increased from 230 to 250 °C, the conversion rate of the catalyst 20Nb-NaOH tended to level up with HB zeolite with the benefits of only EE and DEE products.

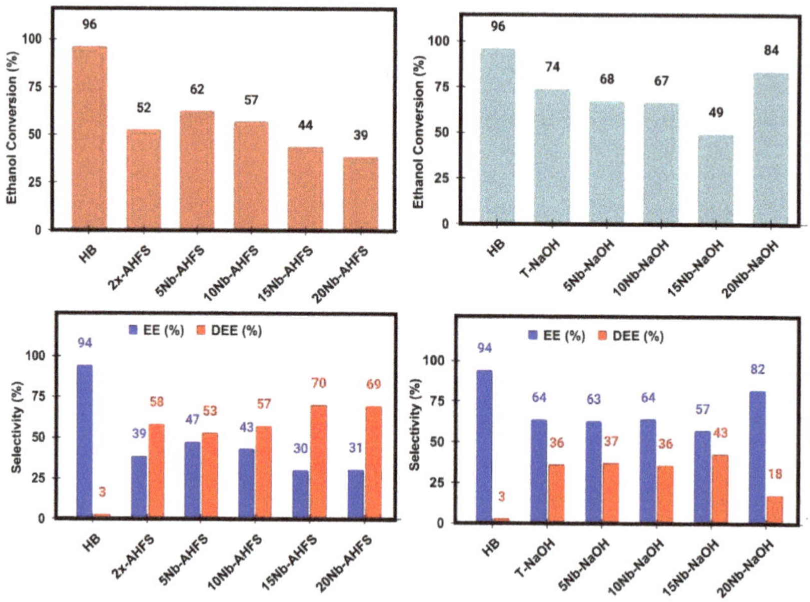

Figure 8. Results of conversion and selectivity for ethanol dehydration at 250 °C using the catalysts of HB, 2x-AHFS, and T-NaOH, and the respective x%Nb_2O_5 supported.

Dehydration reaction results showed that factors such as catalyst morphology and the presence of mesoporosity may enhance mass transfer efficiency, enabling more effective access of alcohol to the active sites, and resulting in higher conversions. The strength and distribution of Brønsted/Lewis sites affected the selectivity for alkene versus ether. The presence of Nb_2O_5 on these catalysts clearly modulates these latter parameters. As we observed earlier [14], the addition of an overlayer of Nb_2O_5 weakens the catalyst compared to the original HB sites. However, selectivity control is clearly dependent on the dealumination treatment. HB treated with 1x-AHFS was more selective for DEE than EE, while the opposite was observed for the 2x-AHFS treatment. The higher efficiency for producing EE with 20Nb-NaOH can be attributed to a combination of increased Brønsted sites from the addition of niobium and better accessibility to these sites due to the mesoporous contribution in this catalyst.

2.8. 1-Propanol Catalytic Dehydration

Propene is one of the fundamental products to support petrochemical processes [19]. Producing propene from renewable sources like propanol can be important for an economy based on lower carbon emissions using biomass [49]. Thus, our catalysts were evaluated in this reaction. Figure 9 presents the results of conversion and selectivity for the dehydration process at 230 °C. The conversion exceeded 88% for all catalysts. For instance, HB converted 90% of 1-propanol, while the treated zeolites showed 96% for 2x-AHFS and 99% with T-NaOH. This indicates that both treatments were efficient in producing more active catalysts than the parent zeolite. The addition of Nb_2O_5, independent of the treatment, maintains the level of conversion above 96%, which demonstrates the efficiency of these catalysts over HB. Actually, the effect on selectivity is much more interesting. HB zeolite showed only 18% selectivity for propene (PP), whereas the 2x-AHFS and T-NaOH had 100 and 89%, respectively. The addition of Nb_2O_5 brought about the same level (99%) of selectivity for PP for treatment with 2x-AHFS and an average of 90% for T-NaOH. Clearly, the main enhancement in the high selectivity for PP was achieved by the dealumination treatment of the parent zeolite (HB).

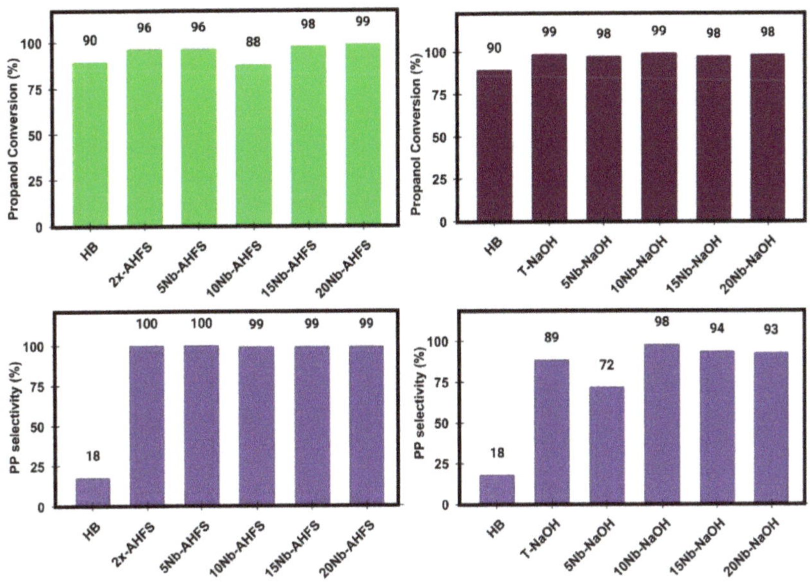

Figure 9. Results of conversion and selectivity for 1-propanol dehydration at 230 °C using the catalysts of HB, 2x-AHFS, and T-NaOH, and the respective x%Nb_2O_5 supported.

One important factor that could explain the higher selectivity for propene in these catalysts may be related to the increased mesoporous area after the dealumination treatments. As we have discussed, this increased area may affect the accessibility of 1-propanol to the Brønsted acid sites of these zeolites, which can better accommodate intermediates based on dimers or trimers of the 1-propanol adsorbed on the acid sites. This is consistent with mechanistic studies in the literature on 1-propanol dehydration over HZSM-5 zeolite [50].

3. Materials and Methods

3.1. Hierarchization of *BEA Zeolite

Zeolite NH_4BEA, obtained from Zeolyst International (CP814E, mole ratio $SiO_2/Al_2O_3 = 25$, Conshohocken, PA, USA), was calcined (8 h at 550 °C) forming a proton sample (HB), which underwent dealumination through two different treatments:

(i) Solid state hierarchization: a fraction of HB was subjected to solid-state dealumination using ammonium hexafluorosilicate (AHFS, 98%, Aldrich, Burlington, MA, USA) with the intention of removing a theoretical percentage of 70 mol% of Al from its crystal lattice by successive (2x) for comparison purposes. The solids were placed in an agate mortar and pestle, and mixed for 10 min, followed by placing the mixture in a desiccator containing a saturated ammonium chloride solution (>99.5%, Sigma-Aldrich, Burlington, MA, USA) at atmospheric pressure. After 24 h, the mixture was heated in a muffle furnace (3 h, 190 °C), followed by washing with ammonium acetate solution and deionized water (Milli-Q), both at room temperature. Finally, the mixture was dried in an oven (24 h, 120 °C) and calcined (8 h, 550 °C) [14].

(ii) Hierarchization in base and acid solutions: the HB zeolite was treated with a 0.2 M sodium hydroxide solution (97%, Aldrich, Burlington, MA, USA), under magnetic stirring at 75 °C, for 4 h. Subsequently, this mixture was washed with deionized water for 1 h, also at 75 °C. Following this step, the resulting material was treated with a 0.5 M hydrochloric acid solution (37%, Aldrich, Burlington, MA, USA) under the same conditions as the base treatment. Finally, the same washing procedure was conducted, and the resulting zeolite was placed in a crucible, dried in an oven for 12 h at 120 °C, and calcined for 8 h at 550 °C.

3.2. Impregnation of Niobium

The insertion of niobium pentoxide into the materials prepared by the methodologies was achieved through aqueous impregnation. The samples were dried at 200 °C for 4 h under vacuum to remove adsorbed water molecules. Subsequently, the mass of the dry materials was measured, and based on these data, the amount of Nb_2O_5 to be used (5, 10, 15, and 20 wt.%) was determined. Niobium ammonium oxalate (99%, CBMM, Araxá, Brazil) was used as the niobium source and was completely dissolved in deionized water. The solutions containing niobium and zeolite were placed under magnetic stirring at 90 °C until the solvent had fully evaporated. Finally, the resulting materials were kept dry in an oven for 12 h at 120 °C, followed by calcination for 8 h at 550 °C. Table 6 presents the nomenclature used to identify the different samples.

Table 6. Nomenclature for the studied *BEA zeolite.

Code	Description
HB	Protonic *BEA zeolite
2x-AHFS	HB dealuminated twice 70 mol%
5Nb-AHFS	HB dealuminated twice 70 mol% and impregnated with 5 wt.% of Nb_2O_5
10Nb-AHFS	HB dealuminated twice 70 mol% and impregnated with 10 wt.% of Nb_2O_5
15Nb-AHFS	HB dealuminated twice 70 mol% and impregnated with 15 wt.% of Nb_2O_5
20Nb-AHFS	HB dealuminated twice 70 mol% and impregnated with 20 wt.% of Nb_2O_5
T-NaOH	HB treated with NaOH and HCl
5Nb-NaOH	HB treated with NaOH and HCl and impregnated with 5 wt.% of Nb_2O_5
10Nb-NaOH	HB treated with NaOH and HCl and impregnated with 10 wt.% of Nb_2O_5
15Nb-NaOH	HB treated with NaOH and HCl and impregnated with 15 wt.% of Nb_2O_5
20Nb-NaOH	HB treated with NaOH and HCl and impregnated with 20 wt.% of Nb_2O_5

3.3. Methods of Characterization

X-ray diffraction (XRD) data were obtained using a powder diffractometer (Panalytical, model Empyrean, Westborough, MA, USA) emitting radiation from a copper tube ($K\alpha$ = 1.5406 Å), at 40 kV and 45 mA, with a scanning rate of 2° per minute (2θ range from 2° to 60°, with a step size of 0.02°). The crystallinity (%C) was obtained by a comparison of the XRD pattern of the standard HB, calculated by integration of the area under peaks (2θ = 5 to 60°), according to Equation (1):

$$\%\,C = \frac{\sum \text{dealuminated zeolite peaks}}{\sum \text{HB peaks}} \times 100. \quad (1)$$

Energy dispersive X-ray fluorescence (EDXRF) spectrometer (Shimadzu, model EDX 720, Kyoto, Japan) was employed to determine the quantities of silicon, aluminum, and niobium. The equipment uses rhodium (Rh) as an X-ray target and operates between 15 and 50 kV. The samples were prepared with polypropylene film and were analyzed under vacuum focused on calculating the total silica/alumina ratio.

All infrared spectra were acquired in a FT-IR spectrometer (Thermo Scientific spectrometer, Nicolet, model 6700, Waltham, MA, USA), with 512 scans and 4 cm^{-1} spectral resolution. They were analyzed in the range of 4000 to 430 cm^{-1}, generally using 0.6 wt.% of the catalyst diluted in dried KBr pellets (>99%, Merck, Rahway, NJ, USA).

The textural data were obtained through gaseous N_2 physisorption at −196 °C in a surface analyzer equipment (Micromeritics, model ASAP 2020C, Norcross, GA, USA). Before the analysis, samples were degassed with evacuation (target pressure of 50 μm Hg) at 300 °C for 4 h. The equations of BET (Brunauer, Emmet, and Teller) in the range $P/P_0 = 0$ to 0.1, t-Plot and BJH (Barrett, Joyner, and Halenda) were used to describe the experimental isotherms.

The average particle diameter was calculated using Equation (2).

$$L(nm) = \frac{6000}{S(m^2/g).\rho(g/cm^3)}. \quad (2)$$

Scanning electron microscopy (SEM) was performed using a scanning electron microscope JEOL (model JSM, Tokyo, Japan) equipped with a secondary electron detector (LED, low energy detector), under high vacuum and a voltage of 15 kV and magnifications ranging from 100 to 10,000 times.

Solid-state nuclear magnetic resonance spectra were acquired with magic angle spinning rotation (MAS NMR) in a spectrometer (Bruker, model Avance III HD Ascend, 14.1 T, 600 MHz for ^{1}H, Karlsruhe, Germany) using 2- or 4-mm CP/MAS probes. Each catalyst was packed inside a zirconia rotor and specific calibration parameters were adopted for each nucleus: (i) ^{27}Al MAS NMR (156.4 MHz); 10 kHz spin rate; 0.4 μs (π/20) pulse duration; 1 s interval between pulses; 2000 acquisitions; external reference: hexa(aqua)aluminum(III) trichloride salt, $[Al(H_2O)_6]Cl_3$ (δ = 0 ppm). (ii) ^{29}Si MAS NMR (119.3 MHz); 10 kHz spin rate; 4.25 μs (π/2) pulse duration; 20 s interval between pulses; 3072 acquisitions; external reference: tetramethylsilane, TMS, $Si(CH_3)_4$ (δ = 0 ppm).

The relative distribution of aluminum atoms in each chemical environment observed in the ^{27}Al spectrum (octahedral–O_h or tetrahedral–T_d) was calculated according to Equation (3), whereas the Si/Al ratio of the zeolitic framework was calculated according to Equation (4) [43,44], which are related to the intensity of the signals referring to the chemical environments Q^4, Q^3, Q^2, and Q^1 after the deconvolution of the ^{29}Si spectrum (using Gaussian Lorentz function, LB = 10, Python, Version: 3.9.7 (default, 16 September 2021, 16:59:28) [MSC v.1916 64 bit (AMD64)], Comments: Enhanced by IPython 8.18.1, Python Software Foundation, Wilmington, DE, USA).

$$\%Al(O_h \text{ or } T_d) = \frac{\text{Signal area of interest}}{\sum \text{Spectrum areas}}, \quad (3)$$

$$\left(\frac{Si}{Al}\right)_{lattice} = \sum_{n=0}^{4} \frac{n}{4} I_{Si(nAl)}, \quad (4)$$

where $I_{Si(nAl)}$ corresponds to the signal intensity related to the environment of Si bonded to n Al atoms. As the signal intensity is proportional to the number of Si atoms in each chemical environment, the total number of Si is proportional to the sum of all chemical environments (numerator). According to Löwenstein's rule, there are no Al-O-Al groups, so each aluminum is surrounded by four Si atoms, so each Si(nAl) group corresponds to ¼ of aluminum atoms (denominator).

Acidity was measured by pyridine adsorption. Before gaseous pyridine (Py) adsorption, each sample (~20 mg) was placed in an aluminum crucible and inserted into a glass tube inside a tubular furnace (Thermolyne, model F21100, Cole-Parmer, Vernon Hills, IL, USA). The catalysts were dehydrated in dried N_2 flow (100 mL min^{-1}) at 300 °C for 1 h. Then, the system was cooled in situ to 150 °C to initiate gaseous Py (>99.8%, Sigma-Aldrich, USA) passage through samples for 1 h. After that, the temperature was held at 150 °C in an N_2 (5.0, White Martins, São Paulo, Brazil) environment for 1 h to remove any physically adsorbed Py. Immediately after cooling the system, each sample was analyzed using thermal analysis and FT-IR (the sample was prepared with a mixture of 10:170 wt.% sample/KBr). Other details can be found elsewhere [14].

3.4. Catalytic Dehydration Reactions

Dehydration reactions of ethanol (99.5%, Dinamica, São Carlos, Brazil) or 1-propanol (99.5%, Sigma-Aldrich, St. Louis, MO, USA) were evaluated by three alcohol injections under 10 mg of catalyst in a pulse microreactor coupled to a gas chromatograph with a flame ionization detector (Shimadzu GC-FID, model 2010, Kyoto, Japan), equipped with a Shimadzu CBP1 PONA column (M50-042, 50 m × 0.15 m × 0.33 μm). Reactions were carried out at temperatures of 230 or 250 °C under these conditions: alcohol injection volume: 0.3 μL; pressure: 100 kPa; total flow: 10.8 mL/min; column flow: 0.1 mL/min; linear velocity: 6.5 cm/s; purge flow: 1 mL/min; split rate: 100; column temperature: 50 °C; flame temperature: 250 °C. The calculations for the conversion of ethanol or 1-propanol, and selectivity for ethylene (EE), dimethyl ether (DME), and propene (PP) were defined by Equations (5)–(8), where n is the number of moles of the reactant.

$$\text{Conversion } (\%) = \frac{n_{\text{ethanol}\,\text{initial}} - n_{\text{ethanol}\,\text{final}}}{n_{\text{ethanol}\,\text{initial}}} \times 100. \tag{5}$$

$$\text{Selectivity } (\%) = \frac{n_{EE} \text{ or } n_{DME}}{n_{\text{ethanol}\,\text{initial}} - n_{\text{ethanol}\,\text{final}}} \times 100. \tag{6}$$

$$\text{Conversion } (\%) = \frac{n_{\text{propanol}_{\text{initial}}} - n_{\text{propanol}_{\text{final}}}}{n_{\text{propanol}_{\text{initial}}}} \times 100. \tag{7}$$

$$\text{Selectivity } (\%) = \frac{n_{PP}}{n_{\text{propanol}_{\text{initial}}} - n_{\text{propanol}_{\text{final}}}} \times 100 \tag{8}$$

4. Conclusions

This work focuses on the hierarchization of *BEA zeolite using two methods: aluminum removal with solid ammonium hexafluorosilicate via two sequential steps (2x-AHFS) and desilication with NaOH followed by hydrochloric acid (T-NaOH). In general, the second treatment (T-NaOH) was more efficient considering the results obtained in the ethanol dehydration reaction. Top–down syntheses involving base and acid treatments are more readily implemented at an industrial scale, allowing interconnectivity and accessibility from the external surface of the zeolite crystal. We presented two beneficial effects of post-synthetic hierarchization of BEA zeolite followed by impregnation of Nb_2O_5 (5, 10, 15, and 20 wt.%) for the dehydration of ethanol and 1-propanol. The hierarchization of the *BEA zeolite by 2x-AHFS and T-NaOH promoted an increase in mesopores and the SiO_2/Al_2O_3 ratio, with only a slight decrease in the average size of crystal domains. This is important since a decrease in size could result in lower selectivity of the zeolite. An increase of 2% in crystallinity was observed for 2x-AHFS, and a decrease of 12% for T-NaOH, compared to HB zeolite. Moreover, ^{27}Al and ^{29}Si MAS NMR results indicated that a more hydrophobic catalyst was formed after the treatment of 2x-AHFS due to an increase of 12% in Q^4 Si environments and a decrease in octahedral Al. On the other hand,

T-NaOH had the highest relative quantity of tetrahedral Al (76%) and the lowest octahedral Al (24%), with a 6% decrease in the Q^4 Si environments, leading to a less hydrophobic catalyst. The gradual addition of Nb_2O_5 tended to balance the number of tetrahedral Al (e.g., 80% of Al (Td) for 20 wt.% Nb on T-NaOH). Based on these combined physicochemical properties, the performance of the catalysts in the dehydration reactions can be explained. For ethanol dehydration at 230 °C, the best catalyst was 20 wt.% Nb_2O_5 supported on T-NaOH zeolite (20Nb-NaOH), which produced 84% conversion with 86% selectivity for ethylene (EE) and 14% diethyl ether (DEE) only. This is advantageous over HB, which showed 93% conversion but produced more than 1% of different by-products. On the other hand, the 20 wt.% Nb_2O_5 supported on 2x-AHFS zeolite (20Nb-AHFS) was the most active for 1-propanol dehydration at 230 °C, with about 99% conversion and selectivity to propene. This performance surpassed that of HB zeolite, which showed 90% conversion but only 18% selectivity for propene.

Supplementary Materials: The following supporting information can be downloaded at https://www.mdpi.com/article/10.3390/catal15040340/s1, Figure S1: FT-IR spectra (in Nujol) for HB, 2x-AHFS, and T-NaOH zeolites; Figure S2: N2 adsorption–desorption isotherms at −196 °C of the catalysts: (A) HB; (B) 2x-AHFS and the respective x%Nb_2O_5 supported; and (C) T-NaOH and the respective x% Nb_2O_5 supported; Figure S3: Trending of some textural properties with the hierarchization method of *BEA zeolite, such as specific surface area (SBET); SiO_2/Al_2O_3 ratio; mesoporous area (SMeso); Figure S4: SEM images and EDX analyses of: (A) HB; (B) 2x-AHFS; (C) T-NaOH; (D) 20Nb-AHFS, and (E) 20Nb-NaOH; Figure S5: 29Si MAS NMR deconvoluted spectra of: (A) HB; (B) 2x-AHFS and the respective x% Nb_2O_5 supported; (C) T-NaOH and the respective x% Nb_2O_5 supported.

Author Contributions: D.d.S.V. conceptualization, data curation, formal analysis, investigation, methodology, writing—review and editing. W.H.R.d.C. data curation, investigation, methodology. A.L.F.F. data curation, investigation, methodology. G.d.F.M. data curation, investigation, methodology. M.R.S. data curation, investigation, methodology. P.T.A.C. formal analysis, investigation, methodology. J.A.D. conceptualization, funding acquisition, investigation, methodology, project administration, writing—review and editing. S.C.L.D. conceptualization, formal analysis, funding acquisition, investigation, methodology, project administration, writing—original draft, writing—review and editing. All authors have read and agreed to the published version of the manuscript.

Funding: Conselho Nacional de Desenvolvimento Científico e Tecnológico, CNPq (Grant Nos. 308693/2022-1, 307413/2021-7); Coordenação de Aperfeiçoamento de Pessoal de Nível Superior, CAPES (Grant no. 001); Decanato de Pesquisa e Inovação (DPI) and Instituto de Química (IQ) from Universidade de Brasília (DPI/IQ/UnB); Fundação de Apoio à Pesquisa do Distrito Federal (FAPDF) (Grant numbers 00193-00001144/2021-60 and 00193-000001176/2021-65); Fundação de Empreendimentos Científicos e Tecnológicos (FINATEC); Financiadora de Estudos e Projetos (FINEP/CTPetro/CTInfra); Petrobras. APC was provided by University of Brasília-Brazil and MDPI.

Data Availability Statement: All data are within the article and the Supplementary Materials and may be available upon request.

Acknowledgments: We would like to thank Richieli Vieira (commercial development coordinator, PQ Silicas Brazil) for providing *BEA zeolite (CP814E*). In addition, we would like to thank Tatiane Oliveira dos Santos from Laboratório Multiusuário de Microscopia de Alta Resolução (LabMic) at IF/UFG-Brazil for SEM/EDX measurements.

Conflicts of Interest: The authors declare no conflicts of interest.

References

1. Gómez-López, P.; Puente-Santiago, A.; Castro-Beltrán, A.; Nascimento, L.A.S.; Balu, A.M.; Luque, L.; Alvarado-Beltrán, C.G. Nanomaterials and catalysis for green chemistry. *Curr. Opin. Green Sustain. Chem.* **2020**, *24*, 48–55. [CrossRef]
2. Li, Y.; Li, L.; Yu, J. Applications of zeolites in sustainable chemistry. *Chem* **2017**, *3*, 928–949. [CrossRef]

3. Mohammed, B.B.; Yamni, K.; Tijani, N.; Alrashdi, A.A.; Zouihri, H.; Dehmani, Y.; Chung, I.; Kim, S.; Lgaz, H. Adsorptive removal of phenol using faujasite-type Y zeolite: Adsorption isotherms, kinetics and grand canonical Monte Carlo simulation studies. *J. Mol. Liq.* **2019**, *296*, 111997. [CrossRef]
4. Li, J.; Corma, A.; Yu, J. Synthesis of new zeolite structures. *Chem. Soc. Rev.* **2015**, *44*, 7112. [CrossRef]
5. Zikrata, O.V.; Larina, O.V.; Balakin, D.Y.; Nychiporuk, Y.M.; Khalakhan, I.; Švegovec, M.; Volavšek, J.; Yaremov, P.S.; Soloviev, S.O.; Orlyk, S.M. Influence of acid-base characteristics of different structural-type zeolites (FER, MFI, FAU, BEA) on their activity and selectivity in isobutanol dehydration. *ChemCatChem* **2024**, *16*, e202400068. [CrossRef]
6. Roth, W.J.; Nachtigall, P.; Morris, R.E.; Čejka, J. Two-dimensional zeolites: Current status and perspectives. *Chem. Rev.* **2014**, *114*, 4807–4837. [CrossRef]
7. Kumar, N.D.; Swaminathan, M.; Swaminathan, M. Review on hierarchically porous BEA and ZSM-5 zeolites and its industrial catalytic applications. *ES Mater. Manuf.* **2024**, *24*, 1151. [CrossRef]
8. Galadima, A.; Muraza, O. Hydrocracking catalysts based on hierarchical zeolites: A recent progress. *J. Ind. Eng. Chem.* **2018**, *61*, 265–280. [CrossRef]
9. Liu, Y.; Wang, Y.; Xiao, P.; Bekhti, S.; Kunkes, E.; Iemhoff, A.; Bottke, N.; Vos, D.E.; Meng, X.; Xiao, F.; et al. Needle-like hierarchical Beta zeolite synthesis via postsynthetic morphology control in the presence of cetyltrimethylammonium bromide for catalytic dehydration of sorbitol. *ACS Appl. Nano Mater.* **2025**, *8*, 1042–1053. [CrossRef]
10. Fernandez, S.; Ostraat, M.L.; Zhang, K. Toward rational design of hierarchical beta zeolites: An overview and beyond. *AIChE J.* **2020**, *66*, e16943. [CrossRef]
11. Kerstens, D.; Smeyers, B.; Waeyenberg, J.V.; Zhang, Q.; Yu, J.; Sels, B.F. State of the art and perspectives of hierarchical zeolites: Practical overview of synthesis methods and use in catalysis. *Adv. Mater.* **2020**, *32*, 2004690. [CrossRef]
12. Hartmann, M.; Machoke, A.G.; Schwieger, W. Catalytic test reactions for the evaluation of hierarchical zeolites. *Chem. Soc. Rev.* **2016**, *45*, 3313–3330. [CrossRef] [PubMed]
13. Bai, R.; Song, Y.; Li, Y.; Yu, J. Creating hierarchical pores in zeolite catalysts. *Trends Chem.* **2019**, *1*, 601–611. [CrossRef]
14. Valadares, D.S.; Clemente, M.C.H.; Freitas, E.F.; Martins, G.A.V.; Dias, J.A.; Dias, S.C.L. Niobium on BEA dealuminated zeolite for high selectivity dehydration reactions of ethanol and xylose into diethyl ether and furfural. *Nanomaterials* **2020**, *10*, 1269. [CrossRef] [PubMed]
15. Groen, J.C.; Peffer, L.A.A.; Moulijn, J.A.; Pérez-Ramírez, J. Mechanism of hierarchical porosity development in MFI zeolites by desilication: The role of aluminum as a pore-directing agent. *Chem. Eur. J.* **2005**, *11*, 4983–4994. [CrossRef]
16. Clemente, M.C.H.; Valadares, D.S.; Lacava, A.B.; Barbosa, L.S.; Martins, G.A.V.; Dias, J.A.; Dias Sílvia, C.L. Catalytic transformation conditions of ethanol on dealuminated BEA zeolites. *J. Braz. Chem. Soc.* **2019**, *30*, 2182–2190. [CrossRef]
17. Groen, J.C.; Abelló, S.; Villaescusa, L.A.; Pérez-Ramírez, J. Mesoporous beta zeolite obtained by desilication. *Microporous Mesoporous Mater.* **2008**, *114*, 93–102. [CrossRef]
18. Ouayloul, L.; Agirrezabal-Telleria, I.; Arias, P.L.; Dumeignil, F.; ElDoukkali, M. Tuning the acid nature of the ZSM-5 surface for selective production of ethylene from ethanol at low temperatures. *Energy Fuels* **2024**, *38*, 4492–4503. [CrossRef]
19. Phung, T.K.; Pham, T.L.M.; Vu, K.B.; Busca, G. (Bio)Propylene production processes: A critical review. *J. Environ. Chem. Eng.* **2021**, *9*, 105673. [CrossRef]
20. Yue, Y.; Guo, X.; Liu, T.; Liu, H.; Wang, T.; Yuan, P.; Zhu, H.; Bai, Z.; Bao, X. Template free synthesis of hierarchical porous zeolite beta with natural kaolin clay as alumina source. *Microporous Mesoporous Mater.* **2020**, *293*, 109772. [CrossRef]
21. Xiong, G.; Yang, H.; Liu, L.; Liu, J. Post-synthesis of Sn-beta zeolite by aerosol method. *RSC Adv.* **2023**, *13*, 4835–4842. [CrossRef] [PubMed]
22. Willhammar, T.; Zou, X. Stacking disorders in zeolites and open-frameworks—Structure elucidation and analysis by electron crystallography and X-ray diffraction. *Z. Kristallogr.* **2013**, *228*, 11–27. [CrossRef]
23. Ordomsky, V.V.; Murzin, V.Y.; Monakhova, Y.V.; Zubavichus, Y.V.; Knyazeva, E.E.; Nesterenko, N.S.; Ivanova, I.I. Nature, strength and accessibility of acid sites in micro/mesoporous catalysts obtained by recrystallization of zeolite BEA. *Microporous Mesoporous Mater.* **2007**, *105*, 101–110. [CrossRef]
24. Li, Q.; Zhang, Y.; Cao, Z.; Gao, W.; Cui, L. Crystallization behavior of zeolite beta from acid-leached metakaolin. *Pet. Sci.* **2010**, *7*, 541–546. [CrossRef]
25. Hassanzadeh-Tabrizi, S.A. Precise calculation of crystallite size of nanomaterials: A review. *J. Alloys Compd.* **2023**, *968*, 171914. [CrossRef]
26. Chalupka, K.; Sadek, R.; Valentin, L.; Millot, Y.; Calers, C.; Nowosielska, M.; Rynkowski, J.; Dzwigaj, S. Dealuminated beta zeolite modified by alkaline earth metals. *J. Chem.* **2018**, *2018*, 7071524. [CrossRef]
27. Ma, Y.; Rigolet, S.; Michelin, L.; Paillaud, J.; Mintova, S.; Khoerunnisa, F.T.; Daou, J.; Ng, E. Facile and fast determination of Si/Al ratio of zeolites using FTIR spectroscopy technique. *Microporous Mesoporous Mater.* **2021**, *311*, 110683. [CrossRef]
28. Dzwigaj, S.; Millot, Y.; Méthivier, C.; Che, M. Incorporation of Nb(V) into BEA zeolite investigated by XRD, NMR, IR, DR UV–vis, and XPS. *Microporous Mesoporous Mater.* **2010**, *130*, 162–166. [CrossRef]

29. Kiricsi, I.; Flego, C.; Pazzuconi, G.; Parker, W.O.; Millini, R.; Perego, C.; Bellussi, G. Progress toward understanding zeolite ß acidity: An IR and 27Al NMR spectroscopic study. *J. Phys. Chem.* **1994**, *98*, 4627–4634. [CrossRef]
30. Sazama, P.; Wichterlová, B.; Sklenák, S.; Parvulescu, V.I.; Candu, N.; Sádovská, G.; Dedeck, J.; Klein, P.; Pashkova, V.; Štastny, P. Acid and redox activity of template-free Al-rich H-BEA* and Fe-BEA* zeolites. *J. Catal.* **2014**, *318*, 22–33. [CrossRef]
31. Nowak, I.; Ziolek, M. Niobium compounds: Preparation, characterization, and application in heterogeneous catalysis. *Chem. Rev.* **1999**, *99*, 3603–3624. [CrossRef] [PubMed]
32. Wachs, I.E.; Jehng, J.M.; Deo, G.; Hu, H.; Arora, N. Redox properties of niobium oxide catalysts. *Catal. Today* **1996**, *28*, 199. [CrossRef]
33. Braga, V.S.; Dias, J.A.; Dias, S.C.L.; Macedo, J.L. Catalyst materials based on Nb_2O_5 supported on silica-alumina: Preparation and structural characterization. *Chem. Mater.* **2005**, *17*, 690–695. [CrossRef]
34. Candu, N.; Fergani, M.E.; Verziu, M.; Cojocaru, B.; Jurca, B.; Apostol, N.; Teodorescu, C.; Parvulescu, V.I.; Coman, S.M. Efficient glucose dehydration to HMF onto Nb-BEA catalysts. *Catal. Today* **2019**, *325*, 109–116. [CrossRef]
35. Santos, L.R.M.; Silva, M.A.P.; Menezes, S.C.; Chinelatto, L.S.; Lam, Y.L. Creation of mesopores and structural re-organization in beta zeolite during alkaline treatment. *Microporous Mesoporous Mater.* **2016**, *226*, 260–266. [CrossRef]
36. Hould, N.D.; Kumar, S.; Tsapatsis, M.; Nikolakis, V.; Lobo, R.F. Structure and colloidal stability of nanosized zeolite beta precursors. *Langmuir* **2010**, *26*, 1260–1270. [CrossRef]
37. Yamamoto, S.; Sugiyama, S.; Matsuoka, O.; Kohmura, K.; Honda, T.; Banno, Y.; Nozoye, H. Dissolution of zeolite in acidic and alkaline aqueous solutions as revealed by afm imaging. *J. Phys. Chem.* **1996**, *100*, 18474–18482. [CrossRef]
38. Hartman, L.; Fogler, H.S. Reaction kinetics and mechanisms of zeolite dissolution in hydrochloric acid. *Ind. Eng. Chem. Res.* **2005**, *44*, 7738–7745. [CrossRef]
39. Shestakova, P.; Martineau, C.; Mavrodinova, V.; Popova, M. Solid state NMR characterization of zeolite beta based drug formulations containing Ag and sulfadiazine. *RSC Adv.* **2015**, *5*, 81957–81964. [CrossRef]
40. Zhao, Z.; Xu, S.; Hu, M.Y.; Bao, X.; Peden, C.H.F.; Hu, J. Investigation of aluminum site changes of dehydrated zeolite H-beta during a rehydration process by high-field solid-state NMR. *J. Phys. Chem. C* **2015**, *119*, 1410–1417. [CrossRef]
41. Hartanto, D.; Yuan, L.S.; Sari, S.M.; Sugiarso, D.; Murwarni, I.K.; Ersam, T.; Prasetyoko, D.; Nur, H. The use of the combination of FTIR, pyridine adsorption, 27Al and 29Si MAS NMR to determine the Brønsted and Lewis acidic sites. *J. Teknol.* **2016**, *78*, 2016. [CrossRef]
42. Parry, E.P. An infrared study of pyridine adsorbed on acidic solids. Characterization of surface acidity. *J. Catal.* **1963**, *2*, 371–379. [CrossRef]
43. Dias, A.P.S.; Rijo, B.; Pereira, M.F.C.; Zăvoianu, R.; Pavel, O.D. Valorization of (bio)ethanol over $MoO_3/(WO_3\text{-}ZrO_2)$ sol-gel-like catalysts. *Reactions* **2024**, *5*, 260–273. [CrossRef]
44. Thangaraj, B.; Monama, W.; Mohiuddin, E.; Mdleleni, M.M. Recent developments in (bio)ethanol conversion to fuels and chemicals over heterogeneous catalysts. *Bioresour. Technol.* **2024**, *409*, 131230. [CrossRef]
45. Zhukova, A.; Chuklina, S.; Fionov, Y.; Vakhrushev, N.; Sazonova, A.; Mikhalenko, I.; Zhukov, D.; Isaikina, O.; Fionov, A.; Il'icheva, A. Enhanced ethanol dehydrogenation over Ni-containing zirconia-alumina catalysts with microwave-assisted synthesis. *Res. Chem. Intermed.* **2024**, *50*, 1331. [CrossRef]
46. Ngcobo, M.; Makgwane, P.R.; Mathe, M.K. A minireview on solid acid catalysts for dehydration of bioethanol to renewable ethylene: An update on catalysts development progress. *Appl. Catal. O* **2024**, *193*, 206976. [CrossRef]
47. Garbarino, G.; Pampararo, G.; Phung, T.K.; Riani, P.; Busca, G. Heterogeneous catalysis in (bio)ethanol conversion to chemicals and fuels: Thermodynamics, catalysis, reaction paths, mechanisms and product selectivities. *Energies* **2020**, *13*, 3587. [CrossRef]
48. Nanda, A.S.F.; Kadja, G.T.M. Bio-based templates for generation hierarchical zeolites: An overview for greener synthesis pathway. *J. Porous Mater.* **2024**, *31*, 1155–1173. [CrossRef]
49. Motte, J.; Nachtergaele, P.; Mahmoud, M.; Vleeming, H.; Thybaut, J.W.; Poissonnier, J.; Dewulf, J. Developing circularity, renewability and efficiency indicators for sustainable resource management: Propanol production as a showcase. *J. Clean. Prod.* **2022**, *379*, 134843. [CrossRef]
50. Zhi, Y.; Shi, H.; Mu, L.; Liu, Y.; Mei, D.; Camaioni, D.M.; Lercher, J.A. Dehydration pathways of 1-propanol on HZSM-5 in the presence and absence of water. *J. Am. Chem. Soc.* **2015**, *137*, 15781–15794. [CrossRef]

Disclaimer/Publisher's Note: The statements, opinions and data contained in all publications are solely those of the individual author(s) and contributor(s) and not of MDPI and/or the editor(s). MDPI and/or the editor(s) disclaim responsibility for any injury to people or property resulting from any ideas, methods, instructions or products referred to in the content.

catalysts

MDPI

Article

Modified Fischer–Tropsch Pathway for CO_2 Hydrogenation to Aromatics: Impact of Si/Al Ratio of H-ZSM-5 Zeolite on Light Aromatics Selectivity

Shaocong Wang [1], Yu Sun [1], Shiyuan Lin [1], Zhongxu Bian [1], Yuanyuan Han [1], Xinze Bi [1], Zhaorui Zhang [1], Xiaojie Liu [1], Dandan Liu [1,*], Yang Wang [1,*] and Mingbo Wu [1,2,*]

[1] State Key Laboratory of Heavy Oil Processing, College of New Energy, China University of Petroleum (East China), Qingdao 266580, China; by1903020418@s.upc.edu.cn (S.W.); s23150055@s.upc.edu.cn (Y.S.); bz23150003@s.upc.edu.cn (S.L.); z23150006@s.upc.edu.cn (Z.B.); b24150013@s.upc.edu.cn (Y.H.); b23150014@s.upc.edu.cn (X.B.); z24030054@s.upc.edu.cn (Z.Z.); b24150012@s.upc.edu.cn (X.L.)

[2] College of Chemical Engineering, Qingdao University of Science & Technology, Qingdao 266042, China

* Correspondence: liudandan@upc.edu.cn (D.L.); wangyang@upc.edu.cn (Y.W.); wumb@upc.edu.cn (M.W.)

Abstract: Despite significant advancements in designing tandem catalysts for CO_2 hydrogenation to aromatics, the role of zeolite acid property in regulating the selectivity of light aromatics (benzene, toluene, and xylene, abbreviated as BTX) remains unclear. Herein, we report H-ZSM-5 zeolite (denoted as HZ-X, where X represents the Si/Al ratio) integrated with a Na-promoted FeCo-based catalyst (NaFeCo) for CO_2 hydrogenation into aromatics via a modified Fischer–Tropsch synthesis pathway. This study systematically modulates the Si/Al ratio of acidic zeolite and examines its critical role in influencing the light aromatics selectivity. The optimized NaFeCo/HZ-50 catalyst achieves a CO_2 conversion of 43% with an aromatics selectivity of 41%, including a BTX fraction of 57% in total aromatics. Multiple characterization techniques (NH_3-TPD, Py/DTBPy-IR, ^{27}Al NMR, etc.) clarify that acidic zeolite HZ-50 exhibits appropriate acid density and lower external surface acid sites, which synergistically boost the efficient aromatics and BTX synthesis while suppressing the undesirable alkylation and isomerization reactions on the external acid sites. This work develops a highly efficient multifunctional catalyst for CO_2 hydrogenation to light aromatics, especially offering guidance for the rational design of acidic zeolite with unique shape-selective functions.

Keywords: CO_2 hydrogenation; multifunctional catalysts; light aromatics synthesis; H-ZSM-5 zeolite; acidic property

1. Introduction

In recent decades, the extensive utilization of fossil energy sources has substantially contributed to massive CO_2 emissions and unprecedented environmental challenges. The utilization of CO_2 as a raw material for valuable chemicals synthesis could provide a non-petroleum pathway for basic chemical supplies and alleviate environmental pressure [1–4]. Aromatics are an important chemical raw material with a wide range of applications. Benzene, toluene, and xylene (BTX), as the most valuable aromatic hydrocarbons, are important raw materials in the polymer industry [5,6]. Compared to the traditional fossil fuel-based aromatics production pathway, the direct transformation of $CO_2 + H_2$ into aromatics via thermal/catalytic strategies holds great potential to realize sustainable development [7–9].

Thermal/catalytic CO_2 hydrogenation to aromatics primarily involves a methanol-mediated route [10–13] and olefin-mediated route [14–18]. In the methanol-mediated

https://doi.org/10.3390/catal15060557

pathway, CO_2 first reacts with hydrogen over a reduced metal oxide catalyst to form methanol. Then, the methanol molecules diffuse into the zeolite channels and are converted into aromatics with the aid of acid sites and the shape-selective catalysis function of zeolite. Even though high aromatics selectivity can be achieved from the methanol-mediated pathway, the thermodynamic mismatch between CO_2 hydrogenation to methanol and the methanol aromatization reaction results in low CO_2 conversion and high undesirable CO selectivity, hindering the high-yield synthesis of aromatics. In the olefin-mediated pathway, the reverse water–gas shift reaction first occurs to generate CO. Subsequently, CO undergoes Fischer–Tropsch synthesis (FTS) with H_2 to form olefins. At last, these olefins undergo aromatization in the channels of H-ZSM-5 zeolite to produce aromatics (Scheme 1). The olefin-mediated pathway (also known as the modified FTS pathway) delivers exceptional CO_2 conversion efficiency and low CO selectivity, offering promising potential for high-yield synthesis of aromatics [19,20]. From the viewpoint of industrial application, the olefin-mediated route, with the advantage of high-yield aromatics synthesis, could decrease the reactor size and investment cost.

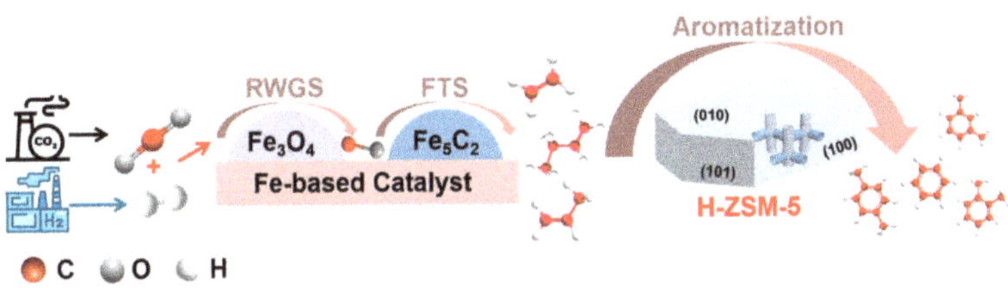

Scheme 1. Reaction process for the hydrogenation of CO_2 to aromatics via the modified FTS pathway.

The design of a multifunctional catalyst that is capable of synergistically catalyzing RWGS, FTS, and aromatization reactions (tandem catalytic system) is the main challenge for the olefin-mediated route. Sun et al. developed a Na-modified ZnFeOx composite catalyst coupled with H-ZSM-5 zeolite for CO_2 hydrogenation to aromatics [21]. This study revealed that Zn acts as a structural accelerator to enhance the dispersion of Fe species, significantly facilitating the exposure of the active sites. In addition, the suitable amount of residual sodium (ca. 4.25 wt%), hierarchical pore structure, and the appropriate density of Brønsted acid sites in the $ZnFeO_x$-nNa/H-ZSM-5 composite catalyst significantly enhanced the aromatic selectivity and catalytic stability. Gascon et al. designed a multifunctional catalyst composed of Fe_2O_3@KO_2 and H-ZSM-5 [22]. The KO_2-doped Fe_2O_3 exhibited excellent olefin production activity, while H-ZSM-5 boosted the aromatization process. Under industrial conditions, the catalyst achieved a CO_2 conversion of 50% and an aromatic space-time yield (STY) of 9.2 mmol g^{-1} h^{-1}. Although a tandem catalytic system based on modified FTS pathways for mixed aromatics (containing large amounts of heavy C_{9+} aromatics) has been extensively studied, the strategy for improving BTX selectivity needs to be further investigated.

The acidic property of zeolite (mainly H-ZSM-5 with MFI topology structure) is the key factor in optimizing the selectivity of BTX [23–25]. The charge imbalance caused by substituting Al^{3+} for Si^{4+} in the skeleton is the origin of the acidity of zeolite [26,27]. The substitution of Al for Si in the zeolite skeleton results in a negative charge that requires a cation to compensate, and when the cation is H^+, a Brønsted acid site is formed [28]. These acid sites are crucial active sites for the aromatization process, promoting C–C bond formation and hydrogen transfer reaction through the proton transfer mechanism [29,30]. Furthermore, the molecular shape-selective effect of microporous in H-ZSM-5 permits only the molecules with kinetic diameters (e.g., BTX) matching the pore dimensions to diffuse through the channels, while bulkier C_{9+} aromatics are theoretically restricted within the channels. However, the formation of C_{9+} aromatics in practical application is mainly due to the alkylation and isomerization initiated by the acid sites at the pore mouth of channels and the external surface. These side reactions significantly reduce BTX selectivity. Therefore, by precisely modulating the spatial distribution of acid sites in H-ZSM-5 and rationally suppressing the side reactions on the external surface, the selectivity of BTX from olefin-mediated CO_2 hydrogenation can be enhanced effectively [31,32]. The current strategies for erasing external surface acid sites of zeolites are typically through epitaxial growth of Silicalite-1 or SiO_2 on the external surface, but these methods involve complicated procedures [13,33]. Recently, a novel H-ZSM-5 zeolite with unique Al distribution characteristics (Al-rich core and Al-deficient surface) has been discovered [34–36]. This special acid site distribution could eliminate the side reactions, offering a new approach for designing multifunctional catalysts in olefin-mediated CO_2 hydrogenation to BTX. Even though it has been shown that the spatial distribution characteristic of Al sites in this type of H-ZSM-5 zeolite could significantly enhance the BTX selectivity, its application in the modified FTS pathway for CO_2 hydrogenation into BTX has not been investigated. In particular, the detailed mechanism by which the Si/Al ratio (Si/Al) affects the product distribution through the modulation of acid strength, acid density, and shape-selective effect has not been systematically revealed.

In this work, we varied the Si/Al ratio of the special H-ZSM-5 zeolite with an Al-rich core and an Al-deficient surface (denoted as HZ-X, where X represents the Si/Al ratio, X = 25, 50, and 150) by adjusting the composition of the synthesized materials. We found a link between zeolite acidity and target product selectivity. NaFeCo/HZ-50 exhibited excellent BTX synthesis performance, which was attributed to its high Bronsted acidity and low external surface acidity. The multifunctional catalyst NaFeCo/HZ-50 achieved 43% CO_2 conversion and 41% aromatics selectivity, with BTX accounting for 57% of the total aromatics. At a high Si/Al ratio (HZ-150), the weak acid sites could not powerfully induce the aromatization of olefins toward aromatics. Conversely, under a low Si/Al ratio (HZ-25), the strong acidity at both internal and external surfaces induced the side reactions, resulting in a marked decline in BTX selectivity.

2. Results

2.1. Physicochemical Property of the H-ZSM-5 Zeolite

The preparation process of H-ZSM-5 zeolites (denoted as HZ-X, where X represents the Si/Al ratio) was illustrated in Figure 1a. The X-ray fluorescence (XRF) spectra verified the Si/Al ratio in HZ-X, as shown in Table S1. The morphology of the prepared HZ-X zeolites with different Si/Al ratios was investigated by scanning electron microscopy (SEM). As shown in Figure 1b–d, the HZ-X zeolite presented a hexagonal morphology. The average length of the zeolite was 20 μm.

The crystal phase structure and crystallinity of HZ-X zeolite with different Si/Al ratios were analyzed by X-ray diffractometer (XRD) patterns. The obtained HZ-X zeolites delivered a typical diffraction peak of the zeolite with MFI topology (PDF#47-0638) (Figure 2a), confirming the successful synthesis of H-ZSM-5 zeolite. As shown in Figure 2b and Table S2, with the increase in the Si/Al ratio, the relative crystallinity of zeolite first increased and then decreased. HZ-50 exhibited the highest relative crystallinity, indicating a more regular and stable crystal structure.

Nitrogen adsorption and desorption isotherms were employed to investigate the impact of the Si/Al ratio on the architecture property of the acidic zeolites. As demonstrated in Figure 2c, the nitrogen adsorption capacity of all zeolites rapidly increased at low relative pressures (P/P_0 = 0–0.2), which is a characteristic of type I isotherms and confirms the microporous-dominated structure. The detailed data are shown in Table S2. As the

Si/Al ratio decreased, the specific surface area of the zeolite gradually increased from 378 $m^2\ g^{-1}$ to 435 $m^2\ g^{-1}$, and the micropore volume progressively rose from 0.161 $cm^3\ g^{-1}$ to 0.185 $cm^3\ g^{-1}$. The pore size distribution plots indicated that the pore size in zeolites all concentrated in the range of 0.5–0.6 nm, further confirming the microporous structure of the acidic zeolites with different Si/Al ratios (Figure 2d).

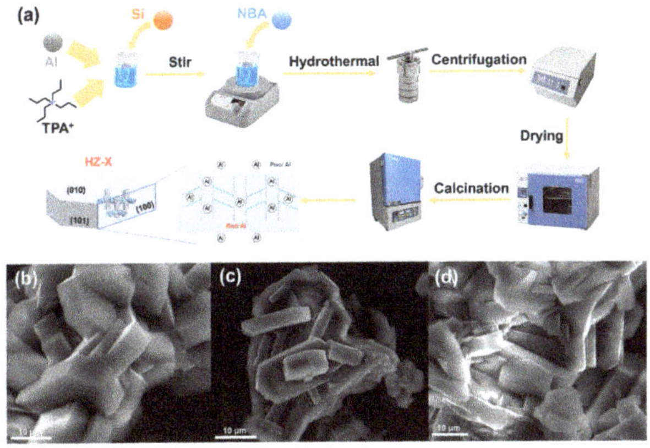

Figure 1. (**a**) Schematic illustration of the fabrication process of HZ-X zeolites. (**b–d**) SEM images of the HZ-X zeolites. (**b**) HZ-25, (**c**) HZ-50, and (**d**) HZ-150.

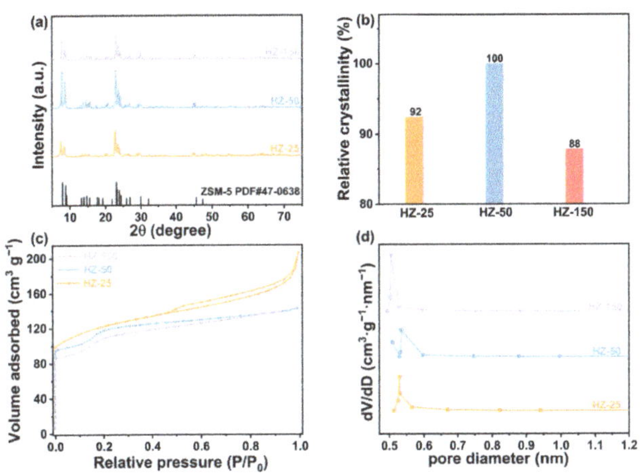

Figure 2. XRD patterns (**a**), relative crystallinity (**b**), nitrogen adsorption-resolution isotherms (**c**), and pore size distribution (**d**) of the HZ-X zeolites with different Si/Al ratios.

2.2. The Catalytic Performance of Aromatics Synthesis from CO_2 Hydrogenation

The catalytic performance of NaFeCo/HZ-X catalysts was evaluated in a fixed-bed reactor at 320 °C and 3 MPa for the hydrogenation of CO_2 to aromatics. NaFeCo and HZ-X zeolites were pelletized into 20–40 mesh particles, respectively, after which they were physically mixed and loaded into the reactor. The detailed results are shown in Figure 3, with specific data in Tables S3 and S4. Increasing the Si/Al ratio of the zeolitic catalyst led to a continuous reduction in CO_2 conversion, dropping from 44.3% to 35.5%. The variation in CO_2 conversion can be attributed to the different driving forces stemming from the HZ-X zeolites with different Si/Al ratios and acid strengths. For the product distribution, as the Si/Al ratio increased, the selectivity of CH_4 and light olefins was improved, while the selectivity of aromatics exhibited a decreasing trend. These results demonstrated that a lower

Si/Al ratio is favorable to enhance the efficiency of olefin transformation and aromatization synthesis. To further compare the aromatization capability of different zeolites, the product distribution of C_{5+} hydrocarbon compounds is presented in Figure 3c. With the Si/Al ratio increased, the selectivity of olefins and iso-paraffins showed a slight enhancement, while the selectivity of n-paraffins experienced a mild decline. We then analyzed the selectivity of the most important products, light aromatics, BTX. Figure 3b demonstrates the BTX selectivity and space-time yield (STY) obtained from the NaFeCo&HZ-X catalysts. It was not difficult to find that when NaFeCo&HZ-50 had the highest selectivity of BTX and STY, NaFeCo&HZ-25 had the second highest performance, and NaFeCo&HZ-150 had the lowest performance. The above results may be because the excessively strong acidity in HZ-25 resulted in the conversion of the produced light aromatics to C9+ heavy aromatics. In contrast, the excessively weak acidity in HZ-150 was not sufficient to support the aromatization reaction. The NaFeCo&HZ-50 catalyst showed the optimal BTX synthesis performance with a BTX selectivity of 23.7% and a BTX STY of 59.8 $mg_{BTX} \cdot g_{FeCo}{}^{-1} \cdot h^{-1}$. The study of the distribution of aromatics fraction (Figure 3d and Table S4) revealed that toluene (T), xylene (X), and C_{9+} heavy aromatics were dominant in the products. When the Si/Al ratio increased from 25 to 150, the proportion of toluene in total aromatics gradually decreased from 22% to 12.8%, and the proportion of ethylbenzene also declined from 5.3% to 4.2%. In contrast, the proportion of C_{9+} heavy aromatics first decreased and then increased, first from 40.7% to 37.2% and then to 47.1%. The above-discussed product distribution characteristics indicated that adjusting the Si/Al ratio can effectively optimize the product distribution. The NaFeCo&HZ-50 catalyst achieved high selectivity of aromatics and BTX, with aromatics selectivity of 41.1% and BTX fraction of 57.7% in total aromatics. Furthermore, benefiting from the higher CO_2 conversion, NaFeCo&HZ-50 also delivered an excellent STY of light aromatics.

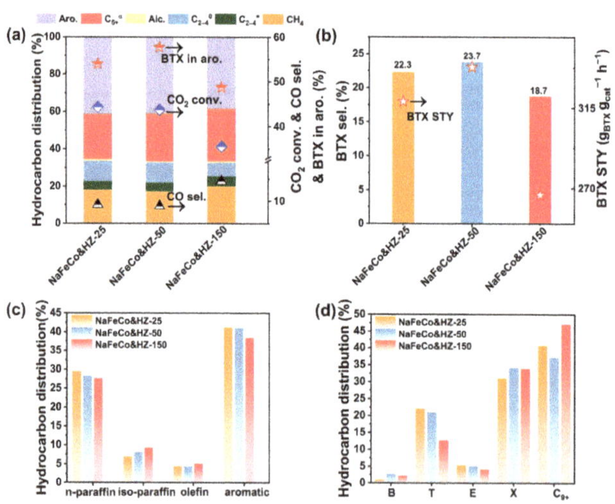

Figure 3. (**a**) Catalytic performance of various catalysts for CO_2 conversion. (**b**) BTX selectivity and STY of BTX from NaFeCo/HZ-X catalysts. (**c**) Product distribution of C_{5+} hydrocarbons. (**d**) Product distribution of various aromatics. $C_{5+}{}^{\alpha}$ denotes the C_{5+} hydrocarbons excluding aromatics. Alc. denotes the alcohol compounds. C_2–$C_4{}^0$ and C_2–$C_4{}^=$ denote the paraffins and olefins of C_2–C_4 hydrocarbons, respectively. Reaction conditions: 320 °C, 3 MPa, (23.75% CO_2, 71.25% H_2, and 5% Ar), GHSV = 4500 mL $g_{cat}{}^{-1}$ h^{-1}, time on stream (TOS) = 24 h, metal oxide/zeolite mass ratio = 1:3, and catalyst weight of 0.8 g.

2.3. Acidic Property of HZ-X Zeolites

In the CO_2 hydrogenation reaction, the synthesis of aromatics, especially BTX, is closely related to the acidic property of zeolites. Optimizing the density of Brønsted acid sites plays a critical role in enhancing aromatics selectivity. Figure 4a presents the ammonia temperature-programmed desorption (NH_3-TPD) profiles of different zeolites, where two distinct desorption peaks appeared at 130 °C and 400 °C, respectively. The peak at 130 °C was associated with the weak acid sites, while the peak at 400 °C corresponded to the medium and strong acid sites. As the Si/Al ratio decreased, the desorption peaks shifted to higher temperatures, indicating an increase in acid strength. The specific density of the acid sites is provided in Table S5. HZ-25 exhibited the highest acid density, reaching a total of 144 μmol g^{-1}, while the acid density of HZ-50 decreased to 54 μmol g^{-1}. To further investigate the acidic property of the zeolites, we performed pyridine infrared (Py-IR) spectroscopy analysis (Figure 4b). The three typical peaks at 1445, 1490, and 1545 cm^{-1} can be assigned to the Lewis acid sites, both Lewis and Brønsted acid sites, and Brønsted acid sites, respectively. In H-ZSM-5 zeolite, the Brønsted acid is associated with the inserted Al^{3+} in a tetrahedral framework as Si–OH–Al in H-ZSM-5 zeolite, and the Lewis acid is generally derived from the Al^{3+} located in the extra-framework [37]. As the Si/Al ratio decreased, the amount of Brønsted acid sites increased, which is associated with the increased Al species content. HZ-25 exhibited the highest Brønsted acid concentration of 0.72 mmol g^{-1}, while HZ-50 exhibited a slightly lower value of 0.52 mmol g^{-1} (Table S6). The variation trends in acid density observed from the Py-IR spectra aligned well with the NH_3-TPD results. It should be noted that the difference in the amount of acid sites determined by NH_3-TPD and Py-IR was due to the different accessibilities of probe molecules with different molecular sizes (pyridine and NH_3) to the acid sites [38]. With a lower Si/Al ratio in zeolite, its acidity gradually enhances, leading to increased CO_2 conversion and aromatics selectivity, whereas BTX selectivity first improved and then decreased.

It has been proven that the acidic characteristic of zeolite possesses a significant correlation with the spatial distribution of Al sites in the skeletal structure [38]. The zeolite framework contains silanol (Si–OH) and silanol–aluminum (Si–OH–Al) groups [39,40]. Si–OH corresponds to weak acid sites, while Si–OH–Al typically corresponds to stronger Brønsted acid sites associated with tetrahedral coordinated Al^{3+} in the framework. Therefore, as the amount of tetrahedral Al increases, the acidity strengthens [41]. To characterize the coordination of Al^{3+} in the zeolite framework and elucidate the association of Al^{3+} with acidic properties, we conducted an analysis of the coordination states using ^{27}Al solid-state magic angle spinning nuclear magnetic resonance (^{27}Al MAS NMR) spectroscopy (Figure 4c). The spectrum exhibited a prominent peak at 55 ppm, which can be diagnosed as frame Al species. In contrast, the peak at 0 ppm corresponded to the non-framework Al species. All zeolite samples exhibited typical tetrahedral coordination framework Al structure characteristics. It was worth noting that the HZ-25 sample exhibited a weak signal of the non-framework Al species due to the high content of Al^{3+}.

The acid sites on the external surface of the zeolites tend to induce the formation of C_{9+} heavy aromatics compounds through the alkylation and isomerization of light aromatics [25,42,43]. To investigate the variation in external surface acidity of zeolites with different Si/Al ratios, the 2,6-di-tert-butylpyridine infrared (DTBPy-IR) spectra were obtained. The bulky DTBPy could primarily only interact with the acid sites on the external surface of the zeolite, thereby avoiding interference from acid sites within the pore interiors. Thus, it is an effective probing molecule for studying the external surface acidity of zeolites. As shown in Figure 4d, it had large peaks at 1600, 1630, 1445, 1490, and 1545 cm^{-1}. The large peak at 1600 cm^{-1} is related to the bending vibration of C=C in the pyridine ring, while the large peak at 1630 cm^{-1} is related to the hydrogen bonding in the pyridine molecule. Since DTBPy also has a pyridine ring structure, three desorption peaks at 1445, 1490, and 1545 cm^{-1} also appeared in the spectra. The peak between 1600 and 1700 cm^{-1} reflected the acidity of the external surface of zeolites. It can be observed that the external surface acidity of the zeolites initially decreased and then increased as the Si/Al ratio increased, with HZ-25 and HZ-150 displaying stronger external surface acidity. In conclusion, moderate Si/Al ratio zeolites featured an Al-rich core and Al-deficient surface, generating spatially distinct acid site distribution (high internal but low surface acid density), which enhances BTX formation efficiency.

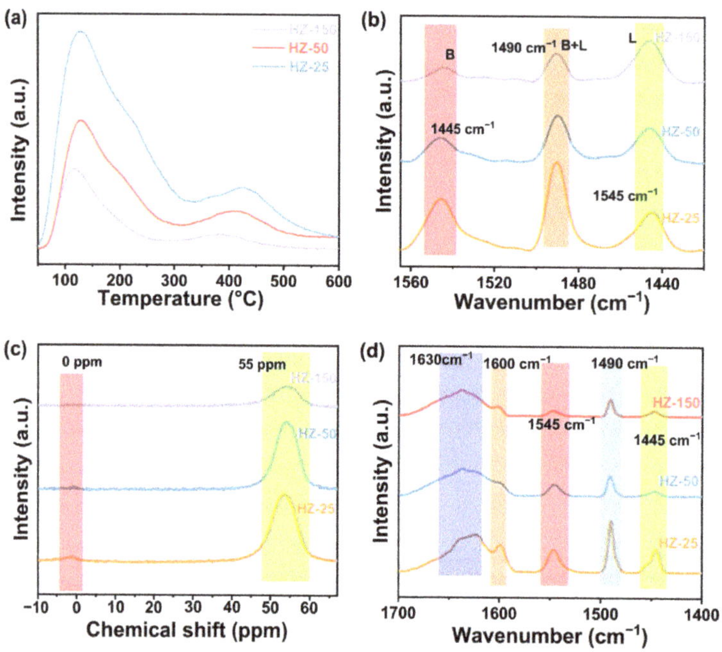

Figure 4. NH_3-TPD profiles (**a**), Py-IR spectra (**b**), ^{27}Al MAS NMR spectra (**c**), and DTBPy-IR spectra (**d**) of all zeolites with different Si/Al ratios. The Py-IR and DTBPy-IR spectra were recorded after removing the weakly adsorbed Py and DTBPy molecules by evacuation at 150 °C.

Through comprehensive analysis, it was found that the Si/Al ratio of zeolite was significantly and positively correlated with the Brønsted acid strength and framework aluminum density (Figure 4). HZ-50 exhibited moderate Al content, with Al exclusively participating in framework formation, which promotes the synthesis of total aromatics. Meanwhile, its lower external surface acid density facilitated the efficient synthesis of light aromatics BTX by avoiding the undesirable side reactions.

2.4. Relationship Between Zeolite Acidic Property and Light Aromatics Synthesis Performance

The relationship between the acidic property of zeolites with different Si/Al ratios and CO_2 hydrogenation performance is shown in Scheme 2. In the modified FTS reaction system, olefin intermediates were first aromatized on acid sites in the pore of the zeolite to form light aromatics. During their diffusion to the external surface, these light aromatics underwent secondary reactions (such as alkylation and isomerization) with the aid of acid sites on the external surface, generating undesirable by-products, such as heavy aromatics or iso-paraffins. The experimental results showed that when the Si/Al ratio was too high, the acid strength was insufficient for the aromatization process. When the Si/Al ratio was too low, a part of the Al might be located on the external surface of the zeolite, which leads to the undesirable side reactions that lower the BTX selectivity. Under the optimal Si/Al ratio, all Al species in the zeolite were involved in the skeleton formation process. It's appropriate acid site density and distribution optimized the aromatization process and minimized the side reactions, therefore achieving the superior BTX synthesis performance. An increase in the Si/Al ratio of zeolites led to a gradual decrease in acid site density, resulting in slower conversion of intermediates and a decreased CO_2 conversion due to hampered driving force in the tandem catalytic system.

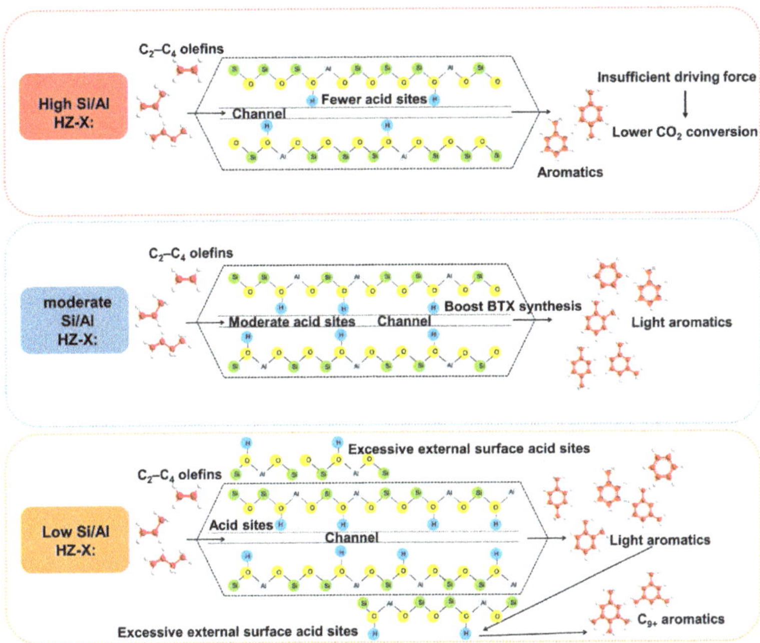

Scheme 2. Relationship between acidic property and catalytic performance of NaFeCo/HZ-X catalysts.

3. Materials and Methods

3.1. Materials and Chemicals

Ferric nitrate nonahydrate [$Fe(NO_3)_3 \cdot 9H_2O$] (Macklin Biochemical Co., Ltd., Shanghai, China), cobalt nitrate hexahydrate [$Co(NO_3)_2 \cdot 6H_2O$] (Macklin Biochemical Co., Ltd., Shanghai, China), sodium meta-aluminate ($NaAlO_2$) (Sinopharm Group Chemical Reagent Co., Ltd., Shanghai, China), tetrapropylammonium bromide ($C_{12}H_{28}BrN$) (Sinopharm Group Chemical Reagent Co., Ltd., Shanghai, China), n-butylamine ($C_4H_{11}N$) (Macklin Biochemical Co., Ltd., Shanghai, China), Urea (CH_4N_2O) (Sinopharm Group Chemical Reagent Co., Ltd., Shanghai, China), anhydrous sodium carbonate (Na_2CO_3) (Sinopharm Group Chemical Reagent Co., Ltd., Shanghai, China), Colloidal silicon oxide LUDOX® HS-40 (SiO_2) (Sigma-Aldrich Life Science & Tech Co., Ltd., Shanghai, China) were of analytical grade and utilized without further purification. The deionized water used for catalyst preparation had a resistivity of 18.2 megohms and was obtained through an ultrapure water system.

3.2. Catalyst Preparation

3.2.1. Synthesis of NaFeCo Catalyst

The FeCo catalyst was synthesized by the urea precipitation method. Typically, $Fe(NO_3)_3{\cdot}9H_2O$ (12.12 g), $Co(NO_3)_2{\cdot}6H_2O$ (2.9103 g), and urea (33.033 g) were dissolved in deionized water (330 mL) to form a homogeneous solution. The solution was then placed in an oil bath, heated to 85 °C, and stirred for 2 h. The solution was then kept at 85 °C for 12 h for aging. The product was collected by centrifugation, washed four times with deionized water, and then dried at 70 °C overnight in an oven. The catalyst was then calcined at 350 °C for 4 h under an air atmosphere at a ramp rate of 5 °C min^{-1} to obtain the FeCo catalyst. Finally, the FeCo catalyst was impregnated with sodium carbonate solution for 3% sodium impregnation, followed by drying in an oven at 70 °C overnight to obtain the final catalyst NaFeCo.

3.2.2. Synthesis of H-ZSM-5 Zeolites

The H-ZSM-5 zeolites with different Si/Al ratios (denoted as HZ-X, where X represents the Si/Al ratio) were synthesized by the hydrothermal method. Typically, deionized water of 91.5 g was added to a 200 mL PTFE bottle. Under the stirring condition, $NaAlO_2$ of 0.08–0.48 g, TPABr of 1.83 g, and $C_4H_{11}N$ of 3.92 g were added into the above solution and magnetically stirred for 15 min to make them dissolve thoroughly. Then, SiO_2 of 22.53 g was slowly added into the above solution drop by drop, and stirring was continued for 30 min. The molar ratio of all the reagents was SiO_2:Al_2O_3:H_2O:TPABr:NBA = 300:1–6:10,160:13.78:107.2. The obtained solution was poured into a stainless steel reactor with PTFE lining and tightly sealed to ensure that the liquid did not leak. The reactor was transferred to a rotary drying oven and fixed on a rotating holder. The hydrothermal temperature was maintained at 180 °C for 48 h, and the rotation speed was 2.0 r min^{-1}. The obtained emulsion was centrifugally washed three times with deionized water, followed by drying overnight at 70 °C to acquire the HZ-X precursor. The precursor was calcinated at 550 °C for 6 h in a muffle furnace with a heating rate of 5 °C min^{-1} to remove the organic template in HZ-X.

3.3. Catalyst Characterization

X-ray diffraction (XRD) analysis of the obtained catalysts was conducted using Cu Kα radiation and a DX-2700BH X-ray diffractometer (Dandonghaoyuan Instrument Co., Ltd., Dandong, China). The relative crystallinity of the zeolites was obtained by comparing the crystal peak areas in the XRD patterns of the zeolites under the same conditions. The morphology of the catalysts was observed via scanning electron microscopy (SEM) using a Regulus 8100 instrument (Hitachi High-Tech Co., Shanghai, China). Further, the specific surface area and pore size distribution of the samples were determined by the JW-BK200 instrument (JingweiGaoBo Co., Ltd., Beijing, China). The specific surface area was calculated using the Brunauer–Emmett–Teller (BET) method. The Horvath–Kawazoe (HK) method was employed to analyze the pore size distribution of microporous. NH_3 temperature-programmed desorption (NH_3-TPD) experiments were carried out on a DCA-1200 instrument (Builder Electronic Technology Co., Ltd., Beijing, China). Prior to the experiment, the catalyst was pretreated at 150 °C for 1 h to eliminate the H_2O present in the microporous zeolite. Subsequently, the sample was saturated with NH_3 at 50 °C, and the physisorbed NH_3 was removed using He flow. Further, NH_3-TPD profiles were recorded under a He flow with a heating rate of 10 °C min^{-1}. The Si/Al ratio of the HZ-X catalyst was evaluated using a PANalytical AXIOS-Petro X-ray fluorescence spectrometer (PANalytical B.V. Co., Netherlands, Beijing, China). Pyridine infrared (Py-IR) and 2,6-di-tert-butylpyridine infrared (DTBPy-IR) spectra were obtained using a Bruker iS50 FTIR spectrometer (Thermo Fisher Scientific Instrument Co., Ltd., Waltham, MA, USA) to distinguish the acid type in HZ-X. Prior to the Py-IR or DTBPy-IR test, the zeolite was pressed into a self-supported wafer and degassed at 400 °C and 10^{-4} Pa for 1 h. After that, inert gas was saturated with pyridine or 2,6-di-tert-butylpyridine and was introduced into the IR cell for 1 h at room temperature. Finally, the catalyst was heated to 150 °C with a heating rate of 10 °C min^{-1}, then desorbed under a helium atmosphere for 30 min, and finally, Py-IR or DTBPy-IR spectra were collected. The distribution of Al in HZ-X zeolites was analyzed using an AVANCE3-400M (Bruker Co., Beijing, China) wide-cavity solid-state Nuclear Magnetic Resonance (NMR) spectrometer. The acidity data for zeolite was calculated using the following equation:

(1) NH_3-TPD:

$$acidity = \frac{V_0 \times w \times A}{m \times A_0 \times V_m}$$

where V_0 is the volume of the quantized ring, w is the volume fraction of NH_3, A_0 and A are the peak areas of the TCD signal peaks measured for the quantized ring and the catalyst, respectively, m is the mass of the catalyst, and V_m is the molar volume at standard conditions.

(2) Py-IR:

$$Brnsted\ acidity = 1.88 \times A_B \times R^2/m$$

$$Lewis\ acidity = 1.42 \times A_L \times R^2/m$$

The quantitative determination of Bronsted and Lewis acids of the catalysts was based on previous studies [44]. where 1.88 and 1.42 are the extinction coefficients of Brønsted acid and Lewis acid, respectively. The peak areas of the characteristic peaks of Brønsted acid and Lewis acid are labeled as A_B/A_L. The radius of the zeolite compression tablet is labeled as R. The mass of the compression tablet is labeled as m.

3.4. Catalytic Activity Evaluation

The catalytic performance of the catalysts was measured in a fixed-bed reactor (ZXBLUE Co., Ltd., Beijing, China) with an internal diameter of 12 mm. As shown in Figure 5, 0.2 g of NaFeCo and 0.6 g of HZ-X were made into 20–40 mesh granules and mixed homogeneously. The mixture was mixed with an equal volume of quartz sand and loaded into the reactor. Prior to the reaction, the catalyst was reduced by pure H_2 at 400 °C for 4 h. After cooling down to room temperature, the reactant gas (23.75% CO_2, 71.25% H_2, and 5% Ar) was fed into the reactor until the pressure reached 3 MPa. At the same time, the temperature of the reactor was increased to 320 °C. The exhaust gas was analyzed by an on-line gas chromatograph (Fuli 9790II, Fuli Analytical Instruments Co., Ltd., Wenling, China) equipped with two detectors (thermal conductivity detector (TCD) and flame ionization detector (FID). The TCD is connected to the TDX-01 (60–80 mesh, 3 mm × 2 m, Ruide-Jingke Instrument Co., Ltd., Zaozhuang, China); Porapak-Q (60–80 mesh, 3 mm × 1 m, Agilent Technologies Co., Ltd., Santa Clara, CA, USA) packed columns for the analysis of Ar, CO, CH_4, and CO_2; and the FID is connected to the HP-PLOT/Q capillary column (0.53 mm × 30 m, Agilent Technologies Co., Ltd., Santa Clara, CA, USA) to detect gas-phase hydrocarbons. An off-line gas chromatograph (Fuli 9790II, Fuli Analytical Instruments Co., Ltd., Wenling, China) equipped with an FID and a capillary column InerCap-5 (30 m × 0.25 mm, GL Sciences, Tokyo, Japan) was employed to analyze the liquid products.

The specific catalytic performance evaluation method is as follows:

(1) CO_2 conversion was calculated according to

$$CO_2\ Conversion = \frac{CO_{2inlet} - CO_{2outlet}}{CO_{2inlet}} \times 100\%$$

in which the mole of CO_2 in the inlet gas is labeled as the CO_2 inlet and that in the outlet gas is labeled as the CO_2 outlet.

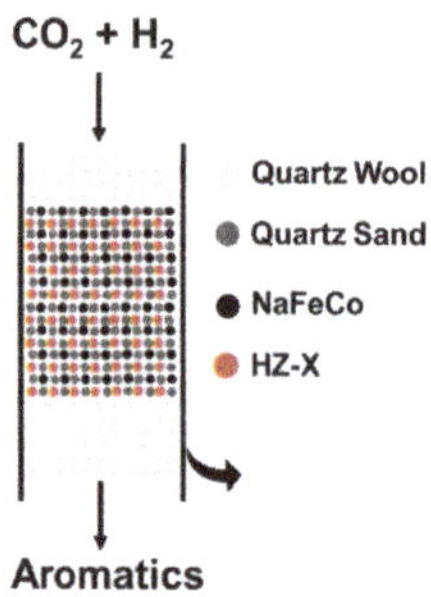

Figure 5. Schematic diagram of catalyst loading in the reactor.

(2) Product selectivity was calculated as the percentage of CO_2 converted into a given product and according to

$$Sel_i = \frac{N_i \times n_i}{\sum_1^i (N_i \times n_i)} \times 100\%$$

where N_i and n_i represent the mole percentage and carbon number of product i, respectively.

(3) CO selectivity for CO_2 hydrogenation was calculated according to

$$CO\ selectivity = \frac{CO_{outlet}}{CO_{2inlet} - CO_{2outlet}} \times 100\%$$

in which the mole of CO in the outlet gas is labeled as the CO outlet. The carbon balances in all catalytic runs were calculated and are all higher than 90%. Typically, the experimental data after the reaction of 24 h were used for discussion.

4. Conclusions

In summary, we synthesized H-ZSM-5 (denoted as HZ-X, where X represents the Si/Al ratio) zeolites with varying Si/Al ratios and evaluated their catalytic performance in CO_2 hydrogenation to aromatics after combining with the NaFeCo catalytic component. The acidic property of the zeolites, particularly the acid density and distribution, plays a crucial role in light aromatics synthesis performance. The structural characteristics of zeolites govern their acid site distribution, and these acid sites serve as the active centers for catalyzing the conversion of CO_2 to aromatics. Therefore, tuning the active sites of zeolites can promote the selective synthesis of the target product. The zeolite with a Si/Al ratio of 50 possessed Al-rich core and Al-deficient surface characteristics. This acidic property synergistically boosted the efficient light aromatics synthesis by suppressing the undesirable alkylation and isomerization reactions on the external acid sites. In contrast, the zeolite with a high Si/Al ratio exhibited insufficient acid site density for the aromatization reaction, while the zeolite with a low Si/Al ratio delivered high heavy aromatics selectivity due to the undesirable side reactions, such as alkylation and isomerization. The optimized NaFeCo/HZ-50 catalyst achieved a CO_2 conversion of 43% with an aromatics selectivity of 41%, including a BTX fraction of 57% in total aromatics. This work demonstrates that the rational design of zeolite acidic property is critical for efficient CO_2 hydrogenation into BTX, offering design guidance for bifunctional catalysts in a targeted synthesis system.

Supplementary Materials: The following supporting information can be downloaded at https://www.mdpi.com/article/10.3390/catal15060557/s1. Table S1: XRF test results of zeolites; Table S2: Texture property of zeolites with different Si/Al ratios; Table S3: Catalytic properties of zeolites with different Si/Al ratios; title; Table S4: Distribution of aromatics in reaction products; Table S5: NH_3-TPD data for zeolites with different Si/Al ratios; Table S6: Py-IR data for zeolites with different Si/Al ratios.

Author Contributions: Conceptualization, S.W. and Y.W.; methodology, S.W. and Y.S.; validation, S.W., X.B. and Y.H.; formal analysis, S.W. and Y.S.; investigation, S.W., S.L., Z.B., Z.Z. and X.L.; resources, D.L., Y.W. and M.W.; writing—original draft preparation, S.W.; writing—review and editing, S.W., Y.S. and Y.W.; visualization, S.W., Y.S. and X.L.; supervision, D.L., Y.W. and M.W.; project administration, Y.W. and M.W.; Funding acquisition, Y.W. and M.W. All authors have read and agreed to the published version of the manuscript.

Funding: This work was financially supported by the National Key Research and Development Program of China (2023YFB4104500, 2023YFB4104502), the National Natural Science Foundation of China (22478436), the Key Research and Development Program of Shandong Province (2024ZLGX08), and the Science and Technology Innovation Project of the Shandong Energy Group Co., Ltd. (SNKJ2023A03).

Data Availability Statement: Data are contained within the article.

Conflicts of Interest: The authors declare that this study received funding from Shandong Energy Group Co., Ltd. The funder was not involved in the study design, collection, analysis, interpretation of data, the writing of this article or the decision to submit it for publication.

References

1. Wang, M.; Luo, J. A coupled electrochemical system for CO_2 capture, conversion and product purification. *eScience* **2023**, *3*, 100155. [CrossRef]
2. Zhang, Z.; Yang, Z.; Liu, L.; Wang, Y.; Kawi, S. Catalytic CO_2 conversion to C1 chemicals over single-atom catalysts. *Adv. Energ. Mater.* **2023**, *13*, 2301852. [CrossRef]
3. Wang, W.; Zeng, C.; Tsubaki, N. Recent advancements and perspectives of the CO_2 hydrogenation reaction. *Green Carbon* **2023**, *1*, 133–145. [CrossRef]
4. Wang, Y.; Sun, J.; Tsubaki, N. Clever Nanomaterials fabrication techniques encounter sustainable C1 catalysis. *Acc. Chem. Res.* **2023**, *56*, 2341–2353. [CrossRef]
5. Zhang, J.; Zhang, M.; Chen, S.; Wang, X.; Zhou, Z.; Wu, Y.; Zhang, T.; Yang, G.; Han, Y.; Tan, Y. Hydrogenation of CO_2 into aromatics over a $ZnCrO_X$-zeolite composite catalyst. *Chem. Commun.* **2019**, *55*, 973–976. [CrossRef]
6. Wei, J.; Yao, R.; Han, Y.; Ge, Q.; Sun, J. Towards the development of the emerging process of CO_2 heterogenous hydrogenation into high-value unsaturated heavy hydrocarbons. *Chem. Soc. Rev.* **2021**, *50*, 10764–10805. [CrossRef]
7. Chang, Z.; Qu, Y.; Gu, Z.; Zhou, L.; Li, R.; Sun, Z.; Xu, M.; Chu, M. Production of aromatic hydrocarbons from catalytic pyrolysis of Huadian oil shale using ZSM-5 zeolites as catalyst. *J. Anal. Appl. Pyrolysis* **2021**, *159*, 104990. [CrossRef]
8. Tsubaki, N.; Wang, Y.; Yang, G.; He, Y. Rational design of novel reaction pathways and tailor-made catalysts for value-added chemicals synthesis from CO_2 hydrogenation. *Bull. Chem. Soc. Jpn.* **2023**, *96*, 291–302. [CrossRef]
9. Wang, Y.; Gao, X.; Wu, M.; Tsubaki, N.J.E. Thermocatalytic hydrogenation of CO_2 into aromatics by tailor-made catalysts: Recent advancements and perspectives. *EcoMat* **2021**, *3*, e12080. [CrossRef]
10. Li, W.; Zhan, G.; Liu, X.; Yue, Y.; Tan, K.B.; Wang, J.; Huang, J.; Li, Q. Assembly of $ZnZrO_X$ and ZSM-5 on hierarchically porous bio-derived SiO_2 platform as bifunctional catalysts for CO_2 hydrogenation to aromatics. *Appli. Catal. B Environ.* **2023**, *330*, 122575. [CrossRef]
11. Wang, W.; He, R.; Wang, Y.; Li, M.; Liu, J.; Liang, J.; Yasuda, S.; Liu, Q.; Wu, M.; Tsubaki, N. Boosting methanol-mediated CO_2 hydrogenation into aromatics by synergistically tailoring oxygen vacancy and acid site properties of multifunctional catalyst. *Chem. Eur. J.* **2023**, *29*, e202301135. [CrossRef] [PubMed]
12. Ni, Y.; Chen, Z.; Fu, Y.; Liu, Y.; Zhu, W.; Liu, Z. Selective conversion of CO_2 and H_2 into aromatics. *Nat. Commun.* **2018**, *9*, 3457. [CrossRef] [PubMed]
13. Wang, Y.; Gao, W.; Kazumi, S.; Li, H.; Yang, G.; Tsubaki, N. Direct and oriented conversion of CO_2 into value-added aromatics. *Chem. Eur. J.* **2019**, *25*, 5149–5153. [CrossRef] [PubMed]

14. Wu, H.; Guo, L.; Wang, X.; Zhou, W.; Chen, F.; Li, D.; Liu, K.; Ai, P.; Wei, Y.; Cai, M.; et al. Spaced-confined capsule catalysts with tunable micro-environments for efficient CO_2 conversion. *AIChE J.* **2024**, *70*, e18445. [CrossRef]
15. Song, G.; Li, M.; Yan, P.; Nawaz, M.A.; Liu, D. High conversion to aromatics via CO_2 FT over a CO-Reduced $CuFe_2O_3$ catalyst integrated with HZSM5. *ACS Catal.* **2020**, *10*, 11268–11279. [CrossRef]
16. Wen, C.; Jin, K.; Lu, L.; Jiang, Q.; Wu, J.; Zhuang, X.; Zhang, X.; Chen, L.; Wang, C.; Ma, L. Insight into the direct conversion of syngas toward aromatics over the Cu promoter Fe-zeolite tandem catalyst. *Fuel* **2023**, *331*, 125855. [CrossRef]
17. Tian, G.; Li, Z.; Zhang, C.; Liu, X.; Fan, X.; Shen, K.; Meng, H.; Wang, N.; Xiong, H.; Zhao, M.; et al. Upgrading CO_2 to sustainable aromatics via perovskite-mediated tandem catalysis. *Nat. Commun.* **2024**, *15*, 3037. [CrossRef]
18. Wang, Y.; Kazumi, S.; Gao, W.; Gao, X.; Li, H.; Guo, X.; Yoneyama, Y.; Yang, G.; Tsubaki, N. Direct conversion of CO_2 to aromatics with high yield via a modified Fischer-Tropsch synthesis pathway. *Appl. Catal. B Environ.* **2020**, *269*, 118792. [CrossRef]
19. Xiang, Y.; Kruse, N. Tuning the catalytic CO hydrogenation to straight- and long-chain aldehydes/alcohols and olefins/paraffins. *Nat. Commun.* **2016**, *7*, 13058. [CrossRef]
20. Galvis, H.M.T.; Bitter, J.H.; Khare, C.B.; Ruitenbeek, M.; Dugulan, A.I.; Jong, K.P. Supported iron nanoparticles as catalysts for sustainable production of lower olefins. *Science* **2012**, *335*, 835–838. [CrossRef]
21. Cui, X.; Gao, P.; Li, S.; Yang, C.; Liu, Z.; Wang, H.; Zhong, L.; Sun, Y. Selective production of aromatics directly from carbon dioxide hydrogenation. *ACS Catal.* **2019**, *9*, 3866–3876. [CrossRef]
22. Ramirez, A.; Chowdhury, A.; Dokania, A.; Cnudde, P.; Caglayan, M.; Yarulina, I.; Abou-Hamad, E.; Gevers, L.; Ould-Chikh, S.; Wispelaere, K.; et al. Effect of zeolite topology and reactor configuration on the direct conversion of CO_2 to light olefins and aromatics. *ACS Catal.* **2019**, *9*, 6320–6334. [CrossRef]
23. Yue, Y.; Tian, J.; Ma, J.; Yang, S.; Li, W.; Huang, J.; Li, Q.; Zhan, G. Regulation of acidity properties of ZSM-5 and proximity between metal oxide and zeolite on bifunctional catalysts for enhanced CO_2 hydrogenation to aromatics. *Appl. Catal. B Environ.* **2024**, *355*, 124158. [CrossRef]
24. Wei, J.; Yao, R.; Ge, Q.; Xu, D.; Fang, C.; Zhang, J.; Xu, H.; Sun, J. Precisely regulating Brønsted acid sites to promote the synthesis of light aromatics via CO_2 hydrogenation. *Appl. Catal. B Environ.* **2021**, *283*, 119648. [CrossRef]
25. Gu, Y.; Liang, J.; Wang, Y.; Huo, K.; Li, M.; Wang, W.; He, R.; Yasuda, S.; Gao, X.; Yang, G.; et al. Tailoring the product distribution of CO_2 hydrogenation via engineering of Al location in zeolite. *Appl. Catal. B Environ.* **2024**, *349*, 123842. [CrossRef]
26. Yang, J.; Gong, K.; Miao, D.; Jiao, F.; Pan, X.; Meng, X.; Xiao, F.; Bao, X. Enhanced aromatic selectivity by the sheet-like ZSM-5 in syngas conversion. *J. Energ. Chem.* **2019**, *35*, 44–48. [CrossRef]
27. Kosari, M.; Lee, K.; Chao, W.; Rimaz, S.; Zhou, S.; Hondo, E.; Xi, S.; Seayad, A.M.; Zeng, H.C.; Borgna, A. Optimizing hollow ZSM-5 spheres (HZSM5) morphology and its intrinsic acidity for hydrogenation of CO_2 to DME with copper-aluminum. *Chem. Eng. J.* **2023**, *470*, 144196. [CrossRef]
28. Zhang, H.; Wang, L.; Xie, Y.; Zhang, S.; Ning, P.; Wang, X. Tailoring mesoporosity and acid sites for enhanced gaseous As_2O_3 adsorption by alkaline-etching ZSM-5 with different Si/Al ratios. *Sep. Purif. Technol.* **2025**, *354*, 129081. [CrossRef]
29. Lukyanov, D.B.; Gnep, N.S.; Guisnet, M.R. Kinetic modeling of ethene and propene aromatization over HZSM-5 and GaHZSM-5. *Ind. Eng. Chem. Res.* **1994**, *33*, 223–234. [CrossRef]
30. Yang, X.; Wang, R.; Yang, J.; Qian, W.; Zhang, Y.; Li, X.; Huang, Y.; Zhang, T.; Chen, D. Exploring the reaction paths in the consecutive Fe-Based FT catalyst-zeolite process for syngas conversion. *ACS Catal.* **2020**, *10*, 3797–3806. [CrossRef]
31. Cheng, C.; Li, G.; Ji, D.; Zhao, Y.; Shen, J. Regulating hierarchical structure and acidity of HZSM-5 for methanol to aromatics via protective desiliconization and external surface modification. *Microporous Mesoporous Mater.* **2021**, *312*, 110784. [CrossRef]
32. Zhang, C.; Hu, K.; Chen, X.; Xu, L.; Deng, C.; Wang, Q.; Gao, R.; Jun, K.; Kim, S.; Zhao, T.; et al. Direct hydrogenation of CO_2 into valuable aromatics over K/Fe-Cu-Al@HZSM-5 tandem catalysts: Effects of zeolite surface acidity on aromatics formation. *Fuel Process. Technol.* **2023**, *248*, 107824. [CrossRef]
33. Wu, X.; Wang, C.; Zhao, S.; Wang, Y.; Zhang, T.; Yao, J.; Gao, W.; Zhang, B.; Arakawa, T.; He, Y.; et al. Dual-engine-driven realizing high-yield synthesis of para-xylene directly from CO_2-containing syngas. *Nat. Commun.* **2024**, *15*, 8064. [CrossRef] [PubMed]
34. Lin, Y.; Gao, W.; Liu, G.; Tsubaki, N. A review of CO_2 hydrogenation to liquid fuels. *ChemSusChem* **2025**, *18*, e202402756.
35. Lin, S.; He, R.; Wang, W.; Wang, Y.; Gu, Y.; Liu, Q.; Wu, M. Highly selective transformation of $CO_2 + H_2$ into para-xylene via a bifunctional catalyst composed of Cr_2O_3 and twin-structured ZSM-5 zeolite. *Catalysts* **2023**, *13*, 1080. [CrossRef]
36. Wang, C.; Zhang, L.; Huang, X.; Zhu, Y.; Li, G.; Gu, Q.; Chen, J.; Ma, L.; Li, X.; He, Q.; et al. Maximizing sinusoidal channels of HZSM-5 for high shape-selectivity to p-xylene. *Nat. Commun.* **2019**, *10*, 4348. [CrossRef]
37. Xin, S.; Wang, Q.; Xu, J.; Chu, Y.; Wang, P.; Feng, N.; Qi, G.; Tébosc, J.; Lafon, O.; Fan, F.; et al. The acidic nature of "NMR-invisible" tri-coordinated framework aluminum species in zeolites. *Chem. Sci.* **2019**, *10*, 10159–10169. [CrossRef]
38. Wang, S.; Wang, P.; Qin, Z.; Chen, Y.; Dong, M.; Li, J.; Zhang, K.; Liu, P.; Wang, J.; Fan, W. Relation of catalytic performance to the aluminum siting of acidic zeolites in the conversion of methanol to olefins, viewed via a comparison between ZSM-5 and ZSM-11. *ACS Catal.* **2018**, *8*, 5485–5505. [CrossRef]

39. Li, H.; Yu, J.; Du, K.; Li, W.; Ding, L.; Chen, W.; Xie, S.; Zhang, Y.; Tang, Y. Synthesis of ZSM-5 zeolite nanosheets with tunable silanol nest contents across an ultra-wide pH range and their catalytic validation. *Angew. Chem. Int. Ed.* **2024**, *63*, e202405092. [CrossRef]
40. Wang, H.; Hou, Y.; Sun, W.; Hu, Q.; Xiong, H.; Wang, T.; Yan, B.; Qian, W. Insight into the effects of water on the ethene to aromatics reaction with HZSM-5. *ACS Catal.* **2020**, *10*, 5288–5298. [CrossRef]
41. Ene, A.B.; Archipov, T.; Roduner, E. Spectroscopic study of the adsorption of benzene on Cu/HZSM5 zeolites. *J. Phys. Chem. C* **2010**, *114*, 14571–14578. [CrossRef]
42. Inagaki, S.; Sato, K.; Hayashi, S.; Tatami, J.; Kubota, Y.; Wakihara, T. Mechanochemical approach for selective deactivation of external surface acidity of ZSM-5 zeolite catalyst. *ACS Appli. Mater. Interfaces* **2015**, *7*, 4488–4493. [CrossRef] [PubMed]
43. Cui, X.; Lyu, H.; Chai, Y.; Liu, B.; Zhao, D.; Liu, C. Selective aromatization of 1-hexene to BTX over core-shell structured Silicalite-1@ZSM-5 catalyst. *Sep. Purif. Technol.* **2024**, *349*, 127881. [CrossRef]
44. Emeis, C. Determination of integrated molar extinction coefficients for infrared absorption bands of pyridine adsorbed on solid acid catalysts. *J. Catal.* **1993**, *147*, 347–354. [CrossRef]

Disclaimer/Publisher's Note: The statements, opinions and data contained in all publications are solely those of the individual author(s) and contributor(s) and not of MDPI and/or the editor(s). MDPI and/or the editor(s) disclaim responsibility for any injury to people or property resulting from any ideas, methods, instructions or products referred to in the content.

MDPI AG
Grosspeteranlage 5
4052 Basel
Switzerland
Tel.: +41 61 683 77 34

Catalysts Editorial Office
E-mail: catalysts@mdpi.com
www.mdpi.com/journal/catalysts

Disclaimer/Publisher's Note: The title and front matter of this reprint are at the discretion of the Guest Editor. The publisher is not responsible for their content or any associated concerns. The statements, opinions and data contained in all individual articles are solely those of the individual Editor and contributors and not of MDPI. MDPI disclaims responsibility for any injury to people or property resulting from any ideas, methods, instructions or products referred to in the content.

www.ingramcontent.com/pod-product-compliance
Lightning Source LLC
Chambersburg PA
CBHW061626210926
51850CB00009B/2448
9783725870004